187
Advances in Polymer Science

Advances in Polymer Science

Recently Published and Forthcoming Volumes

Intrinsic Molecular Mobility and Toughness of Polymers I

Volume Editor: Hans-Henning Kausch

With contributions by
J. L. Halary · H.-H. Kausch · F. Lauprêtre
G. H. Michler · L. Monnerie

 Springer

The series *Advances in Polymer Science* presents critical reviews of the present and future trends in polymer and biopolymer science including chemistry, physical chemistry, physics and material science. It is adressed to all scientists at universities and in industry who wish to keep abreast of advances in the topics covered.
As a rule, contributions are specially commissioned. The editors and publishers will, however, always be pleased to receive suggestions and supplementary information. Papers are accepted for *Advances in Polymer Science* in English.
In references *Advances in Polymer Science* is abbreviated *Adv Polym Sci* and is cited as a journal.

Springer WWW home page: http://www.springeronline.com
Visit the APS content at http://www.springerlink.com/

Library of Congress Control Number: 2005926289

ISSN 0065-3195
ISBN-10 3-540-26155-9 Springer Berlin Heidelberg New York
ISBN-13 978-3-540-26155-1 Springer Berlin Heidelberg New York
DOI 10.1007/b136948

Springer is a part of Springer Science+Business Media

springeronline.com

© Springer-Verlag Berlin Heidelberg 2005
Printed in Germany

Cover design: *Design & Production* GmbH, Heidelberg
Typesetting and Production: LE-TEX Jelonek, Schmidt & Vöckler GbR, Leipzig

Printed on acid-free paper   02/3141 YL – 5 4 3 2 1 0

Advances in Polymer Science
Also Available Electronically

For all customers who have a standing order to Advances in Polymer Science, we offer the electronic version via SpringerLink free of charge. Please contact your librarian who can receive a password or free access to the full articles by registering at:

springerlink.com

If you do not have a subscription, you can still view the tables of contents of the volumes and the abstract of each article by going to the SpringerLink Homepage, clicking on "Browse by Online Libraries", then "Chemical Sciences", and finally choose Advances in Polymer Science.

You will find information about the

– Editorial Board
– Aims and Scope
– Instructions for Authors
– Sample Contribution

at springeronline.com using the search function.

Preface

The enormous length of macromolecules and the low intra- and intermolecular barriers opposing rotation and displacement of molecular groups or of even longer segments are at the origin of the unique visco- and rubber-elastic behaviour of polymer solids. Molecular mobility influences all phases of processing and use of such materials. Thus segregation and phase separation in the melt as well as structure development through crystallization depend on chain dynamics. The same is true for most deformation mechanisms, sample stiffness and ultimate properties such as toughness. Considerable progress has been obtained in the last decade in the understanding of the mutual relationship between the primary molecular parameters chain configuration, architecture and molecular weight (MW) on the one hand, and the response of a loaded entanglement network, the nature of the processes limiting stress transfer and the resulting mode of mechanical breakdown on the other. In view of the large technical importance of mechanical performance it seems to be adequate to review this subject, the *Intrinsic Molecular Mobility and Toughness of Polymers*.

In their introductory contribution Kausch and Michler discuss the elementary, time-dependent molecular deformation mechanisms, the competition between them, and their influence on the different failure modes of thermoplastic polymers (crazing, creep, yielding and flow, fracture through crack propagation). By establishing a *micro-morphological model* of polymer deformation and durability the authors highlight the dual role of segmental jumps and displacements to improve toughness by energy dissipation and relaxation of critical stresses and to influence without exception all damage mechanisms.

The dynamic response of a chain segment to thermo-mechanical excitation strongly depends on in-chain cooperative motions. By combining the powerful techniques of multi-dimensional Nuclear Magnetic Resonance and of dielectric and dynamic mechanical analysis Monnerie, Lauprêtre and Halary have investigated the *intensity and molecular origin of sub-T_g relaxations* and their degree of coupling for five structurally quite different amorphous polymers. Their important findings are reported in two comprehensive reviews treating the effect of chain configuration on segmental mobility and its effect on the toughness of these materials, respectively.

Essential features of the entanglement network and of the morphology of semi-crystalline polymers are determined through the crystallization process.

Chan and Li review homogeneous and heterogeneous nucleation. Using the new hot-stage in-situ AFM technique they particularly investigate the propagation of *founding lamellae*, their branching, interaction and development into lamellar sheaves and spherulites. In her contribution Grein gives a thorough *analysis of the influence of phase structure* (α- and β-crystalline polypropylene) as compared to the effect of elastomeric modifier particles. She concludes that the capacity of a matrix to deform remains an essential requirement for high toughness materials.

Stress cracking environments are known to enhance the mobility in the affected surface regions. Altstädt shows that the rate of fatigue crack propagation at *constant stress intensity factor K* proves to be a sensitive quantitative measure of the influence of active media. He also points to the dual role of segmental mobility, permitting stress relaxation followed by strain hardening or unstable softening, respectively. The complex conditions of *fracture during sliding contact* are reviewed by Chateauminois and Baietto-Duboug. They arrive at the conclusion that the main wear mechanism of glassy polymers, asperity scratching, is strongly controlled by competition between crazing processes and shear yielding. In the final contribution Estevez and van der Giessen present a computational analysis of the fracture of glassy polymers. The *applied cohesive zone model* takes into consideration the three steps of crazing (initiation, thickening and breakdown) and seems to be sufficiently flexible to adapt to future refinements.

The editor wishes to thank all authors for their willingness to cooperate in this joint effort, which so heavily depended on the concourse of their special expertise. It is hoped that the resulting detailed overview will be of help to more fully exploit the large potential offered by polymeric systems. Unfortunately the comprehensive treatment has made it necessary to publish the above, closely related eight contributions in two consecutive volumes of the Advances in Polymer Science, Vols. 187 and 188. However, a common *Subject Index* in both volumes and the reproduction of the two *List of Contents* should make it easy for the reader to find the desired information.

Lausanne, September 2005   *Hans-Henning Kausch*

Contents

Contents of Volume 188

Intrinsic Molecular Mobility and Toughness of Polymers II

Volume Editor: Hans-Henning Kausch
ISBN: 3-540-26162-1

Adv Polym Sci (2005) 187: 1–33
DOI 10.1007/b136954
© Springer-Verlag Berlin Heidelberg 2005
Published online: 13 October 2005

The Effect of Time on Crazing and Fracture

Hans-Henning Kausch[1] (✉) · Goerg H. Michler[2]

[1]Institut des Matériaux, EPFL-Ecublens, 1015 Lausanne, Switzerland
kausch.cully@bluewin.ch

[2]Institut für Werkstoffwissenschaft, FB Ingenieurwissenschaften,
Martin-Luther-Universität Halle-Wittenberg, 06099 Halle/S., Germany
michler@iw.uni-halle.de

Abstract By way of introduction to the subject of this Volume[1] this contribution reviews the effects of the most prominent properties of macromolecules (their enormous length, strong elastic anisotropy in axial and lateral directions and high segmental mobility) and of their characteristic dimensions on the elementary molecular deformation mechanisms of thermoplastic polymers. The competition between these mechanisms has a determinant influence on the different failure modes (crazing, creep, yielding and flow, fracture through crack propagation). The main part is devoted to an analysis of failure in creep. The micro-morphological approach is further developed and compared to criteria derived from visco-elastic theory with representative equations. Small angle X-ray analysis of the formation of fibrillar structures in amorphous polymers SAN and PC identifies three distinct regimes associated to fluid-like behaviour and disentanglement by forced reptation at low and moderate stresses (or high temperatures) and chain-scission dominated craze initiation (at low temperatures and high stresses), respectively. In semi-crystalline polymers similar differences are found: homogeneous creep at high stresses

---

[1] The term *This Volume* refers to the Special Double Volume "Intrinsic Molecular Mobility and Toughness of Polymers" of the Advances in Polymer Science, Vol. 187 and 188 (2005)

mostly involving the plastic deformation and break-up of crystal lamellae as opposed
to the formation of craze-like structures due to the disentanglement of chains at low
stresses. This review focuses on the important dual role of molecular mobility, to be at the
origin of time-dependent properties of polymer materials, especially of their toughness,
and to influence without exception all damage mechanisms which limit the strength and
durability of polymer components.

Keywords Crazing · Creep · Molecular deformation mechanisms · Disentanglement ·
Time-dependent strength · Toughness

Abbreviations

C_∞	characteristic ratio
D_K	equilibrium diameter of a molecular coil
E_a	activation energy
E	Young's modulus
E''	mechanical loss modulus
E'	mechanical storage modulus
D_o	fibril spacing (long period in scattering experiment)
G''	mechanical loss shear modulus
G_o	rubber elastic shear modulus
G_{Id}	dynamic energy release rate
J''	mechanical loss compliance
K, K_c	stress intensity factor
K_{Ic}	critical stress intensity factor
K_{Id}	dynamic stress intensity factor
M_e	entanglement molecular weight
M_w	weight average molecular weight
N_A	Avogadro's number
R	molar gas constant
T	absolute temperature
T_g	glass-rubber transition temperature
T_m	melt temperature
U	activation energy
V	activation volume
a	crack length
h	Plank constant
k	Boltzman constant
mol	mole
ν	Poisson's ratio
p	internal pressure
r_g	radius of gyration
s_{max}	maximum of scattering vector
t_f, t_b	time to failure
t_h	healing time
$\tan \delta$	ratio of loss to storage modulus
ε_a	linear amorphous strain
ε_t	linear total strain
ϕ	torsion angle about a chemical bond
γ	surface tension

Γ	fibrillation energy
ν_e	entanglement density
ρ	density
σ_0	tensile stress
2D	two-dimensional
BPA-PC	bisphenol A polycarbonate
CT	compact tension specimen
DCB	double cantilever beam specimen
DENT	double edge notch tensile specimen
DSC	differential scanning calorimetry
ESIS	European Structural Integrity Society
FNCT	full notch creep test
HDPE	high density polyethylene
iPP	isotactic polypropylene
LEFM	linear elastic fracture mechanics
MDPE	medium density polyethylene
NMR	nuclear magnetic resonance
PC	polycarbonate
PE	polyethylene
PVC	polyvinylchloride
PET	poly(ethylene *tere*-phthalate)
PMMA	poly(methyl methacrylate)
POM	polyoxymethylene
PS	polystyrene
SAN	styrene acrylonitrile
SCG	slow crack growth
SEN	single edge notch
UHMWPE	ultra high molecular weight polyethylene
U-PVC	unplasticized polyvinylchloride
WLF	Willams, Landel, Ferry

1
Introduction

This Volume of the Advances in Polymer Science is essentially devoted to the role of *molecular mobility* in the mechanical behaviour of thermoplastic polymers. Thermoplastics are van-der-Waals solids formed of long, *strong* and highly flexible chains; the strength and toughness of such solids is almost exclusively determined by stress transmission through weak secondary bonds, whose dissociation is easily activated under standard conditions. This geometry and the associated important elastic anisotropy of intra- and intermolecular forces are at the origin of the unique visco-elastic and rubber-elastic response of polymeric materials [1–3]. Most deformation mechanisms and without exception all damaging mechanisms depend on molecular motions and thus on time.

For the purpose of the following discussions it is helpful to depict the four *principal* modes of intra- and inter-molecular elastic interaction between neighbouring atoms of a C – C chain (Fig. 1).

The large stiffnesses of the C – C bond stretching ($740 \, GN/m^2$) and bond angle bending modes ($80 \, GN/m^2$) account for the high stiffness of fully extended, kink-free all-trans chains ($290 \, GN/m^2$ for PE). However, the intra-segmental potential barriers opposing the rotation around C – C bonds are generally so small that in a free chain (as in a diluted solution) such rotations are easily accomplished and transitions from trans to gauche conformations and vice versa occur at high frequency (Fig. 2).

The resulting most probable shape of a chain in solution is a random coil; its size is well characterized by the radius of gyration r_g (defined as the root mean square distance of the segments from the centre of gravity of the coil) or alternatively by the root mean square chain-end-to-end distance $\langle r^2 \rangle^{1/2}$. Many different models exist to relate both quantities to the length of a chain and the rotational freedom of its members. For a freely rotating tetrahedral valence bond model chain formed of n bonds of length a a gaussian distribution of chain-end-to-end distance r with a mean square $\langle r^2 \rangle = 2na^2$ is obtained. In experiment it is generally found that coils are more expanded. This is on the one hand due to short-range *intra*molecular interactions (steric hindrances to bond rotation), on the other due to the dilatating effects of *inter*molecular interactions, solvent quality and excluded volume. Without going into detail of the classical theories of polymers in solution [5] it should be indicated that the latter group of effects can be eliminated when such experiments are performed at the (polymer-solvent-specific) Flory- or Θ temperature. The *characteristic ratio* $C_\infty = \langle r^2 \rangle / na^2$ determined under Θ conditions essentially describes the dilatation due to short-range *intra*molecular interactions. For the polymers studied in this article values of C_∞ between

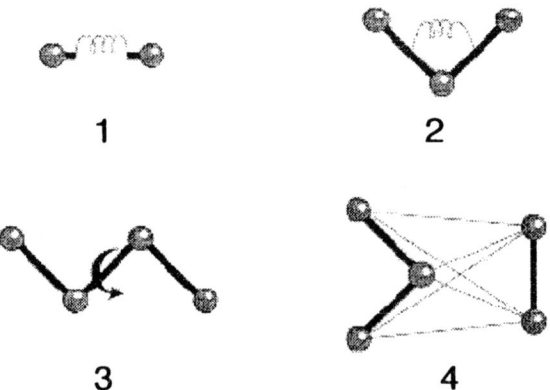

Fig. 1 Anisotropy of the principal elastic interactions in a C–C chain. (1) Bond stretching; (2) Bond angle bending; (3) Bond rotation; (4) Inter-segmental attraction

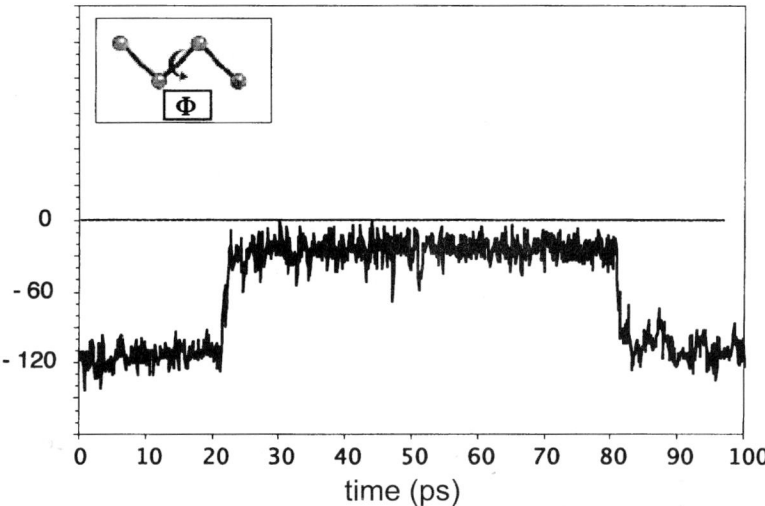

Fig. 2 Time scale involved in a conformational transition in a PMMA chain at ambient temperature obtained by computer simulation (after [4])

2.2 (BPAPC in methylene chloride) and 10.2 (PS in cyclohexane) are obtained. The radius of gyration also increases with C_∞ and is given by $r_g^2 = C_\infty na^2/6$. The large majority of all segments of a given chain are found within a sphere of radius of r_g without filling this volume, however. For technically relevant molecular weights the coil volume $4\pi r_g^3/3$ is more than 10 to 100 times larger than the volume na^3 of an extended chain.

To some extent the above results of chain dimensions in solution remain valid in the solid state. Flory has predicted that a macromolecule in the *melt* well above T_m or T_g should adopt an undisturbed random coil conformation with a radius of gyration identical to that in a solvent under θ-conditions. This hypothesis has been well confirmed by neutron scattering studies of deuterated chains in a hydrogenated matrix. In cooling down a melt the mobility of a chain as a whole (by Rouse diffusion and reptation) is gradually reduced. This means also that the dimensions of a chain, in particular the radius of gyration do not change any more if the *melt* temperature is approaching T_g (or T_m) (for an amorphous polymer Dettenmaier [6] has determined a relaxation time of 32 000 years for a deformed, deuterated PMMA chain of $M_w = 243\,000\,\mathrm{g\,mol^{-1}}$ at $T = 120\,°\mathrm{C}$). The radius of gyration of a chain in an isotropic solid polymer is therefore taken to be given by $C_\infty na^2/6$. The fact that the mass of a given chain only occupies a few percent (or even less) of its coil volume gives rise to an intense interpenetration of the coils. The simulated *structure* of an *amorphous* polymer shown in Fig. 3 gives a good example of the intimate penetration and close space filling of the chains. It was obtained by minimization of the total potential energy with respect to intra-molecular *conformation* and inter-molecular *packing*. These two quantities largely determine the inter-molecular attractive potential and thus the small strain properties (stiffness, hardness, velocity of sound), whereas the otherwise most important parameters molecular weight and density of entanglements hardly influence the *small strain* behaviour. Gentile and Suter [8] give a detailed account of atomistic modelling and of the static and dynamic properties predictable with this approach.

Well above the glass transition temperature the potential barriers formed by the inter-segmental van der Waal's forces are weak, they can be (more or less easily) overcome by thermally excited molecular groups or segments. The polymer behaves fluid-like. However, the rapid displacements of larger chain segments are frozen in at the glass transition temperature. Below T_g only increasingly smaller units remain mobile. An individual, thermally activated event, such as the partial rotation of a pendant group, *contributes* to energy dissipation and/or macroscopic deformation *only if* it interacts with the boundary conditions of the reference cube (external force field). In an unstressed sample such displacements are largely reversible with only a small fraction contributing to loss of free volume and physical aging. In the presence of a stress field, however, a potential difference is created between the displacements of a structural element *in* or *against* stress direction, which leads to anelastic and/or plastic deformation, the major source of energy dissipation [1–3]. However, due to the strong coupling of molecu-

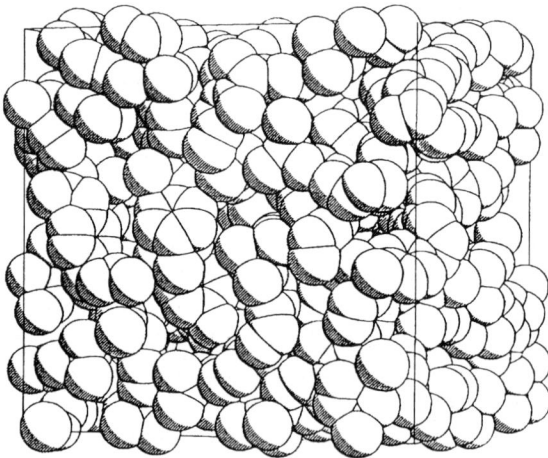

Fig. 3 Computer simulation of a disordered densely packed amorphous polymer (BPAPC, $M = 19\,800$); the density of the simulated structure ($1.2 \times 10^3 \text{ kg/m}^3$) agrees with experimental data. Note that the shown cube of 3 nm edge length contains segments of about 30 different chains thus giving rise to a large number of *topological interactions* or entanglements (Courtesy U. Suter [7])

lar groups within the chain backbone and the repulsion from neighbouring chain segments there are always many (up to 100 and more) backbone atoms involved in the stress-induced displacement of one segment past another (called a *plastic event*) [8]. The apparently homogeneous deformation of a sample under constant load or in steady straining is, therefore, a succession of many discrete local rearrangements (see also Sect. 3.2.2). In recent years the development of methods like multi-dimensional NMR has permitted detailed assignments by studying not only the mobility of a given atom but also the degree of correlation between the motions of any two selected (and marked) atoms on the chain. This important subject, the nature of molecular motions, their degree of cooperativity and their effect on toughness has been investigated in detail by Monnerie, Halary, Lauprêtre et al.; for the first time these authors present in this Volume a comprehensive review of their work [9, 10].

For a semi-crystalline polymer the solidification model of Fischer [11] predicts that the chain dimensions are frozen in at T_m if crystallization occurs sufficiently fast so as to prevent the disentanglement of chains by unravelling and/or their segregation according to their mass. For a series of polyethylene samples Michler [12] has accomplished an instructive comparison between the dimensions of molecular coils and the thickness of crystalline lamellae (Fig. 4).

Low molecular weight lamellae $(8.7\,\text{k})^2$ are highly regular, quite parallel and densely packed with only little amorphous material between them (high degree of crystallinity). As demonstrated in Fig. 4 the diameter of the molecular coils D_k is in this case comparable to the lamellar thickness. A particular chain is likely, therefore, to be incorporated into one single lamella. This means that very few *tie-molecules* connecting different lamellae are formed, which is equivalent to rather low tensile strengths (about 11 MPa for the 10 k material).

The lamellar thickness increases with increasing chain length, the arrangement is less regular, the degree of crystallinity decreases. It is well seen from the coil representation in Fig. 4 that chains with a molecular mass larger than 200 k are necessarily incorporated into different lamellae thus creating a larger number of tie-molecules, which account for efficient *long-range* stress transfer and for larger strains (> 400%) and stresses at break (about 24 MPa). The concentration of tie-chains and their anchoring in the crystalline lamellae are essential parameters in the long-term load bearing capacity of semi-crystalline thermoplastics.

Crystallization is an inherently time-dependent process; the nucleation and growth of crystalline structures, the degree of crystallinity, the phase structure and quality of crystal lamellae, and their connectedness strongly influence the mechanical properties of semi-crystalline polymers. It is for this

---

[2] 8.7 k designates a weight average molecular weight of 8.7 kg mol^{-1}

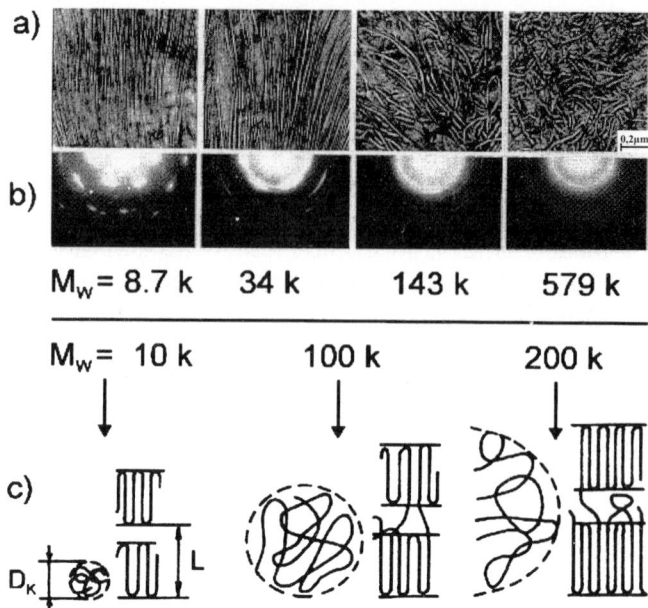

Fig. 4 Influence of molecular weight M_w on (**a**) lamellar arrangement and (**b**) diffraction patterns of polyethylene; (**c**) comparison of the diameter D_K of molecular coils and of the long period L of the lamellar structure (after Michler [12])

reason that these topics are treated in two contributions of this special Volume. The kinetics of crystallization and the effect of molecular variables on lamellar growth are reviewed by Chan and Li [13], who have investigated by in-situ atomic force microscopy the dynamics of nucleation and the growth and space filling of crystalline lamellae of different spherulitic polymers. Grein [14] describes the crystallographic, thermal and mechanical differences between α- and β-polypropylene and identifies their effect on toughness.

Besides chain length (radius of gyration) and chain length distribution it is the *density of entanglements* ν_e, which is particularly important for the long-range transfer of stresses between segments. It can be determined from the standard relations between the shear modulus G_0 of a polymer in the rubber elastic plateau region, density ρ and absolute temperature T, which yields the average molecular weight M_e of a chain section between entanglements:

$$M_e = \rho RT/G_0 \tag{1}$$

from which ν_e is obtained by

$$\nu_e = \rho N_A/M_e \tag{2}$$

with N_A designating Avogadro's number. Both amorphous and semi-crystalline physical networks are generally tougher with increasing density of entanglements ν_e. In his model calculations Brown [15] established a linear relation between ν_e and the energy release rate G_{Ic}. This agrees with the experimental verification of Kausch and Jud [16, 17], who have concluded from their crack healing experiments, that up to saturation the strength G_{Ic} of an interface depends linearly on the number of newly formed entanglements, which is equivalent to the scaling of the critical stress intensity factor K_{Ic} with $(\nu_e)^{1/2}$. The data plotted by Gensler [18] confirm the generally positive linear correlation between critical stress intensity factor K_{Ic} and the square root of ν_e (Fig. 5; however, there are deviations from this linear correlation; their molecular significance will be elaborated by Monnerie, Halary et al. [9, 10] later in this Volume).

In this introduction we have indicated that cohesive energy density, chain length and entanglement density are major strength determining molecular variables of un-oriented thermoplastic homopolymers. In order to fully exploit the potential offered by an adequate combination of intrinsic variables we also have to consider the strong effects of the state of organization (density, phase structure, micro-morphology including the state of orientation and the presence of defects), of the environmental variables (presence of active liquids—including H_2O—or gases—O_2, CO_2 etc) and of stress history. In the following sections we will discuss the major strength-*reducing* mechanisms activated in loading and deformation, whose competition and interaction determine the lifetime of a loaded specimen and the mode of failure.

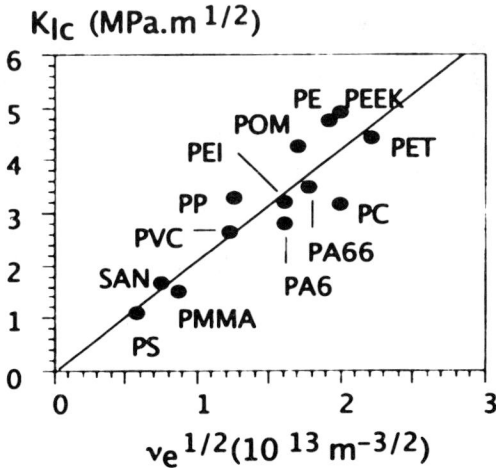

Fig. 5 Critical stress intensity factor K_{Ic} of commercial polymers versus the square root of entanglement density ν_e (after [18])

2
Molecular Micro-Mechanics and Damaging Mechanisms

In the context of this special Volume we are particularly interested in the mutual relations between the molecular structure, the nature and intensity of energy dissipating mechanisms, and ultimate mechanical properties. The latter are among the most important prerequisites for the successful use of a polymer material, no matter whether the mechanical, the optical or some specific functional properties are to be exploited. A thermoplastic construction element will be the stiffer and the stronger the more *homogeneously* —in space and time—large forces can be transferred onto and borne by the chain backbones. However, on a molecular scale polymer materials are necessarily *heterogeneous*. The presence of crystal lamellae separated by amorphous regions and connected by tie-molecules leads to an inhomogeneous distribution of stresses and strains, which strongly influences the mechanical properties. The formation of voids or the presence of defects or cracks has a similar effect. The micromechanical model shown in Fig. 6 may be used to highlight the nature of the deformation and damage mechanisms and their competition.

In any layer-like arrangement of stiff crystalline and soft amorphous regions a strong influence of lamellar orientation with respect to the principal stress direction is observed. The most compliant arrangement is shown in

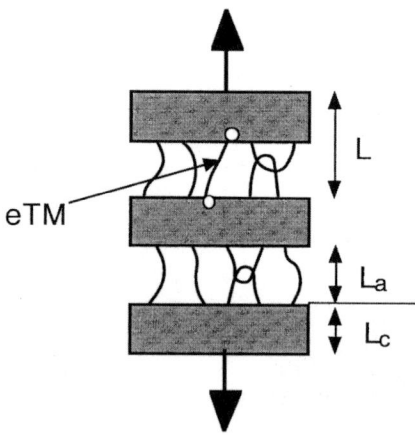

Fig. 6 Micromechanical model of a section of a semi-crystalline polymer with lamellae oriented perpendicular to the principal stress direction showing the long period L and the thicknesses of crystalline (L_c) and amorphous layers (L_a); the latter are composed of loose segments, entangled chains and more or less extended tie molecules. Large forces can be transferred at those *points* (o) where highly extended tie molecules (eTM) enter crystalline lamellae

Fig. 6, where all layers are subjected to the same stress (*Reuss* model). Elastic and anelastic deformation prevail at moderate loading times and small external strains ($\varepsilon_t < 3\%$) with the majority of the deformation born by the generally much weaker amorphous regions; in pre-oriented polyethylene strain concentrations $\varepsilon_a/\varepsilon_t$ of up to 5 have been observed (see [19]). The tensile deformation leads to the stretching of (extended) tie molecules and to the transfer of remarkable stresses to the crystalline lamellae: the pull-out force of a chain from an orthorhombic PE-crystal lamella amounts to 1.4 nN (which is equivalent to 7.5 GPa) [20]. The geometry as shown in Fig. 6 has another mechanically important consequence: the stiff and large crystal lamellae exert strong lateral constraints; the homogeneous stretching of the amorphous regions then must necessarily lead to large hydrostatic tensions. We wish to repeat these two basic facts: the straining of semi-crystalline polymers involves the *stress transfer between crystal lamellae* through the amorphous regions and tie molecules and causes the *dilatation of the amorphous regions.*

The elastic excitation activates different stress relieving mechanisms, which act in competition and at quite different kinetics. The sequence of crystallographic and morphological changes has been extensively investigated in the literature (see [1–3, 19, 21] for further references). There seems to be agreement that *below* the yield point lamellar separation, inter-lamellar shear, and uncorrelated intra-lamellar shear steps dominate in a slow tensile drawing experiment. The axial chain stresses of extended tie molecules are relieved through chain scission as well as through pull-out from crystal lamellae or through the bending, rotation, plastic deformation and/or disintegration of the lamellae to which a particular tie molecule is attached. Not fully extended tie molecules may change their conformation or orientation. Stresses in hydrostatically tensioned amorphous regions are relieved through disentanglement and cavitation. The latter mechanism generally—but not necessarily—coincides with the onset of yielding in steady state deformation. Figure 7a shows the inter-lamellar and/or inter-spherulitic separation and

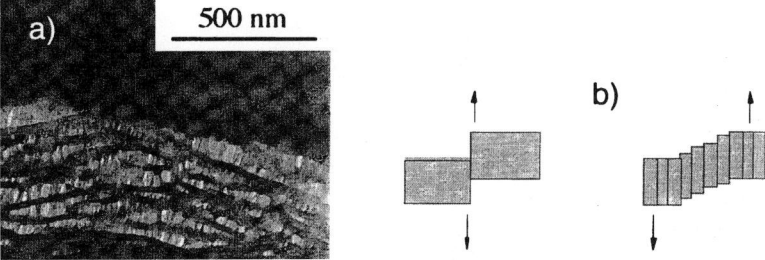

Fig. 7 Beginning of plastic deformation in isotactic polypropylene films (iPP) at room temperature: **a** inter-lamellar separation and voiding; **b** intra-crystalline shear steps (after [18])

voiding at the beginning of plastic deformation of isotactic polypropylene films (iPP) at room temperature. The important post-yield plastic deformation of the then fibrillated structure is mostly born by intra-crystalline shear (Fig. 7b) and de-crystallization accompanied by disentanglement.

The fact that the formations of voids—and subsequently the initiation of crazes—strongly depend on the action of three-dimensional hydrostatic tensile forces is corroborated by two phenomena: *thin layer yielding* and *channel die compression*. Very thin stretched films are in a state of— two-dimensional—plane stress, which favours shear yielding rather than void formation. Kawai et al. [22] were among the first to observe the ductile behaviour of otherwise brittle polymers, namely the plastic drawing of very thin glassy polystyrene lamellae in SBS-block copolymer systems. More recently Adhikari et al. [23] have investigated injection-moulded samples of blends of asymmetric star styrene-butadiene (SB) block copolymers with polystyrene. The morphology of these materials is characterized by alternating, continuous PS and SB layers, whose thickness and regularity depend on blend composition. When stretching the thin PS layers (thickness $D \sim 20$ nm) which exist in a sandwich of pure star copolymer parallel to the direction of orientation the otherwise brittle PS deforms by homogeneous plastic flow up to more than 200% [23]. A transition from ductile to brittle behaviour was clearly observed in the blends; with increasing PS content the thickness of the PS layer increased, at more than 30 nm the lateral constraints became sufficient for void formation leading to the formation of craze-like structures.

The detrimental effect of cavitation has also been demonstrated by Galeski [24], who has compared the behaviour of iPP in tensile drawing and in channel die compression. Cavitation, micro-necking and disentanglement occur during tensile drawing, limiting the strain hardening to a true stress level of below 100 MPa. In channel die compression a rheologically similar deformation is imposed on the polymer network with no cavities formed, however. In this case crystallographic slips lead to progressive local orientation and strain hardening raising the stress level to above 300 MPa (at a strain ratio of 3).

The same molecular mechanisms as in tensile drawing are observed, of course, in constant load experiments. Depending on the stress-time-temperature regime essentially four different failure modes are observed with thermoplastic materials:

1. unstable crack propagation of highly elastically loaded samples;
2. apparently homogeneous, ductile deformation leading eventually to plastic instability;
3. locally heterogeneous deformation: void formation and/or crazing, slow crack growth (SCG) with failure generally occurring at moderate stresses and after longer loading times;
4. thermal aging.

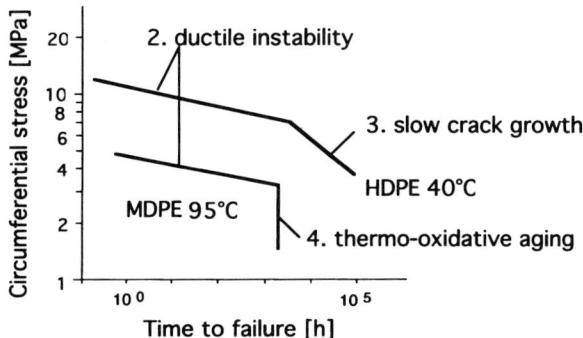

Fig. 8 Stress versus time to failure diagram of internally pressurized HD- and MD-polyethylene pipes; three failure modes are indicated. The circumferential stress σ is calculated according to the standard formula $\sigma = p(d - s)/2s$ (with p internal pressure, d pipe diameter, s pipe wall thickness)

As a representative example, three of the failure modes of internally pressurized polyethylene pipes are indicated in Fig. 8 in the plot of circumferential stress σ_f versus time to failure t_f. (Not shown is the regime of brittle fracture by unstable crack propagation, since it only occurs at much lower temperatures and frequently requires a special crack initiation procedure).

The characteristic change in slope between ductile failure (mode 2) and slow crack growth (mode 3) indicates that (in this material) there is a difference in the kinetics of failure. The associated deformation and damage mechanisms will be discussed in Sect. 3 of this review and in later contributions in this Volume. The almost vertical drop of the time-to-failure curve caused by (thermo-oxidative) aging at higher temperatures (mode 4) can be explained by the total loss of mechanical strength of heavily degraded material (generally after the end of the *induction period*, which is determined by the used stabilization technique).

3
Time-Dependent Deformation and Fracture

3.1
Unstable Brittle Fracture Driven by Elastic Energy

The analysis of brittle fracture is the very domain of Linear Elastic Fracture Mechanics (LEFM). A comprehensive introduction to its fundamentals and the validity of its application to polymers has been given by Williams [25] and more recently by Grellmann and Seidler [26]. The fracture criteria and relevant test procedures elaborated by the ESIS technical committee TC4 can be found in [27].

According to LEFM the stress singularity at the tip of a (sharp) crack of length a can be conveniently described using a *stress intensity factor* $K(a)$; in the crack opening mode $K(a)$ is given by:

$$K_I = f(a/W)\sigma_o\sqrt{\pi a},\qquad\qquad(3)$$

where $f(a/W)$ is a correction function tabulated for standard test specimens taking care of their finite widths W, and σ_o the applied far-field tensile stress [10, 25–27]. Unstable crack propagation occurs if $K(a)$ attains a critical value K_{Ic}. The corresponding energy release rate (in plane strain) is then given by

$$G_{Ic} = \frac{K_{Ic}^2}{E(t)}(1 - \nu^2),\qquad\qquad(4)$$

where $E(t)$ is Young's modulus and $\nu(t)$ the Poison's ratio.

Many variables used and phenomena described by fracture mechanics concepts depend on the history of loading (its rate, form and/or duration) and on the (physical and chemical) environment. Especially time-sensitive are the level of stored and dissipated energy, also in the region away from the crack tip (far field), the stress distribution in a cracked visco-elastic body, the development of a sub-critical defect into a stress-concentrating crack and the assessment of the effective size of it, especially in the presence of micro-yield. The role of time in the execution and analysis of impact and fatigue experiments as well as in dynamic fracture is rather evident. To take care of the specificities of time-dependent, non-linearly deforming materials and of the evident effects of sample plasticity different criteria for crack instability and/or toughness characterization have been developed and appropriate corrections introduced into Eq. 3, which will be discussed in most contributions of this special Double Volume (Vol. 187 and 188).

At this point we wish to look for the effect of time on the fracture behaviour of polymers in the *absence* of notable plastic deformation. The only variables in Eqs. 3 and 4, which explicitly depend on time, are Young's modulus and Poisson's ratio. The resulting implications have been thoroughly investigated by the school of Knauss, they are discussed in the cited literature [25–29] and will be resumed throughout this Volume. The rates of loading and energy transport, the speed of a crack, its acceleration and instantaneous direction influence each other and the critical *dynamic* fracture mechanics variables K_{Id} and G_{Id}. The following examples serve as an illustration of time effects in *elastic* fracture.

An unstable crack is primarily driven by the elastic energy *already stored* in the specimen (and not so much by the energy *introduced into* the specimen by the continued deformation). A pertinent experiment was carried out by Stalder [30], who investigated the impact fracture of PMMA using an instrumented hammer nose and standard notched Charpy specimens with sprayed-on graphite strain gages. With this *double-instrumentation* technique he determined the sequence of events. A transient force $f_h(t)$—the so-called *inertia peak*—is transmitted by the hammer nose to the specimen (in this case for a period of 60 µs). During this period the sample is accelerated and anelastically deformed in bending, but *no* crack is as yet initiated at the notch tip. This occurs only after the external force $f_h(t)$ had already dropped to zero. Crack extension in such a brittle material is, therefore, accomplished *in flight* and at a rate, which rapidly reaches 250 m/s and declines subsequently to a level of 40 m/s. Stalder found that as opposed to PMMA tougher materials (PC, PA6) require a continuous input of energy before and after initiation of a crack. The behaviour of notched and unnotched specimens and the definition and determination of characteristic fracture mechanics parameters for resistance against unstable crack propagation have been summarized by Grellmann et al. [31].

One of the most extreme cases of sustained rapid crack propagation has been investigated by Greig et al. [32]. With a special procedure they introduced a high-speed crack into an internally nitrogen pressurized medium-density polyethylene (MDPE) pipe held at 0 °C. Whereas at low internal pressure the cracks arrested after 1 to 3 m, an undulating crack advanced the full length of the test pipe above a critical pressure p_c. The critical pressure p_c is reasonably well predicted by the Irwin–Corten strain energy analysis [32], which is based on the assumption, that *all* of the stored elastic energy is used to drive the crack:

$$p_c = \frac{s}{D} \sqrt{\left(\frac{8EG_{Id}}{\pi D} \right)}. \tag{5}$$

This expression shows that the critical circumferential *stress* falls with increasing pipe diameter D. This relation and the difficulty to extrude and keep in shape plastic pipes with large wall thicknesses s limit their upper size (at this time the largest external diameter of an extruded (HDPE) pipe is 1400 mm).

The third example of time effects in elastic fracture concerns *accelerated* crack growth. Up to the early 80 s it has been assumed that a unique relation exists between the stress intensity factor K and the rate of crack growth da/dt. This contention had been confirmed in many *steady crack growth* experiments. Using three point bending specimens Chan and Williams [33] obtained for HDPE in water the following relation between da/dt, and the acting stress intensity factor K_c:

$$K_c \sim (da/dt)^n, \tag{6}$$

with n equal to 0.25. This equation implies for a given material a unique relation between K_c and the rate of crack propagation. Whether or not the (dynamic) K_{Id} is also a unique function of $\dot{a}$ has been subject to discussion for many years. In 1987 Takahashi and Arakawa [34] conceived a specially

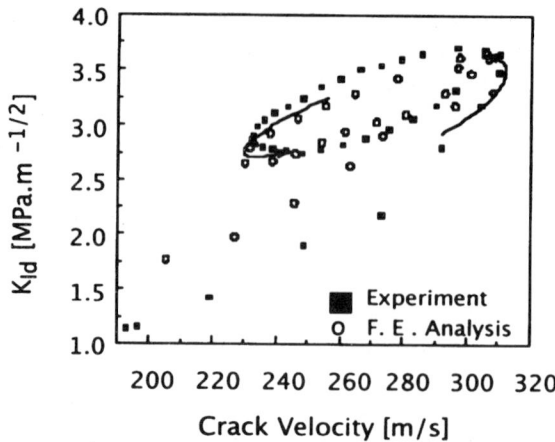

Fig. 9 Dynamic stress intensity factor K_{Id} calculated from a finite element model (o) plotted as a function of crack velocity and compared to experimental data (■) obtained for PMMA (after [34, 35])

shaped SEN tensile specimen, which forced the tip of a running crack to traverse regions of respectively increasing and decreasing stress intensity factor. The crack was accordingly accelerated, decelerated and (possibly) accelerated again. The intriguing data from these caustic experiments clearly reveal, that K_{Id} is *no* unique function of da/dt (Fig. 9).

In the high crack velocity regime *three* different values of K_{Id} can be assigned to *one* rate of crack propagation depending on the state of *crack acceleration*. This behaviour was ascribed to inertia effects associated with crack acceleration and deceleration. Such a hypothesis is corroborated by the computed K_{Id} data (also shown in Fig. 9), which were obtained from a finite element model, taking into consideration the mentioned transient dynamic linear elastic effects [35].

For all further questions of definition, mathematical analysis, limits of applicability, and experimental procedure the reader is referred to the specialized literature. The fracture mechanics of polymers are particularly well treated in [25–27].

3.2
Homogeneous Creep, Flow and Plastic Instability

3.2.1
Definitions

Creep, yielding, and post-yielding plastic deformation (drawing) as well as flow are brought about by the stress-biased deformation and displacement (*jumps*) of molecular groups and chain segments. *Creep* is defined as the time-dependent strain following a step change in stress. It is generally recognized that the *total creep strain* consists of four independent components: an *instantaneous* part, a transient *recoverable* visco-elastic component, which is generally designated as *primary creep*, a *secondary creep phase* characterized by a constant strain rate at constant stress, to which strengthening and weakening processes as well as permanent, non-recoverable flow contribute. At severe conditions progressive sample weakening leads to a phase of accelerating creep rate (tertiary creep).

Although in this review we are not concerned with yielding, we may indicate that the *yield point* is generally defined as the maximum in the engineering stress-strain curve, it frequently—but not necessarily—marks the onset of cavitation and of notable changes in conformation and structural organization, which favour heterogeneous deformation (see Sect. 3.3). Localized yielding at a crack tip or during fibrillation of a homogeneous matrix is generally designated as micro-yielding. For a detailed molecular, visco-elastic and mechanical analysis of polymer plasticity and flow the reader is referred to the cited references [1–3, 36–39].

3.2.2
Molecular Description of Creep Deformation

Deformation and failure in creep is the central subject of this review and a problem of major concern in the application of load bearing polymers. As discussed in the introduction creep is the result of stress-induced segmental *jumps* (displacements and/or conformational changes). The stepwise increase of the shear stress of a reference cube (Fig. 3) leads to an increase of potential energy U over dozens of load increments before a plastic event takes place resulting in an abrupt decrease of U. This is equivalent to saying that an apparently continuous deformation in creep is in reality a succession of very small, normally unresolved steps. In a semi-crystalline polymer with a pronounced lamellar or fibrillar organization deformation steps will also involve the stick-slip motion of such morphological elements with respect to each other. Analyzing the creep of melt crystallized and gel-cast UHMWPE films with a sensitive laser interferometer Mjasnikova et al. [40] have in fact found a large spectrum of creep rates ranging from almost zero to 100 times the average rate.

The relation between the frequency v of local jumps, shear stress τ, temperature T and macroscopic rate of creep $\dot{\varepsilon}$ was well established by Eyring's reaction rate theory [41]. Let us consider that a number of v_{i0} thermally activated structural units attempt per unit time to cross a potential barrier U_i; the net flow v_i of units that will succeed is then given by:

$$v_i = v_{i0} \exp\left(-\frac{U_i}{RT}\right)\left[\exp\frac{V_i\tau_i}{RT} - \exp-\frac{V_i\tau_i}{RT}\right] \tag{7}$$

where V_i is the activation volume (proportional to the *jump width*) and τ_i the (shear) stress acting at the sites i in the jump direction. If there is no unique potential barrier but a distribution of different energy values U_i then the total jump rate is obtained as the sum $\nu = \Sigma \nu_i$. The rate of creep $\dot{\varepsilon}$ can be assumed to be proportional to the jump rate ν. The net flow rate ν and also $\dot{\varepsilon}$ will remain constant if the distribution of stresses τ_i and jump site parameters U_i and V_i does not change during creep.

However, if the creep rate is plotted as a function of time or total strain, it is seen that most materials subjected to a constant force show three phases of non-elastic creep: a first one of decreasing average rate of deformation (*primary creep*), a second one of constant rate, and—if the stress level is sufficiently high and the exposure sufficiently long—a third phase of accelerated deformation (tertiary creep). The *Sherby–Dorn plot* (creep rate versus circumferential strain) of U-PVC for constant hoop stresses of between 37 and 42 MPa at a temperature of 20 °C clearly reveals these characteristic phases (Fig. 10a). In UHMWPE fibres, however, only two creep phases can be observed: a primary one, where misalignments of the ultra-long molecules are gradually eliminated, and a secondary one of irreversible plastic flow with no apparent fibre weakening.

According to the above model the first jumps to be activated are those from a site favourable for a rearrangement (large $V_i\tau_i$, *low* potential barriers U_i). The gradual decrease of creep rate during the first phase indicates that the concentration of favourable sites decreases, in other words, that a jump from a favourable site leads to a less favourable position with a higher U_i (strain hardening).

The fact that the rate of deformation $\dot{\varepsilon}$ is constant during the secondary creep phase, leads to the conclusion that (the majority of) the flow events in this phase do not modify the spectrum of barrier heights significantly as in plastic deformation or true flow. This is rather evident in ultra-high molecular weight fibres, which have high degrees of orientation and crystallinity (> 98%). In un-oriented polymers—such as the studied U-PVC—both, strain *hardening* and *weakening* steps may contribute to secondary creep, but apparently their effects balance each other. The increase in inter-segmental distances (especially in view of dipole-dipole interactions in U-PVC), disentanglement, void formation, and chain scission accompanied or not by craze formation must be considered as weakening steps (see Sect. 3.3).

Rather than modelling the creep behaviour by a *continuous distribution* of potential barriers Wilding and Ward [44] have proposed a model of creep with *two processes acting in parallel*. The first process has a large activation volume; it is operative at low stress levels and does not contribute (much) to permanent flow (this is equivalent to saying that it accounts for the reversible part of creep). The second process has a small activation volume and it is active only at high stresses. Apparently the second process concerns the slip (or even scission) of extended segments, which are responsible for

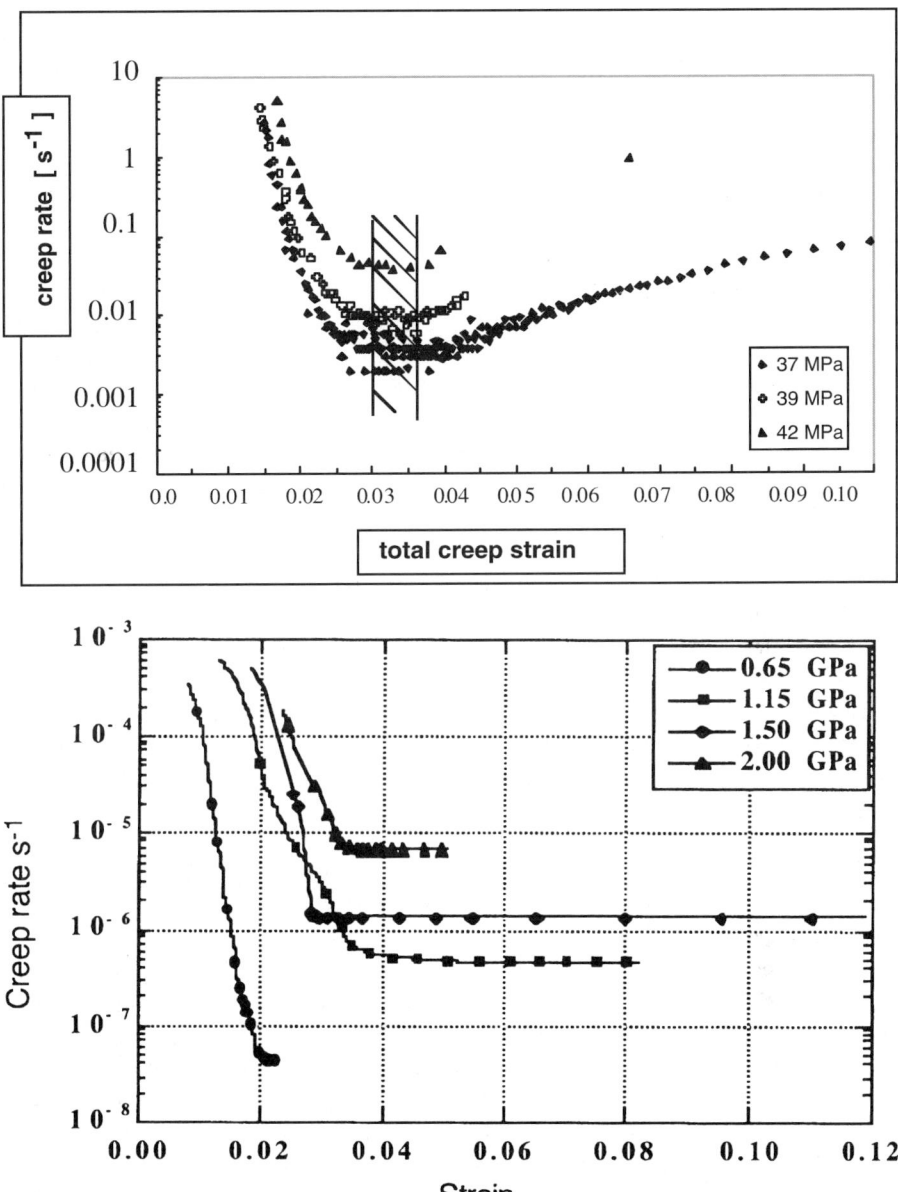

Fig. 10 Sherby–Dorn plots of creep rate versus strain showing the different creep phases **a** U PVC pipes for constant hoop stresses of 37 to 42 MPa at a temperature of 20 °C (60 K below T_g) (using data of Castiglione et al. [42]); **b** Highly oriented UHMWPE fibres at room temperature (using data of Berger et al. [43])

long-range stress transmission within a thermoplastic solid. This also means that load is shifted during creep from non-extended and/or unfavourably oriented segments to the more extended ones (Berger et al. [43] have confirmed this by determining by Raman microscopy the creep-induced changes in the distribution of axial stress acting on main chain C – C bonds). The notion of a network of extended segments responsible for long-range stress transmission has been used for many years to explain the effect of molecular orientation on the stiffness and strength of thermoplastic polymers [16]. The above theoretical concept has been put to industrial use in the last few years by the development of higher temperature resistant resins composed of a bi-modal distribution of chain lengths (also containing appropriate short-chain branches to better control the tie chain concentration and prevent chain pull-out [45, 46]).

3.2.3
Modelling of Creep Deformation

There is strong interest to analytically describe the *time*-dependence of polymer creep in order to extrapolate the deformation behaviour into otherwise inaccessible time-ranges. Several empirical and thermo-dynamical models have been proposed, such as the Andrade or Findley Potential equation [47, 48] or the classical linear and non-linear visco-elastic theories ([36, 37, 49–51]). In the linear viscoelastic range Findley [48] and Schapery [49] successfully represent the (primary) creep compliance $D(t)$ by a potential equation:

$$\varepsilon_V(T, t) = \left[D_0(T) + C(T, \sigma_V)t^n\right]\sigma_V, \tag{8}$$

where ε_V and σ_V are representative (uni-axial or circumferential) sample strain and stress components respectively, and D_0 and C are constants with respect to time. (Gregory et al. [50] have shown that this equation can even be extended into the non-linear stress range if the exponent n is taken to depend on stress). Pilz and Lang [51] use the time-temperature superposition principle to generate a master curve of the relaxation modulus $E(t)$ of externally loaded medium and high density PE-pipes, which is then extrapolated by one decade to obtain the desired creep amplitude after 50 years. Both approaches describe the phase of primary creep dominated by visco elastic molecular displacements; as we have seen above, such rearrangements *strengthen* the material, improve stress transmission, and lead to decreasing creep rates (Phase 1 in Fig. 10). Molecular rearrangements during secondary creep lead on the average to *equivalent states* (which is the definition of *plastic deformation*).

On the other hand the accelerating homogeneous deformation in the tertiary creep phase (Fig. 10a) indicates material *weakening* through the accumulated geometrical and structural changes (reduced cross-section, dilatation, local orientation, disentanglement or decrystallization where applicable, and/or distributed formation of voids, crazes or other damages). A reliable description of tertiary creep through small strain visco-elastic functions is generally impossible (see also Sect. 3.3.4). A further serious obstacle to such an approach is the increasing tendency for localized, heterogeneous deformation. Since the main purpose of modelling creep behaviour is to predict for which period of time a load-bearing element (such as a pipe for instance) will work safely under the given circumstances (stress, temperature), we need to know when (and why) the strengthening primary and the neutral secondary creep steps turn into the weakening tertiary phase. Consideration must be given, of course, to the competing failure modes brittle fracture, slow crack growth, and chemical degradation, which may occur earlier (Fig. 8).

3.2.4
Criteria of Creep Failure

From Eq. 8 three quantities can be identified, which could play a critical role in creep deformation: (i) a maximum admitted level of strain ε_{Vc} (which is related to a minimum admitted relaxation modulus); (ii) a critical strain rate $d\varepsilon_V/dt$; and (iii) the beginning of a notable deviation of creep data from the potential *law*. All three criteria have been used in the literature [42, 51–53].

As an example we will discuss the most recent of work of Castiglione et al. [42], who have investigated the creep of U-PVC. They found that the Findley equation (Eq. 8) applied well to describe the transient phase of creep at RT in the range of applied stresses (from 20 to 44 MPa). The best fit was obtained with a D_0 of 2.91×10^{-4} MPa^{-1} and an exponent n of 0.197 ± 0.004 for internally pressurized pipes, and with a D_0 of 3.65×10^{-4} MPa^{-1} and an exponent n of 0.158 ± 0.002 for tensile specimens. The characteristic, geometry-related differences of the fitting parameters between tubular and tensile specimens—also reported earlier [52]—need not be discussed here, since both types of specimen yielded similar results, provided the true applied stresses were considered. This similarity in behaviour opens up the interesting possibility of substituting the quite demanding experiments using internally pressurized pipes by the much simpler tensile tests [42].

Concerning the above-mentioned critical quantities the authors have in fact established: (i) that irrespective of stress level damage is apparently initiated at a critical creep strain ε_c of 3 to 3.5%; (ii) that a notable deviation of creep data from the potential *law* starts just at this strain level; and (iii) that although the strain rate $d\varepsilon_V/dt$ is a function of stress, the minimum in the Sherby–Dorn plot also occurs (for the tubular specimens) at ε_c. The postulated changes in sample morphology at about the time when the strain values started to deviate from Findley's equation, were in fact seen by these and other authors [42, 52], who detected in U PVC deformed micro-cavities later to be followed by discolouration. Critical creep strain levels of the order of 3 to 3.5% have also been reported for other amorphous and semi-crystalline polymers [53, 54]. The conclusion that cavitation (possibly accompanied by disentanglement) is at the origin of material weakening is also reinforced by the absence of both tertiary creep and defect formation in UHMWPE (Fig. 10b). The experiments of Galeski [24] discussed in Sect. 3 also point in this direction.

It seems to be important to mention that the formation of defects necessarily leads to locally *heterogeneous* deformation, which is the subject of the following section. However, the creep deformation remains macroscopically *homogeneous* if such defects are numerous and spatially well distributed.

3.3
Locally Heterogeneous Deformation: Crazing and Slow Crack Growth

3.3.1
Morphology of Crazes

The formation of voids, crazes and cracks are the most visible signs of heterogeneous deformation. Before discussing in detail their genesis during creep loading, we will briefly describe their morphology. Crazing has been intensively investigated over the last 50 years, with major reviews appearing regularly [55–59]. Referring to the cited references a craze can be described as a well defined, flat deformation zone filled with highly extended, mostly fibrillar matter and interspersed voids, which appears in amorphous and semi-crystalline thermoplastics perpendicular to the main stress axis (Fig. 11).

Since the discovery of the fibrillar microstructure through Kambour [55] in 1968 notable progress has been made concerning the molecular origins of crazing. There are three different phases, craze *nucleation* in those network regions, where sufficient hydrostatic tension meets a low degree of entanglement, craze *extension* through transformation of matrix material into highly oriented craze matter in the *process zone* necessitating an important loss of entanglements, and finally craze failure through *breakdown of stress transmission* (within fibrils or at the process zone). Michler has given a detailed account of microstructure and breakdown mechanisms of crazes in different polymers [1]. It should be noted that the *local strain rate of a growing craze* (i.e. the rate of separation of its borders) can be several orders of magnitude higher than the sample strain rate thus leading to a localization of strain. A single craze generally constitutes a serious defect of a loaded sample, since after its breakdown a crack is formed which further increases the concentration of strain and stress and accelerates sample fracture. On the other hand, the formation of many crazes (*numerous nucleation sites*) having stable fibrils (*large draw ratios* λ) contributes efficiently to material toughness.

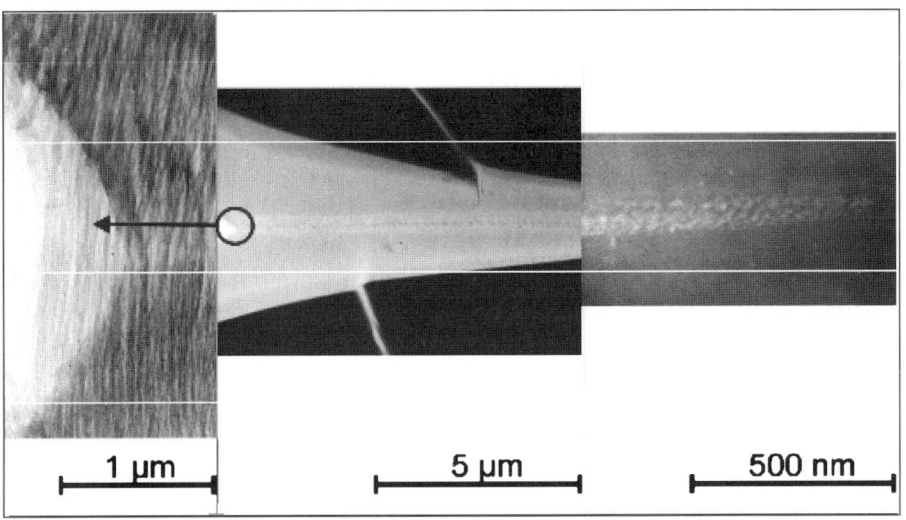

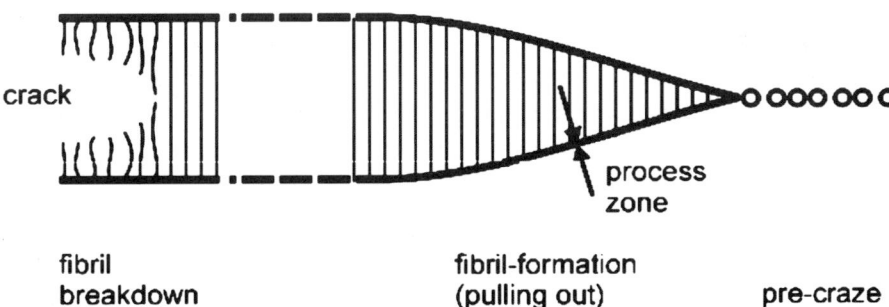

| 1 μm | 5 μm | 500 nm |

fibril breakdown **fibril-formation (pulling out)** **pre-craze**

Fig. 11 Craze in commercial polystyrene showing the characteristic steps: *nucleation* through void formation in a pre-craze zone, *growth* of the fibrillar structure of the widening craze by drawing-in of new matrix material in the process zone, and *final breakdown* of the fibrillar matter transforming a craze into a crack (the crack front is more advanced in the center of the specimen, shielded by a curtain of unbroken fibrils marked by the arrow). The fibril thickness depends—of course—on the molecular variables, the strain rate-stress-temperature regime of the crazing sample and on its treatment (preparation, annealing) and geometry (solid, thin film); for PS typical values of between 2.5 and 30 nm are found [1, 60, 61]

3.3.2
Formation of Spatially Distributed Crazes in Amorphous Polymers under Creep Loading

Section 3.3 is mainly concerned with the molecular mechanisms effective during creep and the competition between crazing and flow. In Sect. 3.2.2 we indicated that during the phase of secondary creep molecular rearrangements take place, which may cause local damage. This includes void formation and craze nucleation. Starke et al. have investigated the microstructure of crazes formed in creep using the small angle X-ray scattering technique [54]. They compared the behaviour of a low entanglement polymer (SAN) with that of bisphenol A polycarbonate known for its high entanglement density (Table 1). Injection moulded tensile specimens were exposed to uni-axial stresses of between 4.5 and 45 MPa up to a total strain of 3.5%. Both materials show the typical creep behaviour discussed in Sect. 3.2.

At a total strain of 3.5% all specimens had developed a partially fibrillar microstructure, which gave rise to characteristic equatorial small angle X-ray scattering diagrams showing a clear interference maximum at a scattering vector s_{max}. The latter is inversely related to the average fibril distance D_0 according to

$$|s_{max}| = s_{max} = 2/(D_0\sqrt{3}) \tag{9}$$

For the studied materials D_0 amounts typically to 28 to 150 nm. A plot of s_{max} as a function of creep stress σ at a given temperature reveals three *distinct regimes* each with a straight slope (Fig. 12); the slope is related to the energy Γ necessary to create the craze fibril surface [60, 61]:

$$s_{max} = \sigma/(4\Gamma\sqrt{3}) \tag{10}$$

It is helpful to recall that the fibrillation energy Γ involves two terms, the van der Waals surface tension γ of the newly created fibril surface and the—mostly dissipated— *extra surface energy* associated with the molecular rearrangements. For chain scission as the main source Kramer [61] has expressed this extra term as $dv_e U/4$, with d the straight-line distance between entanglements and U the activation energy for chain scission. For PC the extra term amounts to 0.20 Jm^{-2}, as compared to the much smaller surface tension γ of 0.045 Jm^2. It is expected, therefore, that the fibrillation energy increases the more chain loading and scission are involved in the process of fibrillation.

For *Regimes I and II* it is possible to determine the fibrillation energy Γ from Fig. 12 using Eq. 10. The corresponding plot (Fig. 13) gives the surprising result that in the small stress regime (I) Γ is constant and amounts to 0.031 Jm^{-2}, which is comparable to but smaller than the surface tension γ of

Table 1 Important parameters of materials used to study craze formation in creep (from [54])

Material	T_g [°C]	M_w [kg mol^{-1}]	M_e [kg mol^{-1}]	v_e [kmol/m^3]	ρ [10^3 kg/m^3]
SAN (76/24)	104.5	200 000	11 600	0.093	1.08
BPAPC	139	131 000	1790	0.672	1.20

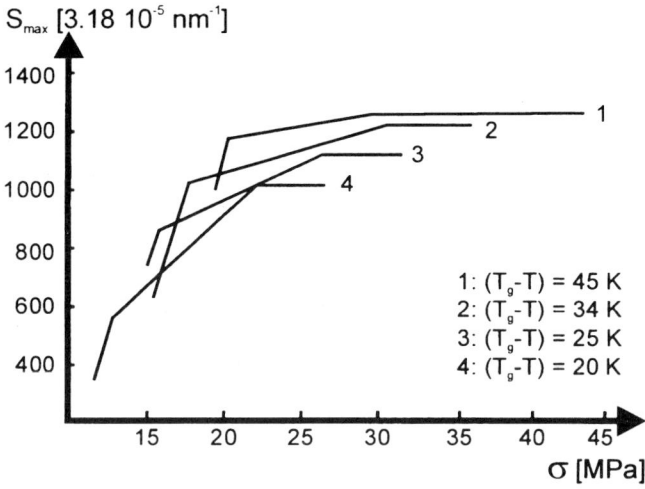

Fig. 12 Scattering vector $|s_{max}|$ of SAN specimens uni-axially deformed to a total creep strain of 3.5% as a function of creep stress σ showing three distinct regimes I, II, III (after [54])

SAN of 0.040 Jm^{-2}. This small value of $\Gamma(I)$ is probably related to the fact that separation of chains at low stress levels occurs in the most favourable sites, the polymer behaves like a visco-elastic *fluid*. As Fig. 12 shows, with increasing stress more such sites are activated and the scattering vector increases. On the other hand, annealing leads to a *de*activation of such sites and to a coarser structure of the formed fibrils [62]. It must be concluded that in this regime no chain scission or forced reptation occur.

The high-entanglement density polymer, PC, shows basically the same behaviour, namely three regions of straight slope. *Regimes I and II* are displaced, however, towards lower values of s_{max} indicating that thicker fibrils are formed at a given stress. The fibrillation energy Γ of the regime-I crazes amounts to 0.042 Jm^{-2}, which compares well with the mentioned surface energy γ of 0.045 Jm^2. These crazes with fibrillar distances D_0 of 100 to 150 nm resemble very much the *intrinsic crazes* discovered by Dettenmaier [63] under slightly different conditions of temperature and stress just 25 years ago. In his seminal investigation he found fibril *diameters* of 95 to 115 nm, which correspond to fibril distances D_0 of 109 to 135 nm.

The fibrillation energy of crazes formed in *Regime II* is strongly temperature dependent; it amounts at *low* temperatures (curve 1, $T = 60\,°C$) to 0.500 Jm^{-2}, which is close to the energy of chain scission. This observation is perfectly in line with the two-network model discussed in Sect. 3.2.2. The restricted molecular motion at low temperatures hinders local stress relaxation, the more extended chains are tightly gripped and increasingly loaded which favours craze nucleation through chain scission and explains the observed

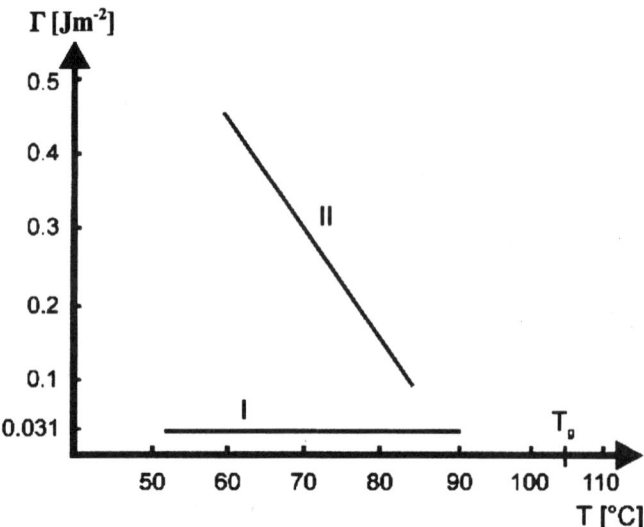

Fig. 13 Influence of temperature on the fibrillation energy Γ of SAN in regimes I and II (after [54])

elevated values of the *extra fibrillation energy*. Crazes formed in regime II show a much finer fibrillated structure, the inter-fibrillar distance D_0 decreases with decreasing temperature; in SAN it approaches a minimum value of 28 nm. The fact that beyond a certain stress D_0 remains constant leads to a horizontal curve of $s_{\max}(\sigma)$ and establishes *Regime III* (Fig. 12). The minimum value of D_0 could be due to thin film ductility, namely to the difficulty in creating voids in the absence of lateral constraints as discussed in Sect. 3.

With increasing temperature slip processes and disentanglement become more important, the fibrillation energy Γ decreases consequently and approaches the van der Waals surface energy γ at 95 °C. Regime II encompasses the whole variety of strongly time- and temperature dependent mechanisms involved in craze initiation as reviewed by Plummer [59] and discussed in Sect. 3. Several aspects of this important subject will be resumed in later presentations in this special Double Volume. Monnerie et al. [9, 10] elucidate the effect of molecular structure on mobility, disentanglement and fibril disintegration, Estevez and van der Giessen [64] model time-effects of craze behaviour through cohesive surfaces, and Altstädt [65] and Chateauminois et al. [66] investigate the influence of molecular *and* environmental parameters on fatigue resistance and friction behaviour respectively.

3.3.3
Creep Deformation of Semi-Crystalline Polymers

Despite the large influence of the presence of crystalline lamellae and tie-molecules on stress transfer—discussed in Sect. 3—there are many common features in the mechanical behaviour of semi-crystalline and amorphous polymers. This can be demonstrated by looking at a medium molecular weight polyacetal (POM). The linear visco-elastic regime in steady state tensile deformation extends up to a tensile strain of 1.4% (at room temperature and strain rates of the order of 10^{-3} s^{-1}), cavitation starts at a strain of 3.5% [67]. There is little *macroscopic* ductility in spite of the high strains to failure (up to 70%), and fracture is qualitatively brittle. In creep POM presents the characteristic three-phase behaviour described in Sect. 3.2.2. Above a temperature of 40 °C samples exposed to a constant load of 25 MPa (about 80% of the ultimate tensile strength) deform macroscopically homogeneously to at least 200% strain (Fig. 14). At this stress the creep strain is essentially borne by the hard visco-elastic bending of bundles of lamellar crystals and substantial internal cavitation; the behaviour of the material in this phase is qualitatively similar to that in steady state tensile deformation.

However, at lower constant loads the rate of crystal plastic deformation decreases and (at 80 °C) disentanglement becomes competitive leading to the development of isolated planar craze-like defects extending perpendicular to the tensile axis (Fig. 15). The ensuing concentration of stress will further localize most of the sample deformation in such creep crazes and lead to a macroscopic ductile-brittle transition—in this material observed at 20 MPa (Fig. 14 [67]).

The most severe of these defects (in terms of material structure, sample geometry and acting stress) will eventually transform into a crack and give

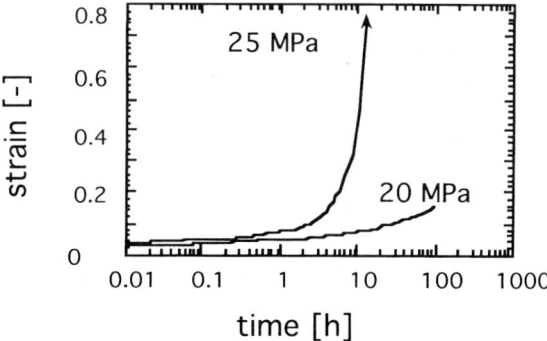

Fig. 14 Creep curves of polyoxymethylene (POM) of medium molecular weight ($M_\mathrm{w} =$ 41 kg/mol) at 80 °C (see text for discussion)

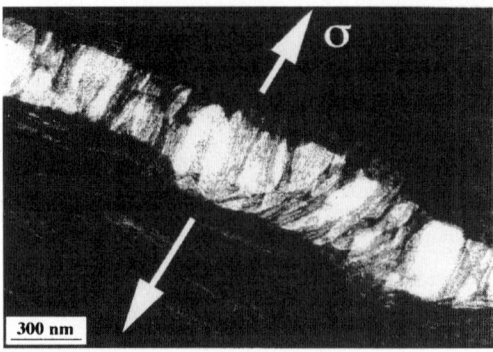

Fig. 15 Craze-like defect formed in the stress-whitened zone in front of the crack in a POM compact tension specimen loaded at 100 °C (from [65])

rise to localized brittle fracture. Just this sequence of events is most conveniently studied using a fracture mechanics specimen.

3.3.4
Slow Crack Growth in Semi-Crystalline Polymers

As indicated in the last paragraph and in Sect. 3 the homogeneous elastic and visco-elastic response of a loaded polymer sample is in competition with heterogeneous, generally localized physical mechanisms (cavitation, disentanglement) and with (chemical) aging. The kinetics of these mechanisms being quite different, the dominant mechanism changes as a function of the stress-time-temperature regime (Figs. 8 and 14). A *quantitative* analysis of the competition between different molecular mechanisms on craze formation in *amorphous* polymers has been given in [1–3] and by Monnerie et al. in this Volume [68]. A major review article by Plummer on the microdeformation and fracture of polyolefins has just appeared in this journal [45], where particular attention has been paid to the experimental methods of fracture mechanics testing and morphological characterization. Most of his observations are of obvious relevance to the topic of this article and can be conveniently discussed using Fig. 16.

Strongly based on the review article by Plummer [45] the following conclusions can be drawn from fracture mechanics studies of slow crack growth:

- Fracture mechanics specimens (CT, DENT, FNCT, DCB) are a convenient means to study creep behaviour and slow crack growth.
- The time to failure t_b of first generation HDPE as a function of K_c shows the typical dependence on stress and the competition between failure modes: *ductile* at high stress intensity K_c, (often yielding across the whole ligament, sometimes accompanied by coarse cavitation in the specimen

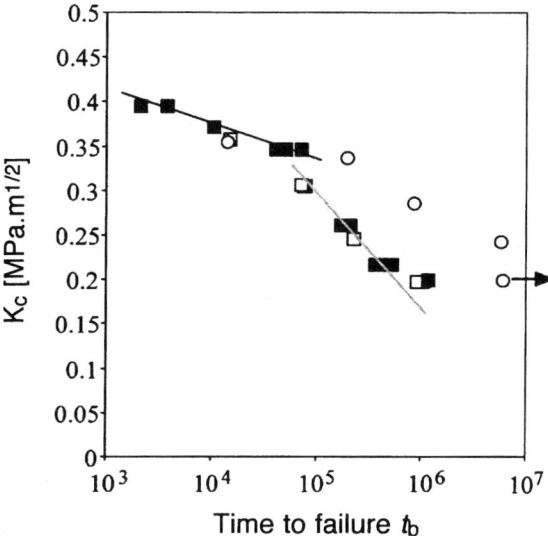

Fig. 16 The initial stress intensity factor K_c plotted as a function of the time to failure t_b for HDPE resins of the first (□■) and third generation (○); the *dashed line* has a slope of −4 (after [45])

interior where triaxial stresses are high), *slow crack growth* (SCG) at moderate K_c with a change in slope.

- In the SCG region failure occurs through formation and extension of a fibrillar crack-tip damage zone (following inter-lamellar cavitation and breakdown of inter-lamellar ligaments); as K_c decreases the fibrillar features become progressively finer.

- Slow crack growth is K-driven with $da/dt \sim (K_c)^{1/n}$, where n amounts to 0.25 for a wide range of PE grades. If t_b is taken to be roughly inversely proportional to da/dt, then the steep slope of $K_c(t_b)$ should be of the order of −4. As seen in Fig. 16, such a line represents the SCG data of the first generation HDPE well. Earlier, Chan and Williams [33] found a similar result (Eq. 6). Brown and Lu [69] related this slope to the rate of fibril extension at the base of the craze preceding the crack, which itself is determined by the *rate of disentanglement* of the chains forming a fibril. This conclusion seems to be well corroborated by the result of the inverse experiment, namely crack healing. Kausch and Jud et al. [16, 17] found through experiment that for amorphous polymers the *interfacial strength* K_c increased as a function of healing time t_h according to $(t_h)^{+1/4}$; their theoretical model explained the exponent of +1/4 from the *rate of formation of new entanglements* in the interfacial region.

- The dominant role of disentanglement in SCG is also underlined by the positive behaviour of bimodal resins, composed of a low and a high molecular weight fraction and where part of the long chains contain small-chain branches [46]. In such resins SCG is largely suppressed (Fig. 16).

- At higher temperatures and in the presence of an active environment the effect of aging and degradation can be preponderant leading to failure at unexpectedly low stresses. This phenomenon is treated in detail by Altstädt in this Volume [65].

3.4
Damage Mechanics

In discussing heterogeneous deformation mention must be made of mathematical models of visco-elastic solids with growing, spatially well distributed damage [71–73]. Such a *damage mechanics approach* assumes that the microscopic damage can be homogenized on a croscopic level and that a pseudo-undisturbed visco-elastic contribution to the compliance can be separated from that of sample weakening. The latter is expressed by a damage parameter S. Rate effects in *traction* of asphalt concrete [72], glass-fibre reinforced PA 66 [73] and semi-crystalline polyolefins [74] have been successfully analysed this way – but an experimental verification in *creep* seems to be lacking so far.

4
Conclusion

In this review the dual role of molecular mobility has been investigated. On the one hand important properties of thermoplastic polymers (low processing temperatures, ductility, increased toughness) depend on it. On the other hand without exception all damage mechanisms, which limit strength and durability, are related to molecular motions. A micro-mechanical model of an entangled physical network has been presented and used to point out the intimate interpenetration of chain molecules in the solid state, to identify the roles of the four major disintegrating mechanisms chain scission, segmental slip, disentanglement and flow in mechanical loading, and to discuss the effect of time and the importance of hydrostatic tensile stresses. The competition of the above mechanisms determines the mode of failure of amorphous and semi-crystalline thermoplastics. Leaving the failure modes at very low (brittle fracture) and high temperatures (flow, chemical degradation) out of consideration it was shown how these four mechanisms intervene in the more *moderate failure modes* of macroscopically homogeneous creep, craze formation and slow crack growth. From an analysis of the rarely investigated craze formation under creep conditions several conclusions could be derived: in SAN at small stresses the fibril formation energy Γ has a temperature independent value of $0.031\,\mathrm{Jm}^{-2}$ (called *Regime I*), which is explained by fluid-like behaviour of the stressed network. *In Regime II* the fibrillation energy is strongly temperature dependent, it decreases from about $0.50\,\mathrm{Jm}^{-2}$ at low temperatures, where chain scission dominates, to the level of the van der Waals surface tension $\gamma = 0.040\,\mathrm{Jm}^{-2}$ at 10 K below the glass transition temperature of 104.5 °C. However, in a large part of the stress-time-temperature regime craze extension as well as slow crack growth are controlled by disentanglement through thermally activated localized molecular displacements (*forced reptation*). This process is apparently *K-driven*, i.e. it depends on the level of stress intensity K at the active site. This contention is supported by a comparison of the kinetics of formation and dissolution of entanglements observed in three independent experiments, which all show the same $t^{1/4}$-dependence on time: the increase of interfacial strength through the formation of new entanglements according to the relation $K_c \sim (t_h)^{+1/4}$ as postulated by Kausch et al. [16, 17], the increase of the rate of crack growth $K_c \sim (\mathrm{d}a/\mathrm{d}t)^{+1/4}$ as found by Chan et al. [13], and the dependence of the time to fracture t_b on $K_c \sim (t_b)^{-4}$ as given by Plummer [45] and shown in Fig. 16. This interpretation corresponds well to the Kausch model of craze initiation [70].

The treated phenomena and investigations summarized above are at the centre of the topic of this special two-volume edition of the Advances in Polymer Science on "Intrinsic molecular mobility and toughness of polymers". However, little has been said about the mutual relations between molecular configuration, polymer morphology, nature, intensity and cooperativity of sub-T_g relaxation mechanisms, and mode of solicitation on the one hand and their effect on disentanglement, craze formation, yield behaviour and ultimate strength on the other. These subjects will be critically evaluated in the subsequent presentations of this Volume.

Acknowledgements The authors gratefully acknowledge fruitful discussions with M. Fischer, St. Antoni, B. Möginger, Rheinbach, A. Pavan, Milano, C.J.G. Plummer, Lausanne, and A. Rodrigez, Stuttgart.

References

1. Michler GH (1992) Kunststoff-Mikromechanik. Carl Hanser, Munich
2. Materials Science and Technology: A Comprehensive Treatment (1993) Cahn RW, Haasen P, Kramer EJ (eds) Structure and Properties of Polymers, Thomas EL, vol-Ed, vol 12. Wiley, New York
3. Kausch H-H, Heymans N, Plummer CJ, Decroly P (2001) Matériaux Polymères: Propriétés Mécaniques et Physiques, Principes de Mise en Oeuvre. Presses Polytechniques et Universitaires Romandes, Lausanne
4. Lousteaux B (2002) PhD Thesis, University Pierre and Marie Curie, Paris
5. Flory P (1953) Principles of Polymer Chemistry. Cornell University Press, Ithaca
6. Dettenmaier M (1986) personal communication based on the experiments published in Macromolecules 19:773
7. Suter U, p 63 in [3]

8. Gentile FT, Suter UW Amorphous polymer microstructure, Chap 2 in [2]
9. Monnerie L, Lauprêtre F, Halary JL (2005) Investigation of Solid-State Transitions in Linear and Crosslinked Amorphous Polymers. Adv Polym Sci 187:35
10. Monnerie L, Halary JL, Kausch H-H (2005) Deformation, Yield and Fracture of Amorphous Polymers: Relation to the Secondary Transitions. Adv Polym Sci 187:215
11. Fischer EW, Dettenmaier M (1978) J Noncryst Solids 31:11
12. Michler GH (1992) p 216 in [1]
13. Chan C-M, Li L (2005) in volume 188
14. Grein C (2005) in volume 188
15. Brown HR (1991) Macromolecules 24:2752
16. Kausch H-H (1987) Polymer Fracture, 2nd ed. Springer, Heidelberg Berlin New York
17. Jud K, Kausch H-H, Williams JG (1981) J Materials Sci 16:204
18. Gensler R (1998) Thesis No 1863, Ecole Polytechnique Fédérale de Lausanne, Lausanne, Switzerland
19. Kausch H-H, Gensler R, Grein C, Plummer CJG, Scaramuzzino P (1999) J Macromol Sci B38:803
20. Kausch H-H, Langbein D (1973) Journal of Polymer Science: Polymer Physics Edition 11:1201
21. Michler GH, Godehardt R (2000) Cryst Res Technol 35:863
22. Kawai H, Hashimoto T, Miyoshi K, Uno H, Fujimura MJ (1980) Macromol Sci Polym Phys 17:427
23. Adhikar R, Michler GH, Goerlitz S, Knoll K (2004) J Appl Polym Sci 92:1208
24. Galeski A (2004) 11. Int Conf Polymeric Materials 2004, Halle, p C06
25. Williams JG (1984) Fracture mechanics of Polymers. Ellis Horwood, Chichester
26. Grellmann W, Seidler S (2001) Deformation and fracture behaviour of polymers. Springer, Berlin Heidelberg New York
27. Williams JG, Moore DR, Pavan A (2001) Fracture Mechanics Testing Methods for Polymers, Adhesives and Composites. Elsevier, Amsterdam
28. Knauss WG (1970) Int J Fracture 6:7
29. Bradley W, Cantwell WJ, Kausch H-H (1998) Mech Time-Dependent Materials 1:241
30. Stalder B cited after [16], p 237
31. Grellmann W, Seidler S, Hesse W (2001) p 71 in [26]
32. Greig JM, Leevers PS, Yayla P (1992) Engineering Fracture Mechanics 42:663
33. Chan MKV, Williams JG (1983) Polymer 24:234
34. Takahashi K, Arakawa K (1987) Experimental Mechanics 27:195
35. Ferrer JB, Fond C, Arakawa K, Takahashi K, Béguelin P, Kausch H-H (1997) Int J Fracture 87:L77
36. Harward RN (1973) The Physics of Glassy Polymers. Appl Sci Publ, London
37. Ferry DJ (1960) Viscoelastic properties of polymers. Wiley, New York
38. Crist B (1993) Plastic deformation of polymers, Chap 10 in [2]
39. Ward IM, Hadley DW (1993) Mechanical properties of solid polymers. Wiley, Chichester
40. Myasnikova LP, Marikhin VA, Ivan'kova EM, Yakushev PN (1999) J Macrom Science-Physics 38B:859
41. Eyring H (1936) J Chem Phys 4:283
42. Castiglione C, Verzanini D, Pavan A (2004) Plastic Pipes XII, Session 10b
43. Berger L, Kausch H-H (2003) Polymer 44:5877
44. Wilding MA, Ward IM (1978) Polymer 19:969
45. Plummer CJG (2004) Adv in Polymer Si 169:75
46. Damen J, Schramm D, Anderson K (2004) Plastic Pipes XII, Session 10b

47. da Andrade ENC (1910) Proc Royal Soc London 84:1
48. Findley WN (1987) Polym Eng Sci 27:582
49. Schapery RA (1997) Mechanics of Time-Dependent Materials 1:209
50. Gregory A, Vogel D, Béguelin PH, Gensler R, Kausch H-H (1999) Mechanics of Time-Dependent Materials 3:71
51. Pilz G, Lang RW (2001) Proceedings of Plastics Pipes XI, p 903
52. Niklas H, Kausch von Schmeling HH (1963) Kunststoffe 53:886
53. Janson LE, Bergstrom G, Backman M, Blomster T, Plastic Pipes XII, Session 10b
54. Starke JU, Schulze G, Michler GH (1997) Acta Polymer 48:92
55. Kambour RP (1968) Appl Polymer Symposia 7:215
56. Rabinowitz S, Beardmore P (1969) CRC Critical Reviews 1:1
57. Kausch H-H (ed) (1983) Crazing in Polymers Advances in Polymer Science, 52/53. Springer, Berlin Heidelberg New York
58. Kausch H-H (ed) (1990) Crazing in Polymers, vol II. Advances in Polymer Science, 91/92. Springer, Berlin Heidelberg New York
59. Plummer CJG (1997) Current Trends in Polymer Sci 2:125
60. Paredes E, Fischer EW (1979) Makromol Chem 180:2707
61. Kramer EJ (1983) p 1 in [57]
62. Michler GH, Section B 1.4, p 193 in [26]
63. Dettenmaier M (1983) p 57 in [57]
64. Estevez R, van der Giessen E (2005) Adv Polym Sci 188:195
65. Altstädt V (2005) Adv Polym Sci 188:105
66. Chateauminois A, Baietto-Dubourg MC (2005) Adv Polym Sci 188:153
67. Scaramuzzino P (1998) Thesis No 1818, Ecole Polytechnique Fédérale de Lausanne, Lausanne, Switzerland
68. Monnerie L, Halary JL, Kausch H-H (2005) Deformation, Yield and Fracture of Amorphous Polymers: Relation to the Secondary Transitions, Sect 2.3 of [10]. Adv Polym Sci 187:215
69. Brown N, Lu X (1995) Polymer 36:543
70. Kausch H-H (1987) p 347 in [16]
71. Schapery RA (1990) J Mech Phys Solids 38:215
72. Park SW, Kim YR, Schapery RA (1996) Mech Mater 24:241
73. Rodrigez A, Eyerer P (2005) 10. Tagung Deformation und Bruchverhalten von Kunststoffen, Merseburg/Germany, June 15–17
74. Strobl G (2005) Symposium "Deformation Mechanisms in Polymers", Halle/Germany, May 19–20

Adv Polym Sci (2005) 187: 35–213
DOI 10.1007/b136955
© Springer-Verlag Berlin Heidelberg 2005
Published online: 13 October 2005

Investigation of Solid-State Transitions in Linear and Crosslinked Amorphous Polymers

Lucien Monnerie[1] (✉) · Françoise Lauprêtre[2] · Jean Louis Halary[1]

[1]Laboratoire PCSM, Ecole Supérieure de Physique et de Chimie Industrielles de la Ville de Paris, UMR 7615, 10 rue Vauquelin, 75231 Paris cedex 05, France
l.monnerie@noos.fr, jean-louis.halary@espci.fr

[2]Laboratoire de Recherche sur les Polymères, UMR 7581, 2 à 8 rue Henri Dunant, 94320 Thiais, France
francoise.laupretre@glvt-cnrs.fr

Abstract This paper deals with the determination of the main characteristics and the assignment of the molecular motions leading to the solid-state transitions observed in amorphous polymers at temperatures lower than the glass–rubber transition. First, the specific features of these secondary transitions (β, γ, δ, ...) are briefly described. Then, the behaviour of various polymer systems is analysed: poly(cycloalkyl methacrylates) and their intracycle motions, poly(ethylene *tere*-phthalate) and bisphenol A polycarbonate β transition and the effect of small molecule antiplasticisers, aryl-alkyl polyamide transitions, the β transition of poly(methyl methacrylate) and its maleimide as well as glutarimide random copolymers. Additionally linear polymers, aryl-aliphatic epoxy systems, with or without antiplasticisers, illustrate the case of crosslinked amorphous polymers. Whereas all the systems considered were investigated by dynamic mechanical measurements and ^{13}C solid-state nuclear magnetic resonance experiments, some systems were also studied by other techniques such as dielectric relaxation and molecular modelling. Poly(methyl methacrylate) and bisphenol A polycarbonate constitute the best examples of the level of description of the molecular motions involved in the β transition and, in particular, the nature and extent of the cooperativity which develops in the high-temperature part of the transition. This is achieved by combining all the experimental and modelling techniques presently available.

Keywords Secondary transitions · Poly(cycloalkyl methacrylate) ·
Poly(ethylene *tere*-phthalate) · Bisphenol A polycarbonate · Aryl-aliphatic polyamide ·
Aryl-aliphatic epoxy resins · Poly(methyl methacrylate) ·
Methyl methacrylate-*co*-N-cyclohexyl maleimide copolymer ·
Methyl methacrylate-*co*-N-methyl glutarimide copolymer

Abbreviations

°C	Degrees Celsius
C_g^1, C_g^2	Coefficients of Williams, Landel, Ferry equation associated with the glass–rubber transition temperature
E_a	Activation energy
E''	Mechanical loss modulus
E'	Mechanical storage modulus
G''	Mechanical loss shear modulus
Hz	Hertz
J	Joule
J''	Mechanical loss compliance
K	Degrees Kelvin
R	Perfect gas constant

T	Absolute temperature
T_1	Spin-lattice relaxation time
$T_{1\rho}$	Spin-lattice relaxation time in the rotating frame
T_2	Transverse relaxation time
T_g	Glass–rubber transition temperature
f	Frequency
h	Plank constant
k	Boltzman constant
m''	Dielectric loss modulus
mol	Mole
$t_{1/2}$	Cross-polarisation time
t_m	NMR mixing time
$\tan \delta$	Ratio of loss to storage modulus
ΔG^*	Transition Gibbs energy
ΔH_a	Activation enthalpy
$\langle \Delta H^2 \rangle$	Second moment of NMR line width
ΔS_a	Activation entropy
ε'	Dielectric constant
ε''	Dielectric loss
ϕ	Torsion angle about a chemical bond
μ	Dipole moment
σ_{ij}	Chemical shift anisotropy tensor component
τ	Correlation time
ω	NMR resonance frequency

Acronyms

2D	Two dimensional
AP	Antiplasticiser additive
Ar-Al-PA	Aryl-aliphatic copolyamide
BPA-PC	Bisphenol A polycarbonate
CMI	Cyclohexyl maleimide unit
$CMIM_x$	Methyl methacrylate-*co*-*N*-cyclohexyl maleimide copolymer containing x molar percent of the latter comonomer
CP	Cross polarisation
DD	Dipolar decoupling
DGEBA	Diglycidyl ether of bisphenol A
DGERO	Diglycidyl ether of resorcinol
DMHMDA	Dimethyl hexamethylene diamine
DMT	Dimethyl *tere*-phthalate
DPC	Diphenyl carbonate
DPP	2,2′Diphenyl propane
DSC	Differential scanning calorimetry
EPPHAA	Hydroxypropyl ether of epoxyphenoxypropane and hydroxyacetanilide
HA	Hexylamine
HMDA	hexamethylene diamine
HPE	hydroxypropyl ether
MAS	Magic angle spinning
MD	Molecular dynamics
MGI	*N*-Methyl glutarimide
$MGIM_x$	Methyl methacrylate-*co*-*N*-methyl glutarimide copolymer containing x molar percent of the latter comonomer

MT_yI_{1-y} Copolyamide whose repeat unit contains 1,5-diamino-2-methyl pentane, y and $1 - y$ *tere*- and *iso*-phthalic moities

NMR   Nuclear magnetic resonance

PC   Polycarbonate

PET   Poly(ethylene *tere*-phthalate)

PMMA   Poly(methyl methacrylate)

REDOR   Rotational-echo double resonance NMR

TMBPA-PC   Tetramethyl bisphenol A polycarbonate

TPDE   Tetrachlorophthalic dimethyl ester

WLF   Willams, Landel, Ferry

xT_yI_{1-y} Copolyamide whose repeat unit contains x lactam-12 sequences, y and $1 - y$ *tere*- and *iso*-phthalic moieties, respectively, and 3,3′-dimethylcyclohexyl methane unit in the regular order

1
Introduction

The occurrence of molecular motions in solid polymer systems is accompanied by a decrease of the mechanical modulus whose amplitude depends on the size of the moving molecular units. Furthermore, in the temperature and frequency ranges where these motions happen, they lead to a specific energy dissipation and give rise to "secondary transitions". The involved motions are more localised than the ones which develop at higher temperatures in the glass–rubber transition. However, in addition to the energy dissipation they generate, they have an influence on mechanical behaviours such as plastic deformation and brittleness, as illustrated in another paper in this volume [1].

It is worth noticing that some small molecule additives, called "antiplasticisers", are able to significantly decrease the amplitude of the β secondary transition (the first transition appearing at a lower temperature than the glass–rubber transition) and, consequently, to affect the material properties.

Any motion occurring within any polymer system leads to a change of the dynamic mechanical behaviour, in particular its mechanical loss. This makes the dynamic mechanical measurements the most appropriate technique for studying solid-state transitions. However, in order to assign the molecular motions involved in the considered transition from only the dynamic mechanical results, it is necessary to perform systematic studies on a large series of compounds with gradual modification of their chemical structure. Such an approach has been used in some cases, but it requires lots of effort in synthesising the various compounds.

For polymers containing polar groups, dielectric relaxation is a convenient technique for checking how much these groups take part in the investigated transition. Such a technique has been used, indeed, for quite a long time. Nevertheless, the information obtained exclusively concerns the polar groups.

Among the spectroscopic techniques, solid-state NMR applied to ^{13}C, either in natural abundance or on labelled polymers, and to ^{2}H-labelled polymers, constitutes a selective investigative tool. Indeed, it applies to both polar and non-polar polymers, and the specific response of most of the carbon groups allows one to examine which groups are involved and what kind of motions these groups undergo. The very large set of NMR parameters that are sensitive to motions offers a large range of investigation.

In spite of the satisfactory assignment of the groups involved and the types of motions that they perform, some questions still remain concerning the cooperativity of some of these motions or the intra- or intermolecular origin of the activation energy associated with these processes. The only approach for getting such information involves molecular modelling, not only of the glassy state structure of the considered polymer, but also of the forced motion of specific groups within the polymer chains.

In this paper, after describing the main characteristics of solid-state transitions in Sect. 2, a series of polymers with different chemical structures is considered in Sects. 3 to 8. The first example, poly(cycloalkyl methacrylates), Sect. 3, deals with the motions occurring within the alkyl cycles. In Sect. 4, the β transition of poly(ethylene tere-phthalate) is analysed by combining mechanical, dielectric and NMR approaches. Furthermore, the effect of small molecule additives (antiplasticisers), which have an influence on this transition, provides an additional way for going further in the understanding of the nature of the cooperativity. In Sect. 5, bisphenol A (and/or tetramethyl bisphenol A) polycarbonate is analysed using the whole set of techniques as well as atomistic modelling. The effect of small molecule antiplasticisers is also considered. More complex chemical structures, undergoing several solid-state transitions, are addressed in Sect. 6 with the aryl-aliphatic polyamides. All the systems considered in Sects. 4, 5, 6 correspond to linear polymer chains with aromatic or aliphatic rings in their chain backbone and without side chains. Still keeping this kind of chemical structure, 7 concerns the effect of crosslinking on the β transition through the case of aryl-aliphatic epoxy networks. The various experimental techniques, as well as the effect of antiplasticisers, yield a detailed description of the motions of the different chemical groups and an understanding of the cooperativity that develops when increasing the crosslink density. The last section, 8, is dedicated to linear polymers in which the β transition originates from motions of a side group: poly(methyl methacrylate) and its random copolymers methyl methacrylate-co-N-cyclohexyl maleimide on one side and methyl methacrylate-co-N-methyl glutarimide on the other side. In addition to the experimental techniques, atomistic modelling has been performed on all three systems. It is worth mentioning that, in the cases of bisphenol A polycarbonate and poly(methyl methacrylate), the level of description of the motions and cooperativity which occurs in the high temperature part of the transition, represents the highest achievement reached up to now.

2
Main Characteristics of Solid-State Transitions

When considering amorphous polymers, an essential characteristic concerns the glass–rubber transition temperature. It is usual to denote as T_g the temperature of occurrence determined by techniques which operate on samples at rest (thermal measurements, spectroscopic techniques, dielectric relaxation). When dealing with techniques involving a mechanical solicitation, like dynamic mechanical measurements, the temperature corresponding to the glass–rubber transition is denoted as T_α. In the case of the dynamic mechanical measurements, T_α is chosen as the temperature at which the mechanical loss, E'' or G'', goes through a maximum. Thus, T_α measured at 1 Hz is shifted upwards by 10 °C relative to T_g measured by differential scanning calorimetry at a heating rate of 20 °C min^{-1}.

It is well known that the glass–rubber transition is characterised by the gradual development of cooperative segmental motions, when approaching the glass–rubber transition from higher temperatures. As a consequence, the temperature dependence of the frequency of the segmental motions, at temperature T and frequency f_T, is described by the Willams, Landel, Ferry (WLF) expression:

$$\log(f_T/f_{T_g}) = [C_g^1(T - T_g)]/[C_g^2 + (T - T_g)], \qquad (1)$$

where the index g refers to the glass–rubber transition temperature T_g, C_g^1 is related to the free volume fraction at the glass–rubber transition temperature, and C_g^2 is related to the thermal expansion of the free volume.

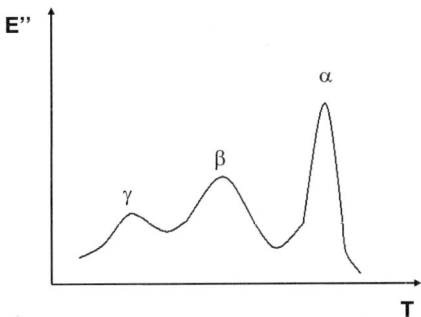

Fig. 1 Typical example of the temperature dependence of the mechanical loss modulus, E'', showing α, β and γ transitions

For amorphous polymers, the glass–rubber transition is usually referred as the *a* transition, the solid state transitions (frequently called sub-T_g or secondary transitions) being designated by β, γ, δ, ... A typical example is shown in Fig. 1 with the temperature dependence of the mechanical loss modulus, E'', which exhibits several peaks corresponding to the various transitions.

2.1
Activation Energy

In the case of any transition occurring at a temperature lower than the glass–rubber transition temperature, T_g, the frequency of the involved process has a temperature dependence which obeys an Arrhenius law:

$$f_T = f_0 \exp(- E_\alpha/RT),\tag{2}$$

where E_α is the activation energy of the process under consideration.

2.2
Relaxation Map

In order to represent the temperature dependence of the process frequencies corresponding to the various transitions, it is usual to consider a relaxation map, as it is called, in which the logarithm of the frequency is plotted as a function of $1/T$, where T is the absolute temperature. A typical example is shown in Fig. 2 for a polymer exhibiting two solid state transitions (β and γ), in addition to the α transition. It is worth pointing out that the lower the transition temperature, the smaller the activation energy.

The relaxation map in Fig. 2 clearly shows that for a measurement performed at a low frequency, f_1, the various transitions appear at temperatures well separated from each other. At a higher frequency, f_2, the transitions appear, firstly, at higher temperatures and, secondly, closer to each other, in such a way that at high frequency an overlapping (merging) of the transitions occurs.

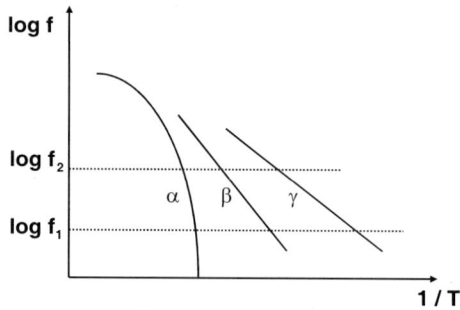

Fig. 2 Example of a relaxation map showing α, β and γ transitions

It is worth pointing out that the relaxation map is quite useful for comparing results obtained by techniques operating at various frequencies, like dynamic mechanical measurements, dielectric relaxation, NMR, etc.

2.3
Activation Entropy

In order to go further in the characterisation of the processes involved in the secondary transitions, Starkweather [2–4] applied the transition state theory. It leads to expressing the motional frequency as:

$$f_T = (kT/2\pi h) \exp(- \Delta G^*/RT), \tag{3}$$

where h is the Planck constant and ΔG^* the transition Gibbs energy:

$$\Delta G^* = \Delta H_a - T \Delta S_a, \tag{4}$$

ΔH_a and ΔS_a are the activation enthalpy and entropy, respectively. By plotting $\ln(f/T)$ versus $1/T$, one gets directly ΔH_a and ΔS_a. Interestingly, Starkweather differentiates between:

- Simple transitions, characterised by low values of the activation entropy $(0–30 \, \mathrm{J \, K^{-1} \, mol^{-1}})$ and associated with motions of small chemical sequences
- Complex transitions, characterised by high values of the activation entropy $(80 \, \mathrm{J \, K^{-1} \, mol^{-1}}$ or more) and corresponding to cooperative motions of neighbouring groups

2.4
Simple or Multiple Motional Processes

When dealing with a secondary transition, for example β, it is interesting to check whether it involves a single motional process or several different motional processes, gradually developing at increasingly higher temperatures through the β transition temperature range. There are two possible approaches for such a check.

The first approach consists in determining the activation energy, E_a and entropy, ΔS_a, for different parts of the transition, typically the low-temperature part, the central part or the maximum of the peak, and the high-temperature part. When E_a and ΔS_a values do not significantly change with the considered part of the transition, a single process is involved. In contrast, when higher and higher values of E_a and ΔS_a are obtained when going from the low- to the high-temperature parts, it means that several motional processes are involved, the extent of their cooperativity depending on the ΔS_a values obtained.

The second approach deals with the set of curves obtained by plotting, as a function of $1/T$, the measured quantity (for example, the mechanical loss modulus, E'') determined at different frequencies. If, after choosing a ref-

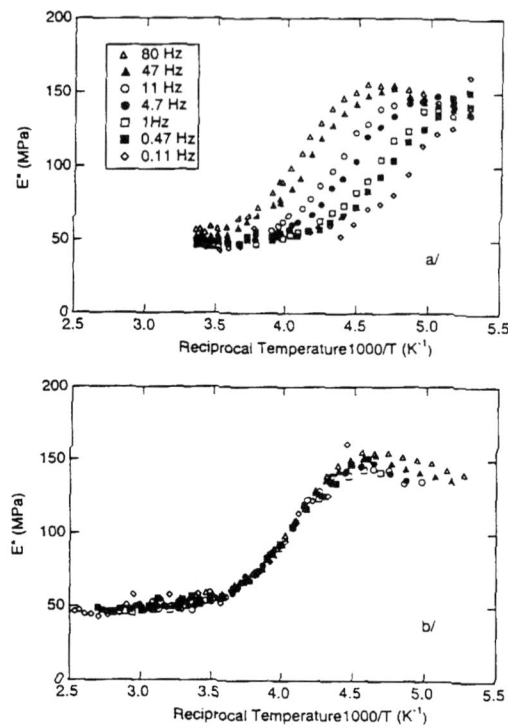

Fig. 3 Typical example where, over the temperature range of a transition (β in this case), a "master curve" can be constructed

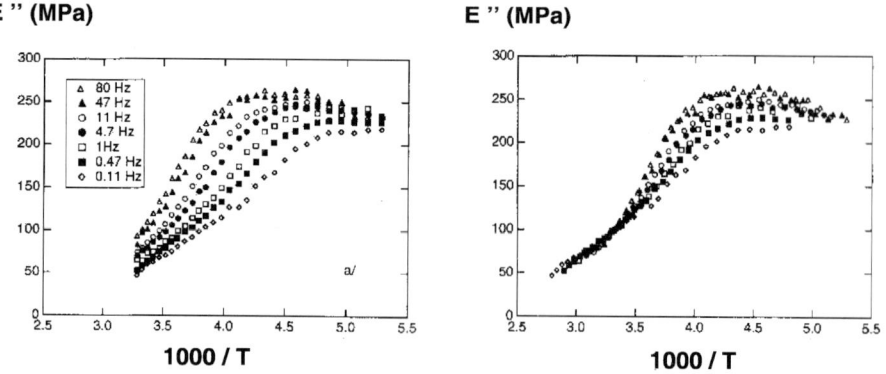

Fig. 4 Typical examples where, over part of the temperature range of the transition (β in this case), a "master curve" cannot be obtained

erence frequency, all the other curves can be translated along the $1/T$ axis to yield a single curve, the "master curve", running over the temperature range corresponding to the considered transition (Fig. 3), a single motional process is involved in this transition. In contrast if, in a particular region of the explored temperature range, a "master curve" cannot be constructed (Fig. 4), it means that other motional processes are involved in this particular region of the considered transition. As regards the distribution of processes, it can be either narrow or broad, depending on the system and transition.

2.5
Conclusion on the General Features of Solid-State Transitions

It is clear from the described characteristics that the secondary transitions have specific features, different from those of the glass–rubber transition. Furthermore, to determine these characteristics, as well as the existence of single or multiple motional processes, measurements have to be performed at various temperatures and frequencies.

The importance of these various characteristics is illustrated in the following sections, where different types of polymer systems are considered and analysed.

3
Poly(cycloalkyl methacrylates)

A series of poly(cycloalkyl methacrylates) has been investigated by dynamic mechanical analysis and solid-state ^{13}C NMR in order to characterise the

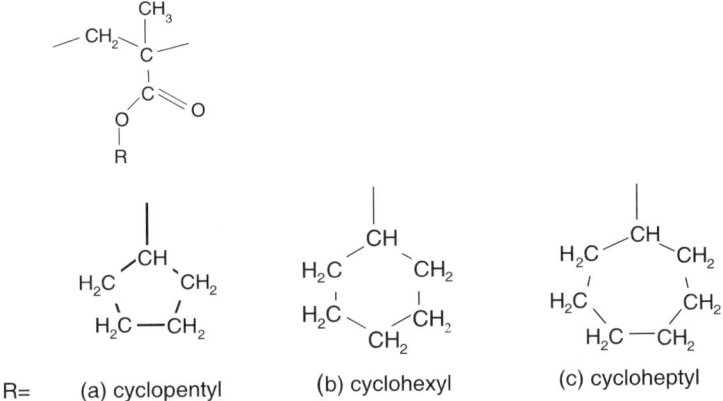

Fig. 5 Chemical structures of the poly(cycloalkyl methacrylates)

motions undergone by the cycloalkyl groups and involved in a solid-state transition. This corresponds to a quite simple case and constitutes the first example in which solid-state ^{13}C NMR has been used to precisely identify the motions involved in a solid-state transition of polymers. The chemical structures of the considered polymers are shown in Fig. 5.

3.1
Dynamic Mechanical Analysis

Under viscoelastic measurements poly(cycloalkyl methacrylates) show a loss maximum (designated γ), located in the very low temperature range ($T < -60$ °C), as illustrated in Fig. 6 in the case of poly(cyclohexyl methacrylate). Such a series of polymers has been extensively studied by Heijboer in his Ph.D. thesis [5], by performing viscoelastic studies at 1 Hz (sometimes 180 kHz) as a function of temperature and exploring quite a large number of cycloalkyls, either substituted or not. In cyclopentyl, cyclohexyl, cycloheptyl derivatives, the γ transition was shown to occur at ca. -185 °C (180 Hz), -80 °C (1 Hz), -180 °C (1 Hz), respectively. The associated activation energies, E_a, are 13, 47, 26 kJ mol^{-1} for the cyclopentyl, cyclohexyl, cycloheptyl derivatives, respectively.

These γ transitions were assigned to motions within the alkyl cycles. In the specific case of poly(cyclohexyl methacrylate), in order to identify the pre-

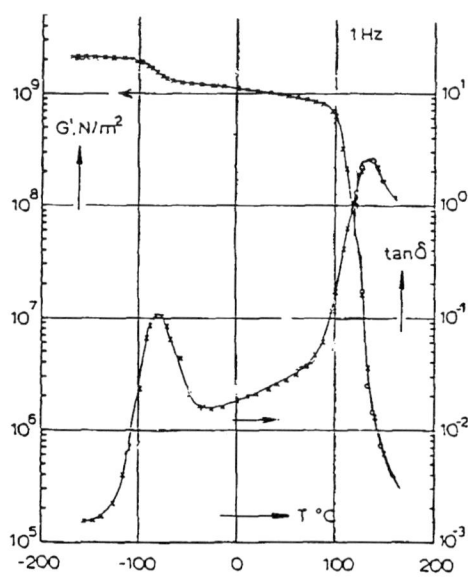

Fig. 6 Temperature dependence of the mechanical modulus, G', and loss tangent, tan δ, at 1 Hz, for poly(cyclohexyl methacrylate) (from [5])

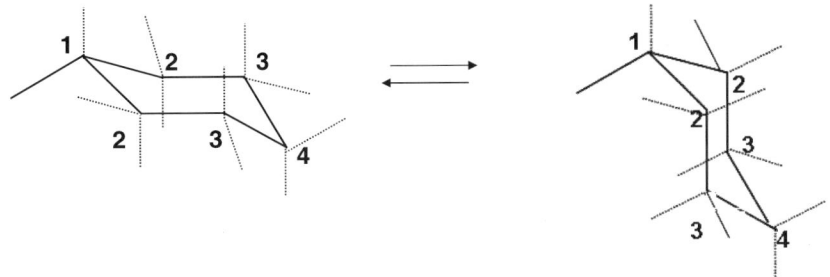

Fig. 7 Chair–chair inversion of a cyclohexyl ring with carbon 1 in a fixed position (from [7])

cise motion involved, Heijboer studied a large series of polymers modified by various substitutions of the cyclohexyl ring, which was quite a painful task. He concluded that this transition originates from a chair–chair inversion of the cyclohexyl side-chain (Fig. 7) between unequivalent conformations. Conformational energy calculations [5] performed on an isolated repeat unit of poly(cyclohexyl methacrylate) indicate a $2.9 \, \text{kJ} \, \text{mol}^{-1}$ energy difference between the two conformers.

Later on, when high-resolution solid-state ^{13}C NMR became available, these questions concerning motions within the rings of poly(cycloalkyl methacrylates) and the assignment of the specific motion occurring in the case of poly(cyclohexyl methacrylate) were revisited [6].

3.2
Solid-State NMR

12-MHz MAS CP DD (magic-angle spinning with cross-polarisation and ^{1}H dipolar decoupling) ^{13}C NMR spectra performed at room temperature for the three poly(cycloalkyl methacrylates) (cyclopentyl, cyclohexyl, cycloheptyl) are shown in Fig. 8.

The spectrum of poly(cyclopentyl methacrylate) (Fig. 8a) consists of seven lines which have been identified from left to right, in order of increasing magnetic field, as:

- A line due to the carboxyl carbon
- A line due to the CH group of the side ring
- A broad line arising from the main-chain CH_2 carbons
- A line due to the main-chain quaternary carbon
- Two lines arising from the cyclopentyl CH_2 carbons
- A shoulder due to the methyl group

The spectra of poly(cyclohexyl methacrylate) and poly(cycloheptyl methacrylate) (Fig. 8b and c) are identical with those of poly(cyclopentyl methacrylate)

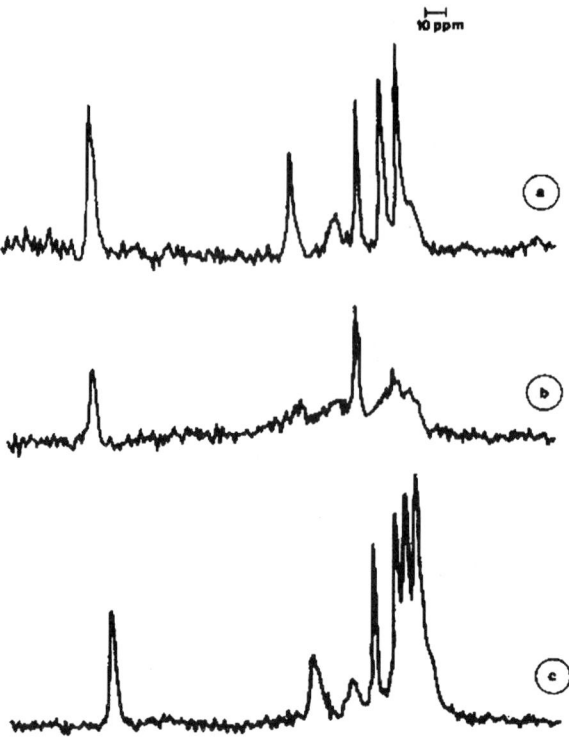

Fig. 8 12-MHz MAS CP DD ^{13}C NMR spectra of **a** solid poly(cyclopentyl methacrylate), **b** poly(cyclohexyl methacrylate) and **c** poly(cycloheptyl methacrylate) (from [6])

except for the side-ring CH and CH$_2$ carbon lines. In the case of the cyclo-heptyl derivative (Fig. 8c), four sharp peaks are observed. In contrast, for poly(cyclohexyl methacrylate) (Fig. 8b), the resolution has disappeared in the side-ring CH and CH$_2$ regions. This broadening of aliphatic protonated car-bon lines is a clear indication of modulation of the carbon–proton dipolar coupling by slow motions of the side ring, with correlation times in the neigh-bourhood of 10^{-6} s.

The data obtained from dynamic mechanical measurements lead to the following values of the correlation times at 25 °C: $\tau = 3 \times 10^{-12}$ s for poly(cyclopentyl methacrylate), $\tau = 3 \times 10^{-6}$ s for poly(cyclohexyl methacry-late), $\tau = 5 \times 10^{-12}$ s for poly(cycloheptyl methacrylate).

Thus, at room temperature, the molecular motions responsible for the mechanical losses of poly(cyclopentyl methacrylate) and poly(cycloheptyl methacrylate) are very rapid processes, having too-high a frequency to broaden the line widths of MAS CP DD ^{13}C NMR spectra, in agreement with the absence of observable motional broadening in the spectra of these two polymers (Fig. 8a and c).

In contrast, in the case of poly(cyclohexyl methacrylate), the calculated correlation time is much longer. Its value agrees nicely with a strong line broadening of the ring carbon NMR peaks.

Complementary NMR measurements, such as rises of carbon polarisation in a spin-lock experiment and determination of ^{13}C spin-lattice relaxation times in the rotating frame, $T_{1\rho}(^{13}C)$, support these conclusions about the correlation times of the side-ring CH and CH_2 motions in the various poly(cycloalkyl methacrylates).

In order to identify more precisely the type of motion occurring in the cyclohexyl side group, a variable-temperature MAS CP DD ^{13}C NMR study has been performed on solid poly(cyclohexyl methacrylate) [7]. The spectra obtained at various temperatures are shown in Fig. 9.

At a low temperature (– 50 °C) the three cyclohexyl CH_2 peaks are narrow and well resolved. They are in sharp contrast with the room temperature spectrum, which is so broadened that the resolution has completely disappeared. Between these two extremes, intermediate features are observed at the other temperatures.

The transverse relaxation times T_{2m} were determined for the ring CH and CH_2 carbons at various temperatures.

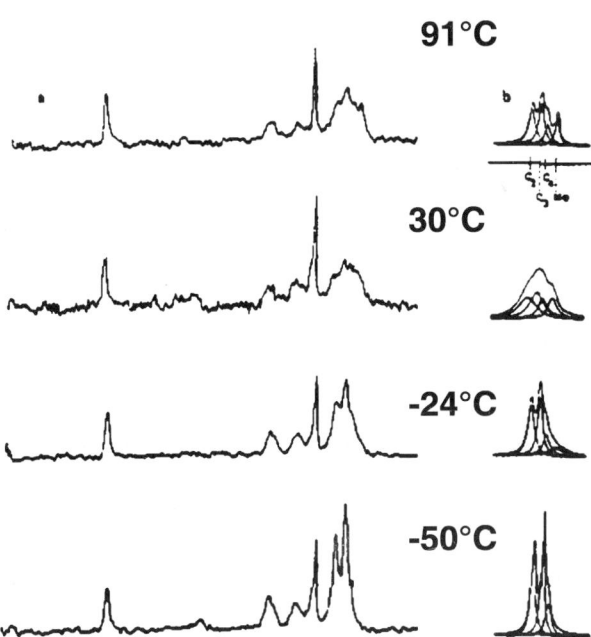

Fig. 9 Experimental 12-MHz MAS CP DD ^{13}C NMR spectra of solid poly(cyclohexyl methacrylate) at different temperatures (from [7])

Figure 10 shows the evolution of T_{2m} for the CH_2 ring carbon C_2 (the carbon code number is given in Fig. 7). The T_{2m} minimum is observed around 20 °C, which indicates that at this temperature the correlation times of the ring motion are of the order of 5×10^{-6} s. Very similar results are observed for the CH_2 ring carbon C_3. The line of the C_4 ring carbon seems to be less broadened than those of carbons C_2 and C_3. Moreover, the widths of the main-chain CH_2 carbon and of the C_1 ring carbon lines do not seem to be temperature-dependent over the range of temperature investigated. All these data indicate that the dynamic process involved is a ring motion that mainly affects the C_2-H and C_3-H internuclear vectors.

These conclusions may be interpreted in terms of a model in which the polymer main chain is fixed while the cyclohexyl ring, which is rigidly bound to the carboxyl moiety, experiences chair–chair inversions as shown in Fig. 7, in agreement with the assignment proposed by Heijboer [5]. A deeper analysis of the variable-temperature NMR results leads to an estimate of 14.6 kJ mol^{-1} for the energy difference, ΔE, between the two conformers and of 47.4 kJ mol^{-1} for the energy barrier. The higher value of ΔE obtained by NMR, relative to the 2.9 kJ mol^{-1} value deduced from conformational energy calculations on an isolated molecule, may reflect the role of intermolecular interactions due to molecular packing.

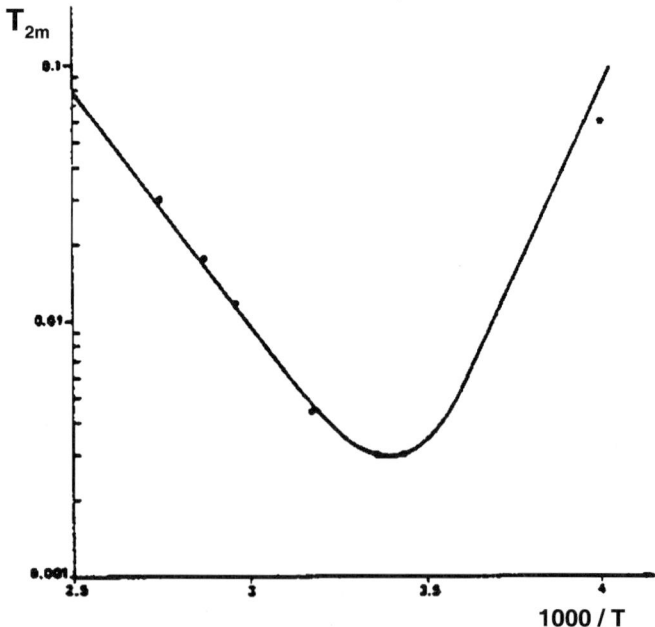

Fig. 10 Transverse relaxation times, T_{2m}, of the methylene ring carbon C_2 as a function of $10^3/T$, where T is the absolute temperature (from [7])

This example dealing with the precise identification of the ring motions involved in the γ transition of poly(cyclohexyl methacrylate) clearly illustrates how convenient and powerful the MAS CP ^{13}C NMR is for studying solid-state transitions of polymers. It is worth noting that, owing to the very low temperature at which the γ transition occurs, the involved motions are very localised.

4
Poly(ethylene *tere*-phthalate)

The solid-state relaxations of poly(ethylene *tere*-phthalate) (PET) were investigated many years ago and the research in this area up to 1964 has been very well reviewed [8]. The chemical structure of PET is:

$$\left[-CH_2-O-\underset{O}{\overset{}{C}}-\left\langle\bigcirc\right\rangle-\underset{O}{\overset{}{C}}-O-CH_2- \right]_n$$

Various attributions for the β relaxation peak, located at about $-70\,°C$ at 1 Hz, have been proposed over the years, involving motion of either the aliphatic part of the chain, or the carboxyl groups, or the phenyl rings, or the restricted rotation of the glycol residue. More recently [9], using selectively deuterated PET samples, the β relaxation has been attributed to the motion of the phenyl rings, whereas the ethylene glycol units do not appear to contribute. Later work [10, 11] has confirmed the importance of ring flips in PET below the glass transition.

Such a question has been recently revisited for pure PET and PET blended with antiplasticiser small molecules [12, 13].

4.1
Pure PET

The results obtained on pure amorphous PET by using the various techniques are presented first.

4.1.1
Dynamic Mechanical Analysis

The dynamic mechanical loss tangent, tan δ, determined [13] for PET at four different frequencies, over a range of temperature between -150 and $75\,°C$ is shown in Fig. 11.

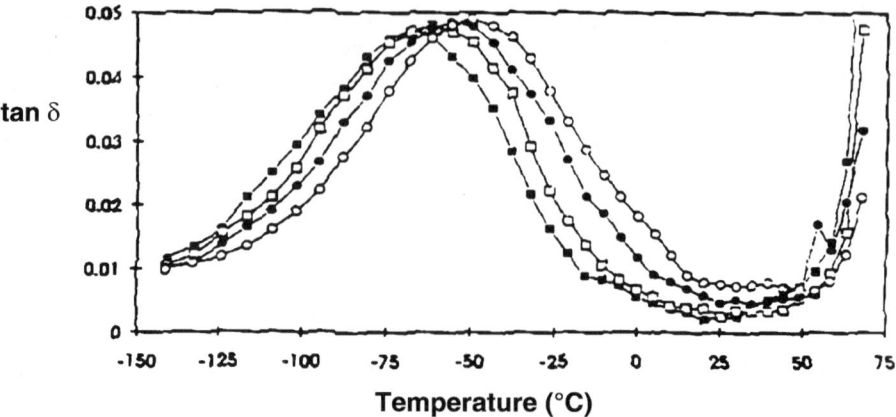

Fig. 11 Dynamic mechanical relaxation processes in PET observed over a range of temperatures at 1 Hz (■), 3 Hz (□), 10 Hz (●) and 30 Hz (○) (from [13])

At 1 Hz, the main secondary relaxation peak is the β peak, that appears at approximately – 70 °C. Furthermore, the peak has an asymmetrical shape, with a high-temperature side broader than the low-temperature side. The activation energy for the β peak is 70 ± 8 kJ mol^{-1} (Table 1).

The associated activation entropy, ΔS_a, has been determined by using the Starkweather expression (Eq. 3 in Sect. 2.3). Considering the asymmetry

Table 1 Activation energies for the β transition peak in PET and PET/additive from mechanical and dielectric measurements

Polymer/additive blend	Activation energy mechanical [kJ]	Activation energy dielectric [kJ]
PET	70 ± 8	56 ± 10
PET/DMT	57 ± 9	53 ± 14
PET/TPDE	54 ± 9	55 ± 10

Table 2 Activation entropy, ΔS_a, at different positions in the β peak of PET (from [13])

Position in the β peak	Mechanical peak entropy [J K^{-1} mol^{-1}]	Dielectric peak entropy [J K^{-1} mol^{-1}]
Low temperature side	46	52 ± 10
Centre of relaxation peak	106	53 ± 10
High temperature side	129	49 ± 10

of the β peak, activation entropies corresponding to the low- and high-temperature sides have been calculated in addition to that for the centre of the relaxation peak. The values, reported in Table 2, show that the degree of cooperativity is significantly higher for the relaxation process involved in the high-temperature side of the peak than for the low-temperature side.

4.1.2
Dielectric Analysis

Dielectric measurements on PET [13], over a range of five different frequencies between 1 Hz and 10 kHz, at temperatures between − 120 and 80 °C, are shown in Figs. 12 and 13 for the dielectric constant, ε', the loss tangent, tan δ, respectively.

The β peak occurs at − 70 °C at 1 Hz. As the only significant dipoles in PET that are mobile are the carboxyl groups, the dielectric β relaxation can be unambiguously attributed to the motion of these groups. Furthermore, the dielectric β peak is quite symmetrical and can be fitted by a Gaussian curve.

The associated activation energy (Table 1) is 56 ± 10 kJ, a value significantly lower than the one derived from mechanical measurements (Table 1). In the same way, the activation entropy (Table 2) of 53 ± 10 J K^{-1} mol^{-1} corresponding to the centre of the relaxation peak is lower that the value obtained from mechanical measurements. The symmetrical character of the dielectric β peak is also reflected in the constant value of ΔS_a over the temperature range (Table 2).

From another point of view, the dielectric increment, $\Delta \varepsilon'$, of the β relaxation process is 0.52, which is much smaller than the one calculated for

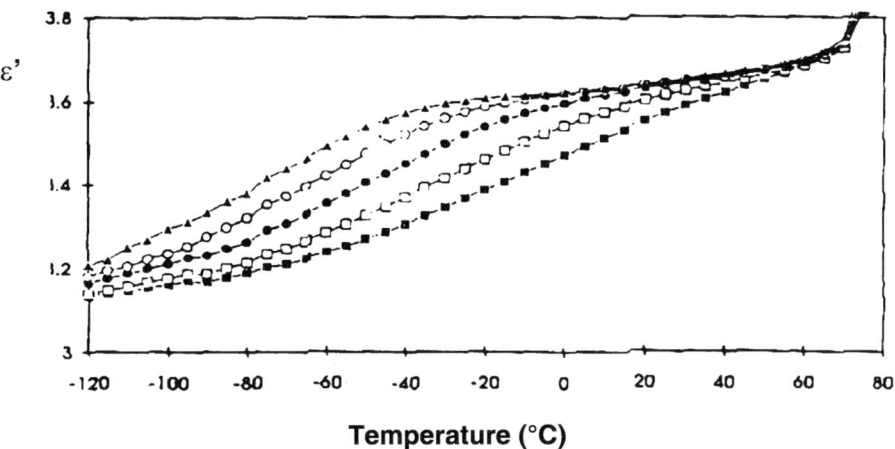

Fig. 12 Increase in the dielectric permitivity of PET that is observed during the relaxation process at 10 kHz (■), 1 kHz (□), 100 Hz (●), 10 Hz (○) and 1 Hz (▲) (from [13])

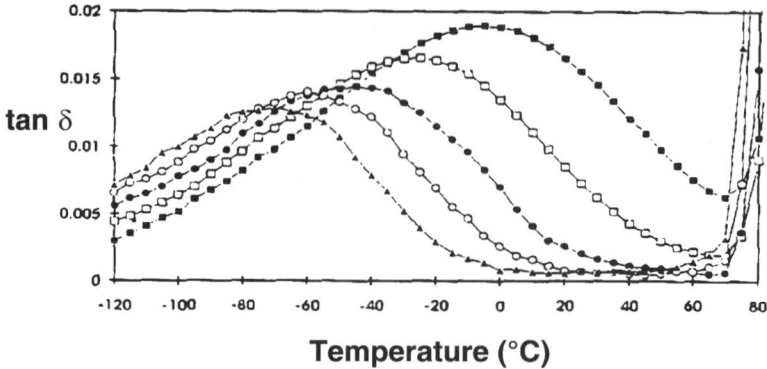

Temperature (°C)

Fig. 13 Dielectric tan δ peaks in PET associated with the β relaxation process observed at 10 kHz (■), 1 kHz (□), 100 Hz (●), 10 Hz (○) and 1 Hz (▲) (from [13])

carboxyl groups undergoing free rotation [13], i.e. 4.52, suggesting that either approximately 10% of the carboxyl groups are involved in such a process or that they undergo a restricted motion.

4.1.3
Solid-State NMR Analysis

In order to get more detailed information about the motions associated with the β relaxation in PET and to understand the differences observed between mechanical and dielectric data, ^{13}C NMR was used, as well as ^{2}H NMR on PET samples selectively deuterated either on the phenyl rings or on the ethylene glycol units [12]. Due to the higher frequency range corresponding to NMR experiments (10^5 Hz), the extrapolation of the dielectric results leads to the occurrence of the motions involved in the β relaxation around 25 °C, which is effectively observed.

First of all, these investigations confirm that there is no detectable mobility of the ethylene groups in PET below the glass transition temperature.

As regards the phenyl ring motions, interesting features are obtained from measurements of ^{13}C chemical shift anisotropy of protonated and unprotonated aromatic carbons. The chemical shift parameters are orthogonal to each other, with σ_{11} and σ_{22} in the phenyl plane and σ_{33} bisecting the phenyl plane.

A relevant quantity is ($\sigma_{33}-\sigma_{11}$), for its value is reduced when the molecule becomes mobile, leading to an averaging of σ_{11} and σ_{33}. Figure 14 shows the variation of ($\sigma_{33}-\sigma_{11}$) as a function of temperature between 20 and 80 °C for protonated and unprotonated aromatic carbons.

It can be seen in Fig. 14 that, as the temperature increases, there is a decrease in the chemical shift anisotropy ($\sigma_{33}-\sigma_{11}$) of protonated carbons, indicating an increase in the mobility of the aromatic rings at higher tem-

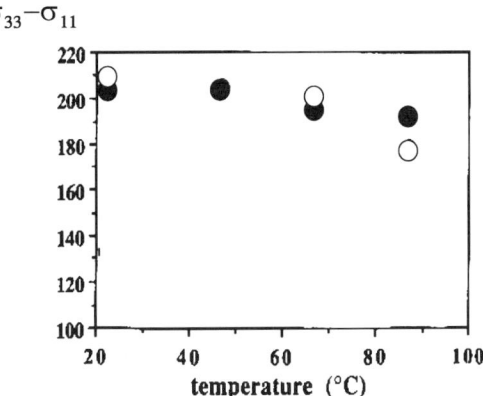

Fig. 14 Chemical shift anisotropy ($\sigma_{33}-\sigma_{11}$) for the protonated ($\bigcirc$) aromatic carbons and unprotonated ($\bullet$) aromatic carbons in PET (from [12])

peratures. The chemical shift anisotropy change with increasing temperature is much smaller for unprotonated aromatic carbons than for protonated aromatic ones. In order to determine the molecular motions that are responsible for the decrease of ($\sigma_{33}-\sigma_{11}$), a model considering either phenyl ring flips or small angle oscillations was developed [12].

As seen from Fig. 15, in the case of unprotonated aromatic *para* carbons, a phenyl ring π-flip does not change the orientation of the chemical shift anisotropy tensor components and, thus, this π-flip does not affect the NMR response of such *para* carbons, whereas phenyl ring oscillations do.

In contrast, for protonated *ortho* and *meta* aromatic carbons, both phenyl ring π-flips and oscillations affect the orientation of the chemical shift anisotropy tensor components and, consequently, their NMR response. The change in the amplitude of the oscillations with temperature for protonated and unprotonated aromatic carbons is shown in Fig. 16. For the two types of carbons the amplitudes of the oscillations are quite similar.

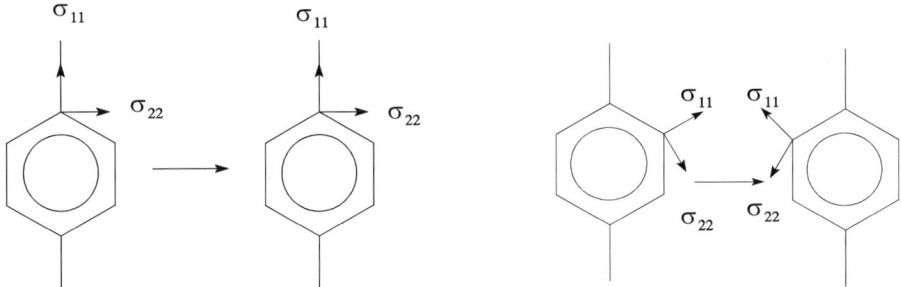

Fig. 15 Effect of a phenyl ring π-flip on the orientation of the chemical shift anisotropy tensor components for unprotonated and protonated aromatic carbons

In the case of the aromatic protonated carbons, cross-polarisation experiments can be performed. The $t_{1/2}$ cross-polarisation time (defined as the contact time required to produce half the maximum value of magnetisation that is possible by cross-polarisation) significantly increases with temperature, as shown in Fig. 17.

In contrast to the chemical shift anisotropy, $t_{1/2}$ is sensitive to both oscillations and ring flips and, after deduction of the oscillation contribution, allows one to calculate the percentage of ring flips as a function of temperature, as shown in Fig. 18.

The occurrence of ring flips with increasing temperature is confirmed by ^{2}H NMR on PET with selectively deuterated phenyl rings [12].

Concerning the carboxyl groups, the change with temperature of the chemical shift anisotropy (Fig. 19) reflects an increase of the amplitude of

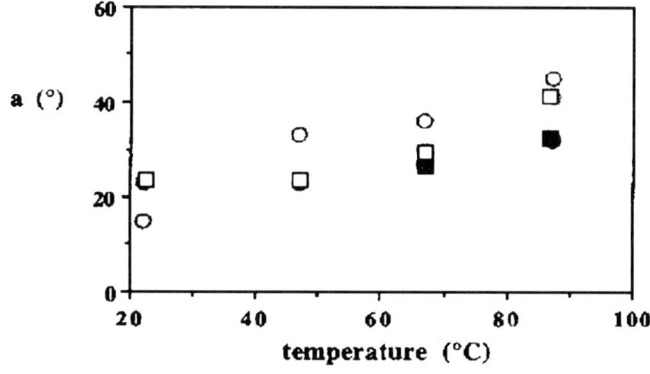

Fig. 16 Amplitude of the oscillations, a, of the phenyl rings and carboxyl groups: ■ protonated carbons, □ unprotonated carbons and ○ carboxyl carbons (from [12])

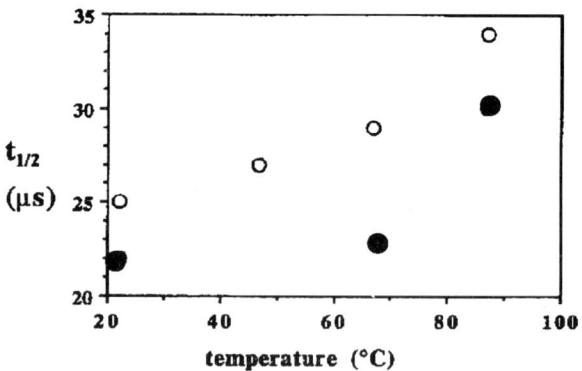

Fig. 17 Cross-polarisation contact times, $t_{1/2}$, for the aromatic carbons in the phenyl rings, in both the pure PET and the PET/TPDE blend: ○ aromatic carbons in PET and ● aromatic carbons in the blend that contains 10% TPDE (from [12])

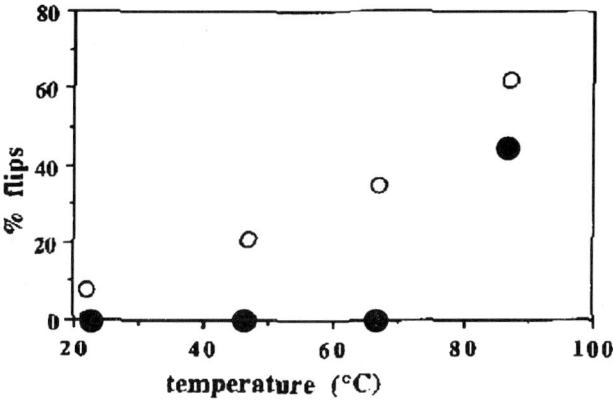

Fig. 18 Estimate of the number of phenyl rings that are flipping in ○ PET and ● the PET blend that contains 10% TPDE (from [12])

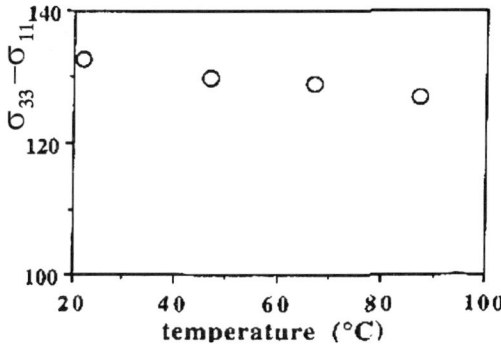

Fig. 19 Chemical shift anisotropy results obtained from the carboxyl groups in PET (from [12])

the small angle oscillations (Fig. 16). It is worth noticing that oscillation amplitudes are similar in the carboxyl groups and the phenyl rings, reaching oscillations of approximately $\pm 40°$ during the β relaxation process. Unfortunately, as the $t_{1/2}$ data of carboxyl groups cannot be easily interpreted in terms of motions, it has not been possible to confirm by NMR that the carboxyl groups undergo flip motions.

4.2
Effect of Small Molecule Antiplasticiser

In the case of polycarbonates, it has been observed that by adding miscible low molecular weight additives, with specific chemical structures, it is possible to increase the yield stress of the polymer, as well as to reduce the local molecular motions that are responsible for the secondary relaxation processes of the polymer [14–17]. These compounds are known as antiplasticisers (due to their effect on the yield stress), in spite of the fact that they lead to a decrease of the glass transition temperature of the polymer, as do the usual plasticisers.

Some antiplasticisers exist for PET [13] and it is interesting to study their effect on the β relaxation in PET, using the same investigation tools as the ones applied to pure PET.

The antiplasticisers considered here are:

- Dimethyl *tere*-phthalate (DMT)

$$H_3C-O-\underset{\underset{O}{\|}}{C}-\text{\Large\textcircled{}}-\underset{\underset{O}{\|}}{C}-O-CH_3$$

- Tetrachlorophthalic dimethyl ester (TPDE)

Other compounds [13] have very similar effects.

4.2.1
Dynamic Mechanical Analysis

As an example, the effect of DMT additive on the β relaxation of PET [13] at 1 Hz is shown in Fig. 20.

It is worth noting that an amount of DMT as small as 2 wt % affects the temperature and the amplitude of the β peak. Saturation seems to be approached from 10 to20% of additive. It can be seen that the additive suppresses the high-temperature side of the β peak considerably more than the low-temperature side. This would appear to support the statement that the β peak consists of more than one relaxation process, as evidenced by NMR studies and that the antiplasticiser has suppressed the relaxation processes that occur on the high-temperature side of the β peak.

At the same concentration (10 wt %) TPDE has the same effect as DMT (Fig. 21).

The shift of the β peak from – 70 to – 80 °C is associated with a decrease of the activation energy (Table 1), as well as the activation entropy (Table 3). The latter result indicates that the cooperativity of the processes involved in the remaining mechanical peak is close to the one associated with the

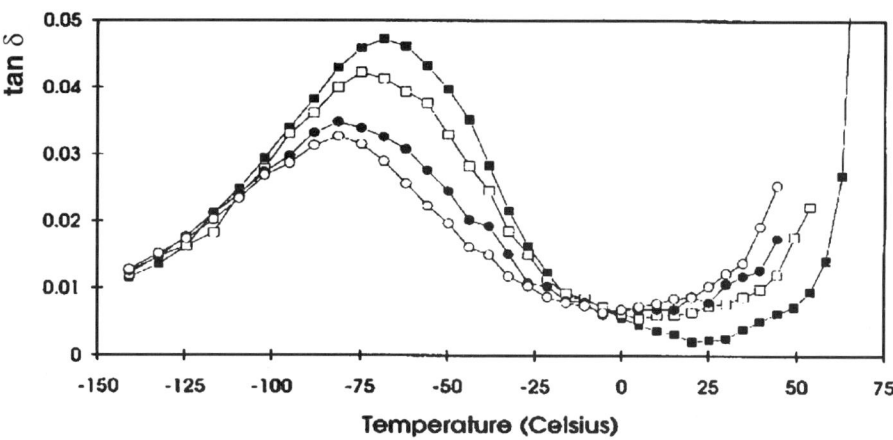

Fig. 20 Effect of different amounts of DMT additive on the mechanical relaxation processes in PET at 1 Hz: ■ pure PET, □ 2% DMT, ● 10% DMT and ○ 20% DMT (from [13])

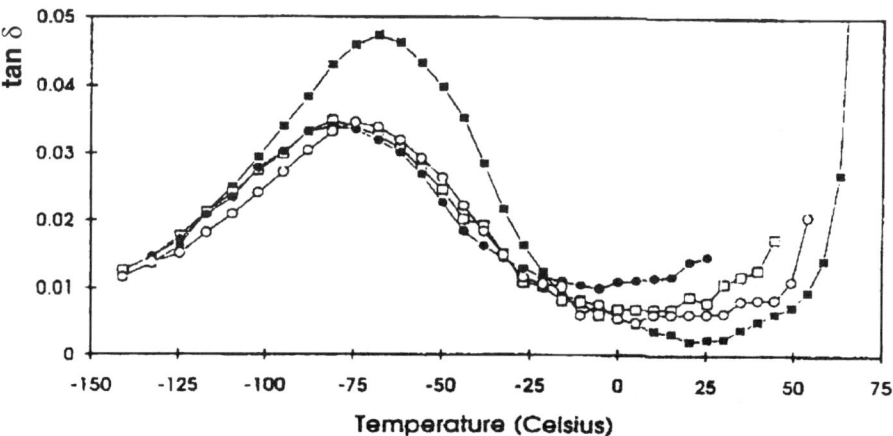

Fig. 21 Effect of different additives on the mechanical relaxation peak of PET at 1 Hz: ■ pure PET, □ 10% DMT, ● 10% TPDE and ○ 20% TPDE (from [13])

low-temperature side of the β peak in pure PET (Table 2) and significantly lower than the cooperativity of the processes corresponding to the high-temperature side of the β peak in pure PET (Table 2).

To examine in more detail the relaxation process suppressed by the additives, the relaxation peaks that were obtained from the PET/DMT blends have been subtracted from those obtained from pure PET. The result (Fig. 22) is a quite symmetrical peak, centred at $-60\,^{\circ}\mathrm{C}$ for 1 Hz, which can be perfectly fitted with a Gaussian curve at each frequency, as shown in Fig. 22. The activation energy of the suppressed relaxation processes is approximately $70\,\mathrm{kJ\,mol^{-1}}$.

Table 3   Activation entropy, ΔS_a, for the β peak in different PET/additive blends (from [13])

Polymer/additive blend	Mechanical peak entropy [J K^{-1} mol^{-1}]	Dielectric peak entropy [J K^{-1} mol^{-1}]
PET	106	53 ± 10
PET/DMT	57	40 ± 11
PET/TPDE	37	51 ± 10

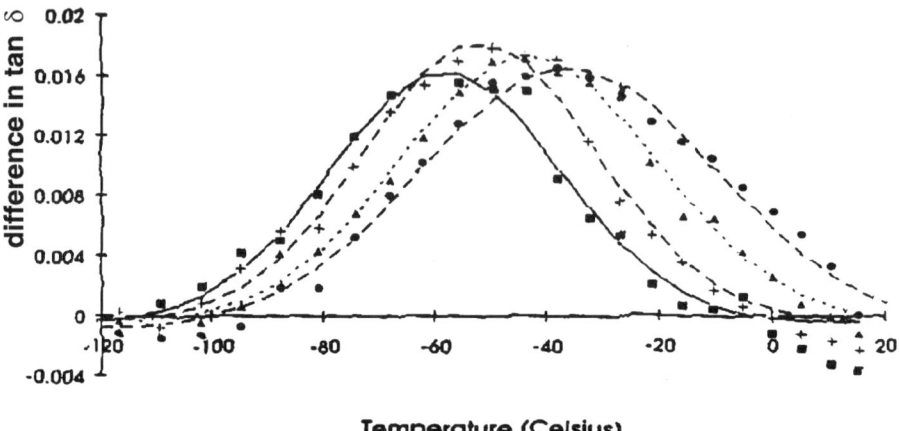

Fig. 22 Gaussian peaks that have been fitted to the suppressed relaxation peaks in PET at: ■ 1 Hz, + 3 Hz, ▲ 10 Hz and ● 30 Hz (from [13])

4.2.2
Dielectric Analysis

The tan δ loss curves obtained at 1 Hz for the PET blends with the DMT and TPDE additives [13] are shown in Fig. 23. In contrast to what happens in the dynamic mechanical experiments, the additives lead to only a small shift of the curves relative to the case of pure PET and to the same peak amplitude as for pure PET. Furthermore, the activation energies derived for the β peak obtained from dielectric measurements are the same as the ones for pure PET (Table 1) and the activation entropies are in the same range (Table 2).

In order to confirm that the processes dielectrically active in pure PET and PET/additive blends correspond to the low-temperature side of the mechanical β peak of pure PET and remain mechanically active in PET/additive blends, the shapes of these two tan δ curves have been compared in Fig. 24. The observed similarity allows one to attribute the processes involved in the low-temperature side of the mechanical β peak in pure PET to motions of the carboxyl groups.

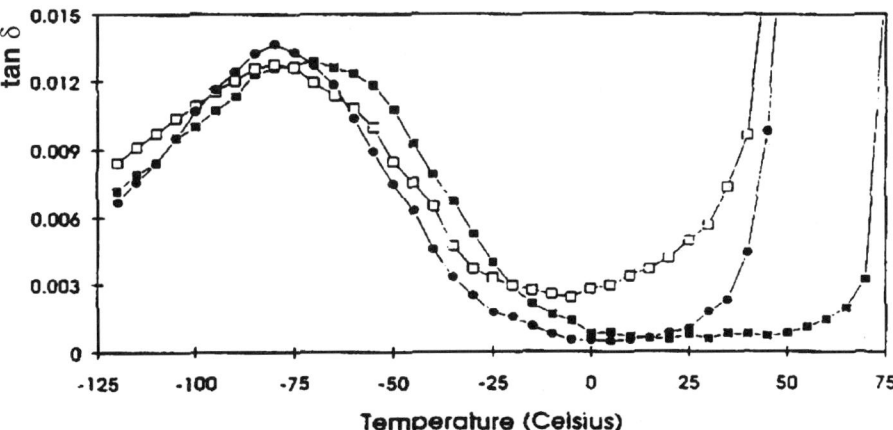

Fig. 23 Effect of different additives on the dielectric relaxation peaks in PET at 1 Hz: ■ pure PET, □ PET with 10% DMT and ● PET with 10% TPDE (from [13])

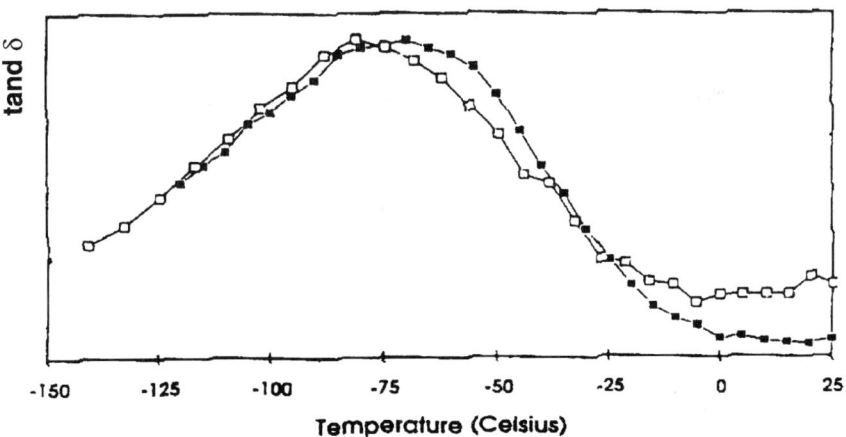

Fig. 24 Similarity between the dielectric relaxation peak in PET (■) and the mechanical peak in a polymer/additive blend containing 10% DMT additive (□) at 1 Hz (from [13])

4.2.3
Solid-State NMR Investigation

To identify the processes that are suppressed by the antiplasticisers, NMR studies are required [12]. The $t_{1/2}$ contact times in PET/TPDE blends, shown in Fig. 17, are lower than those in pure PET. They do not change with temperature until 70 °C. As $t_{1/2}$ is mostly sensitive to ring flips, the number of ring flips in the polymer has been determined for PET/TPDE blends (Fig. 18). Within the accuracy of the experiments, no ring flips are observed between 25 and 70 °C.

4.3
Conclusion

The investigation of pure PET and PET/additive blends by combining dynamic mechanical analysis, dielectric relaxation and solid-state NMR techniques, leads to a clear attribution of the molecular processes involved in the β relaxation of PET, as well as an understanding of the effect of an antiplasticiser additive:

- There are no motions of the aliphatic groups occurring below the glass transition temperature.
- The low-temperature side of the mechanical β peak originates from the carboxyl groups and phenyl rings undergoing small angle oscillations at amplitudes from $\pm 10°$ to $\pm 40°$.
- The high-temperature side of the mechanical β peak reflects the phenyl ring π-flips performed by 10–30% of the phenyl groups. These phenyl ring motions have a much higher cooperativity than those of the carboxyl groups. The phenyl ring oscillations are still present, but in this temperature range their frequency is too high to lead to a contribution in the mechanical response.
- The antiplasticiser additives do not affect the motions of the carboxyl groups, but they hinder the phenyl ring flips.

5
Bisphenol A (and/or Tetramethyl Bisphenol A) Polycarbonate

Bisphenol A polycarbonate, BPA-PC (Fig. 25), has an unusual toughness among the amorphous thermoplastics, which has led to considerable industrial development. As early as the 1960s studies [8] were performed to understand the molecular origin of such a behaviour and to relate it to the secondary transition occurring around – 100 °C.

In order to investigate the molecular motions involved in this transition and, in particular, the nature of the associated cooperativity, it is interesting to consider not only BPA-PC, but also tetramethyl bisphenol A polycarbonate, TMBPA-PC, (Fig. 25) as well as copolymers of BPA and TMBPA carbonates, and compatible blends of BPA-PC and TMBPA-PC.

Furthermore, the effect of miscible small molecule additives, antiplasticisers, on the secondary transition, is worth analysing in order to reach a deeper understanding of the involved molecular motions.

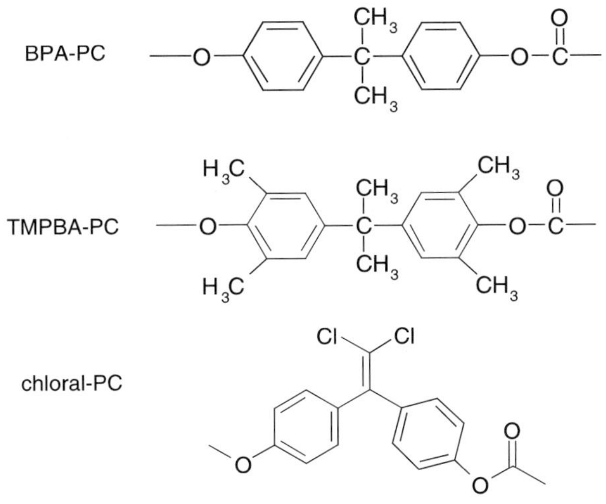

BPA-PC

TMPBA-PC

chloral-PC

Fig. 25 Chemical structures of BPA-PC, TMPBA-PC and chloral-PC

5.1
Dielectric Analysis

The temperature dependence of the dielectric loss, ε'', at 10 Hz, is shown in Fig. 26 for BPA-PC.

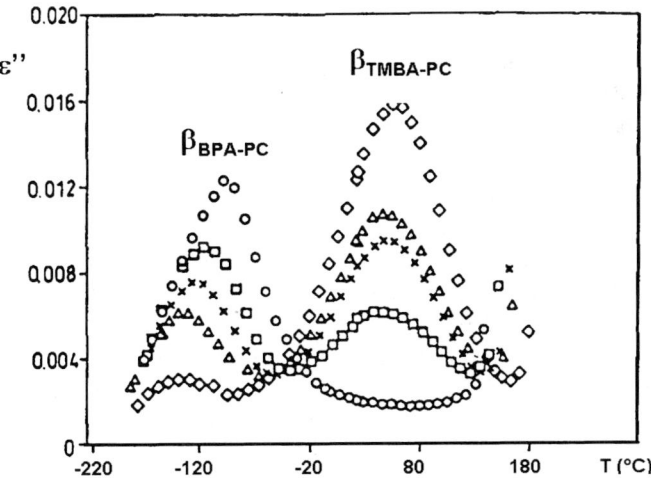

Fig. 26 Temperature dependence of the dielectric loss, ε'', at 10 Hz, for BPA-PC, TMBPA-PC and their compatible blends. The TMBPA-PC volume fraction in the samples is: ◯ 0, □ 0.22, × 0.38, △ 0.53 and ◇ 0.82 (from [19])

The peak observed around – 100 °C corresponds to the β transition of BPA-PC (it is sometimes denoted as γ). As the only significant dipoles in BPA-PC that are mobile are the carbonate groups, the dielectric β relaxation can be unambiguously attributed to the motion of these groups.

The frequency analysis of this β peak has been performed [18] and is shown in Fig. 27 in the temperature range – 112 to – 38 °C. In addition to the expected shift of the frequency at the peak maximum with increasing temperature, there was a large increase of ε''_{max}, as well as a narrowing of the peak. Such a behaviour indicates that the dielectric β relaxation does not correspond to a unique process, but contains different processes with various activation energies, which gradually merge when the frequency is increased.

The activation energy associated with the peak maximum is 30 kJ mol^{-1} (using a broader frequency range, a value of 54 ± 2 kJ mol^{-1} has been estimated [19]).

As mentioned, the dielectric β peak is related to carbonate motions; however, it is important to check whether the adjacent phenyl groups are involved or not in such motions. Concerning this feature, it is interesting to consider the dielectric behaviour of TMBPA-PC. The temperature dependence of the

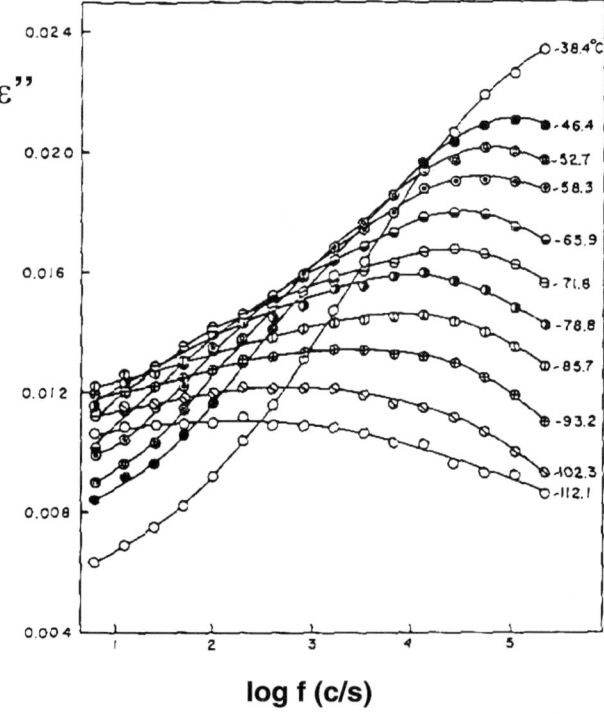

log f (c/s)

Fig. 27 Dielectric loss, ε'', versus logarithm of frequency, f, for BPA-PC in the β transition region (from [18])

dielectric loss, ε'', at 10 Hz, is shown in Fig. 26 for TMBPA-PC [20]. The dielectric β relaxation appears at $+60\,^{\circ}$C, which corresponds to a shift of $+160\,^{\circ}$C consecutive to the introduction of the methyl group on the phenyl rings in *ortho* positions to the carbonate (Fig. 25). Such a behaviour supports the idea that the carbonate motion is coupled to the ring motion. Indeed, if it were a hindering of carbonate motion due to steric hindrance of the methyl groups on the phenyl rings, what should happen would be the disappearance of the dielectric β relaxation instead of a shift of its temperature occurrence. Indeed, in spite of the lower density of TMBPA-PC (1.083 kg m^{-3} instead of 1.208 kg m^{-3} for BPA-PC [19]), a larger free volume is required in the case of TMBPA-PC in order to perform the phenyl ring motion required for getting the carbonate motion responsible for the dielectric β relaxation. Thus, $+60\,^{\circ}$C has to be reached, instead of $-100\,^{\circ}$C for BPA-PC. The activation energy associated with the dielectric β relaxation of TMBPA-PC is 95 ± 2 kJ mol^{-1}.

This result clearly shows the intramolecular cooperativity associated with the carbonate motion.

Another feature is the occurrence of an intermolecular cooperativity. A first approach involved the dielectric behaviour of compatible blends of BPA-PC and TMBPA-PC [19]. The results are shown in Fig. 26. It appears that the TMBPA-PC β peaks in the blends have a shape similar to that of the homopolymer. In the same way, the peak temperatures are close to that of the homopolymer; a slight shift to lower temperatures happens when increasing BPA-PC content ($17\pm 2\,^{\circ}$C at a BPA-PC concentration of 80%). For the BPA-PC β peaks, a different behaviour is observed. Firstly, the low-temperature side of the peak is unchanged by increasing the TMBPA-PC content. In contrast, the high-temperature side gradually disappears, leading to a downward shift of the peak maximum temperature (40 $^{\circ}$C for a TMBPA-PC concentration of 90%). Thus, in the high-temperature part of the dielectric β relaxation, there is an intermolecular cooperativity that is hindered by the motionless (in this temperature range) TMBPA-PC chains present in the surroundings.

5.2
Dynamic Mechanical Analysis

The temperature dependence of the dynamic loss modulus, E'' and $\tan\delta$ at 1 Hz [21] is shown in Fig. 28 for BPA-PC.

A secondary transition, with a peak maximum around $-100\,^{\circ}$C, is clearly evidenced. It is denoted here as β, but some authors call it γ. Indeed, another mechanical transition (referred to as β by these authors) appears in some samples around 80 $^{\circ}$C, but it is not characteristic of BPA-PC for it disappears by annealing. This has been attributed to relaxation of packing defects generated by rapid quenching [22–24]. This transition will not be consid-

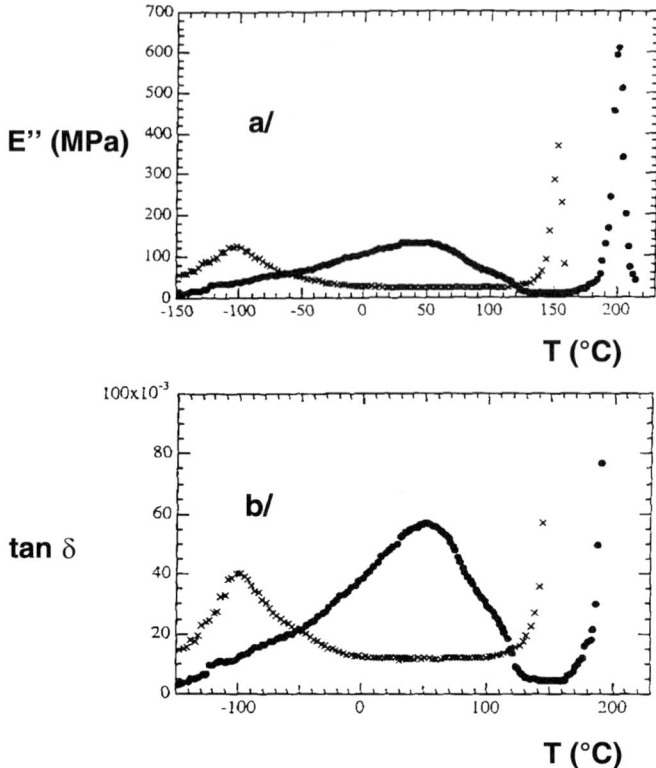

Fig. 28 Temperature dependence of (**a**) the dynamic loss modulus, E'', and (**b**) $\tan \delta$ at 1 Hz for BPA-PC ($\times$) and TMBPA-PC ($\bullet$) (from [21])

ered hereafter, the only transition under concern in this section will be the β transition located around $-100\,^{\circ}\text{C}$ at 1 Hz.

The associated activation energy is $60\,\text{kJ mol}^{-1}$ and the activation entropy, determined according to the Starkweather's treatment (Sect. 2.3), is equal to $110\,\text{J K}^{-1}\,\text{mol}^{-1}$, which corresponds to cooperative motions.

The effect of chemical structure modifications on this β transition (temperature and activation energy) has been reviewed in [25]. It appears that replacing the carbonate group by a formal linkage, $-\text{O}-\text{CH}_2-\text{O}-$, does not affect the transition characteristics. In the same way, replacing the isopropylidene unit, located between the phenyl rings (Fig. 25), by either more flexible or more rigid, bulky groups does not lead to any change of the β transition characteristics. In contrast, introducing substituents such as Cl, Br, or CH_3 on the phenyl rings in *ortho* position relative to the carbonate link, strongly shifts the transition temperature towards higher temperatures, as reported in Table 4; the effect increases from CH_3 to Cl and Br.

Table 4 Effect of various substituents of phenyl groups on the characteristics of the β transition (from [25])

Structure	$T\beta$ [°C]	E_a [kJ mol^{-1}]
$-O-\langle\text{C}_6\text{H}_4\rangle-\text{C}(CH_3)_2-\langle\text{C}_6\text{H}_4\rangle-O-\overset{O}{\underset{\|}{C}}-$	-100	57
same, with Cl on second ring	5	105
same, with two Cl on second ring	50	
Cl-substituted first ring, Cl on second ring	40	118
Cl on first ring, CH$_3$ on second ring	39	92
two Cl on first ring, two Cl on second ring	112	96
two Br on first ring, two Br on second ring	120	
two H$_3$C on first ring, two CH$_3$ on second ring	40	80

In the case of TMBPA-PC, the temperature dependence of E'' and $\tan\delta$ [21] is shown in Fig. 28. It leads to an activation energy of 80 kJ mol^{-1} and an activation entropy of 20 J K^{-1} mol^{-1}, indicating very localised motions, contrasting with the case of BPA-PC.

These results clearly prove the participation of the phenyl rings in the β transition.

In order to analyse the extent of cooperativity of the motions involved in the β transition of BPA-PC, quite a large series of copolymers has been synthesised. The (BPA-C)$_n$ blocks (denoted B$_n$) and (TMBPA-C)$_n$ blocks (denoted T$_n$) have been studied by dynamic mechanical measurements.

The results for the alternate copolymer B$_1$T$_1$ [26] are shown in Fig. 29. Quite a broad β transition is observed, from -150 to $90\,°C$, with a maximum around $0\,°C$ and a weak shoulder at about $-100\,°C$. Such a behaviour is an unambiguous indication of the coupling between the BPA-C and TMBPA-C units. The low temperature shoulder slightly increases in intensity when considering longer blocks such as B$_6$T$_6$ and B$_9$T$_9$.

The other approach [27] was to interrupt the length of a comonomer block by inserting one unit of the other comonomer. The results for the series B$_x$T and BT$_y$ are shown in Figs. 30 and 31, respectively.

From the behaviour of the B$_x$T copolymers, the authors concluded that β blocks of seven to nine units are required for a BPA-PC-like β transition. Such a conclusion can be questioned since the tan δ curves for B$_3$T and B$_5$T copolymers show a clear shoulder in the $-100\,°C$ region, suggesting that in these copolymers the motions involved in the β transition of BPA-PC can occur within the B blocks. The larger tan δ peak around -25 to $0\,°C$ reflects the motions encountered in the alternate BT copolymer. However, it is worth noting that a deeper analysis of the whole set of results would require consideration of the dynamic mechanical loss compliance, J'', as done in Sects. 6, 7 and 8.

As regards the BT$_y$ series (Fig. 31), a shift of the tan δ peak maximum from 0 to $25\,°C$ is observed between BT$_1$ and BT$_3$, but it does not change further when increasing the T$_y$ block.

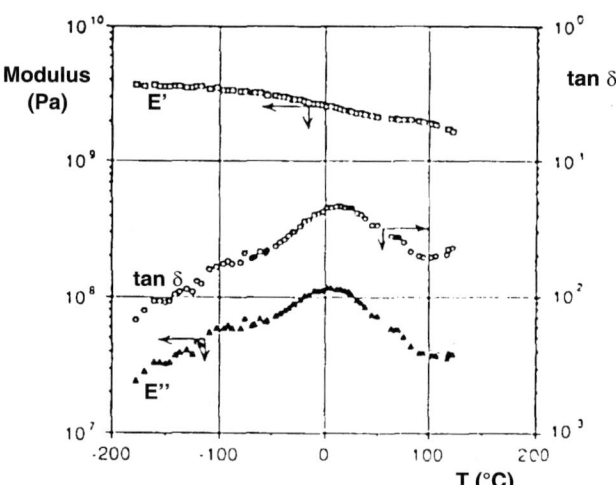

Fig. 29 Temperature dependence of the dynamic mechanical behaviour, at 11 Hz, for the alternating copolymer B$_1$T$_1$ (from [26])

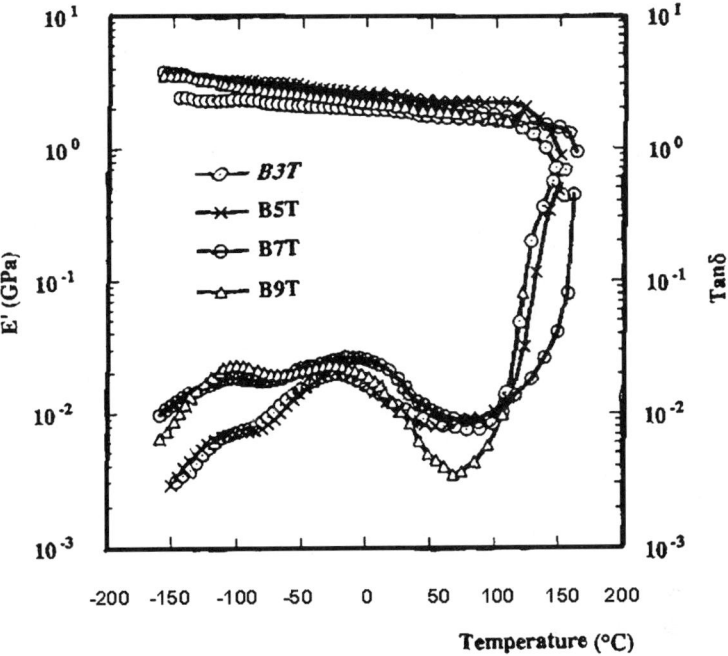

Fig. 30 Temperature dependence of the dynamic mechanical behaviour for the B_xT copolymer series (from [27])

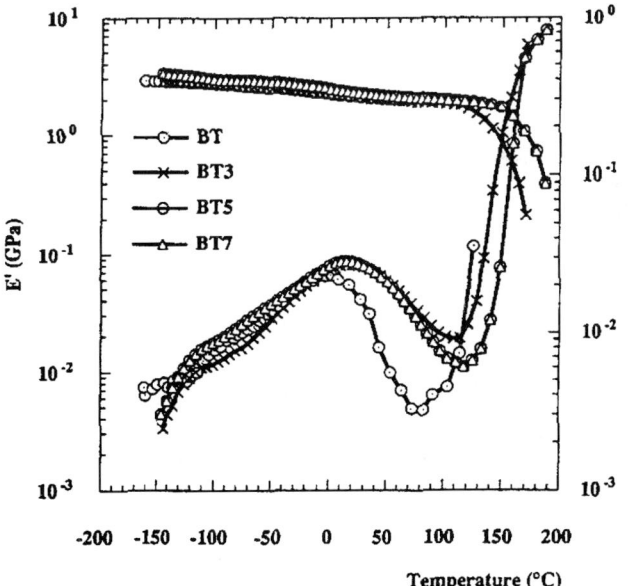

Fig. 31 Temperature dependence of the dynamic mechanical behaviour for the BT_y copolymer series (from [27])

The copolymer approach is interesting, but it does not lead to a direct answer concerning the intra- and intermolecular contributions to the cooperative nature of the β transition motions of BPA-PC. This discussion is postponed until later on in this section, after considering the information provided by the whole set of experimental and atomistic modelling investigations.

To analyse the occurrence of an intermolecular cooperativity, the compatible blends of BPA-PC and TMBPA-PC offer quite a useful opportunity [17, 26, 28]. The dynamic mechanical results [28] are shown in Fig. 32. At first, two β peaks are observed. They are located in the temperature range of each component. However, a more precise examination of the peak positions shows that the blend composition affects the two peaks differently (Fig. 33). Indeed, the TMBPA-PC peak is unchanged, whereas a downward temperature shift is observed for the BPA-PC peak with increasing TMBPA-PC content. Actually, the temperature shift of the maximum does not correspond to a shift of the whole low-temperature peak, but it results from the disappearance of the high-temperature peak when increasing the TMBPA-PC content, as clearly shown in Fig. 34.

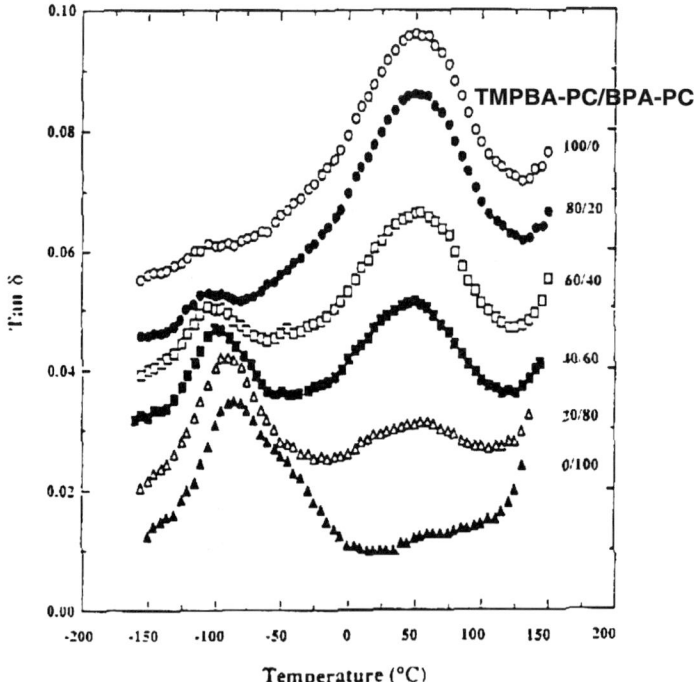

Fig. 32 Temperature dependence of tan δ, at 3 Hz, for TMBPA-PC and BPA-PC blends. As TMBPA-PC content increases, each curve is shifted by 0.01 to get a clearer picture (from [28])

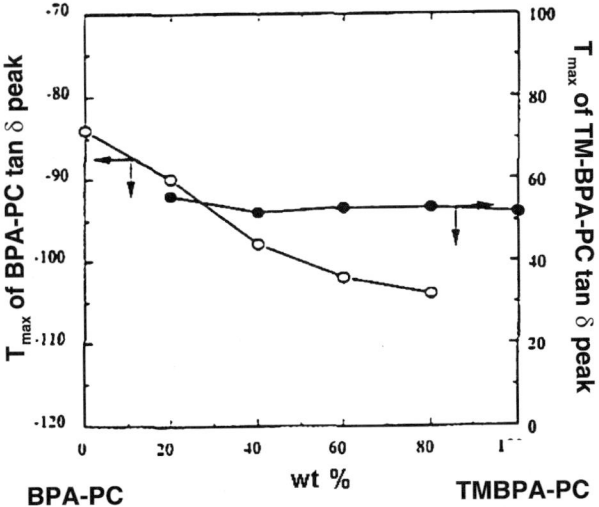

Fig. 33 Temperature shift of the β peak $\tan\delta$ maximum for TMBPA-PC and BPA-PC blends (from [28])

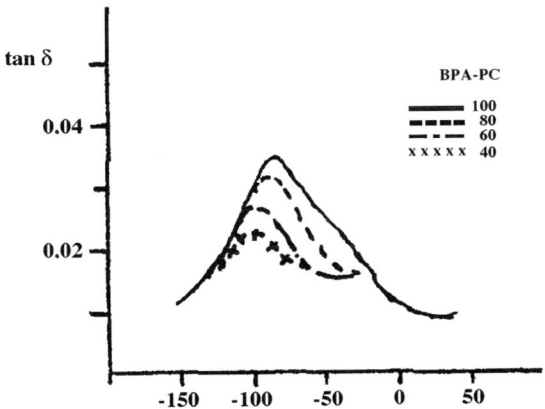

Fig. 34 Low-temperature $\tan\delta$ peak for the TMBPA-PC and BPA-PC blends

Thus, these results demonstrate that there is an intermolecular cooperativity involved in the β transition of BPA-PC. More precisely, this cooperativity concerns the high-temperature part of the β transition, whereas it does not exist for the low-temperature part.

5.3
Solid-State NMR

Many studies have been performed, mostly in the 1980s, on BPA-PC with the various solid-state ^{1}H, ^{2}H and ^{13}C NMR techniques.

5.3.1
Proton NMR

^{1}H NMR applied to bulk polymers can yield information on molecular dynamics. Indeed, molecular reorientations occurring at a rate comparable to the ^{1}H NMR line width ($\sim 10^5$ Hz) will cause line narrowing and decrease the second moment, $\langle \Delta H^2 \rangle$.

In the case of BPA-PC (Fig. 25), two types of protons are present: the methyl and phenyl protons. The temperature dependence of the ^{1}H NMR line width [18] shows two distinct components between -190 and $-150\,^\circ$C (Fig. 35). The broader component is attributed to methyl protons and the other to phenyl protons. For the latter component a moderate narrowing is observed at about $-80\,^\circ$C.

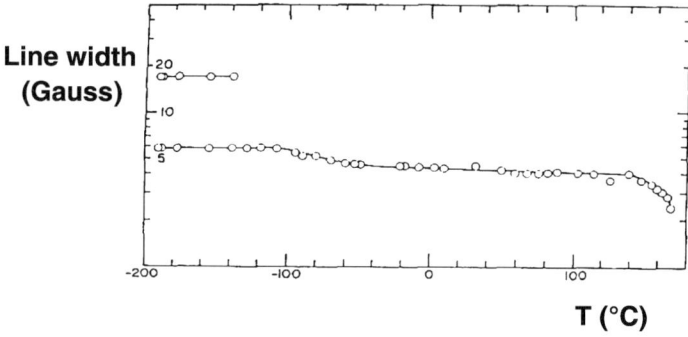

Fig. 35 ^{1}H NMR line width versus temperature for BPA-PC (from [18])

5.3.1.1
Methyl Motion

A more quantitative approach has been performed [29] through the temperature dependence of the second moment of BPA-PC (Fig. 36). The sharp decrease of $\langle \Delta H^2 \rangle$ observed between -170 and $-100\,^\circ$C corresponds to the occurrence of methyl rotation and the second moment values have been quantitatively accounted for by considering the various intra- and intermolecular contributions, leading to an activation energy of 8 kJ mol^{-1}, which agrees with values obtained by NMR for methyl group rotation in low molecular weight compounds. The origin of the additional decrease of $\langle \Delta H^2 \rangle$ at higher temperatures was not identified.

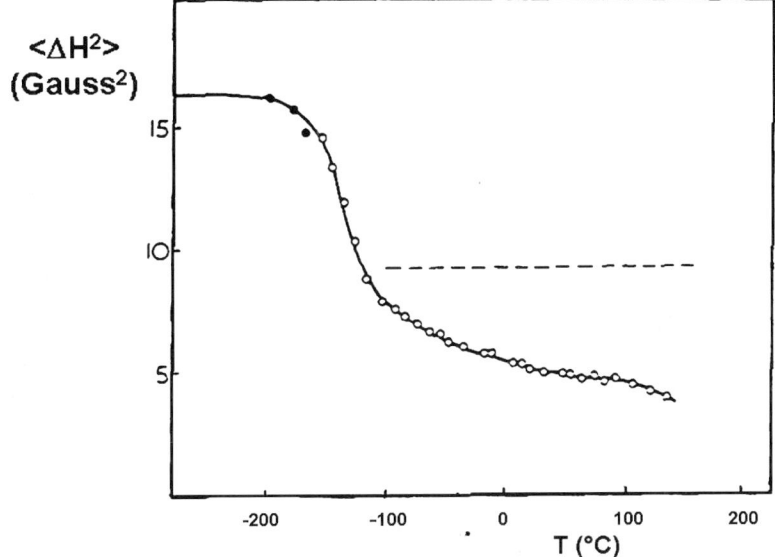

Fig. 36 Temperature dependence of the ^{1}H second moment for BPA-PC. The *broken line* at 9.2 Gauss^2 represents the second moment with rotating methyl groups (from [29])

5.3.1.2
Phenyl Ring Motion

In order to bypass the difficulty arising from the two types of protons existing in BPA-PC, ^{1}H NMR has been performed [30] on chloral polycarbonate, whose structure is shown in Fig. 25.

The choice of chloral-PC is appropriate because (i) it shows a β transition at the same temperature as BPA-PC, and (ii) NMR measurements performed on CDCl$_3$ solutions [31] lead to the same dynamics of segmental and phenyl ring motions as for BPA-PC.

The temperature dependence of $\langle \Delta H^2 \rangle$ for chloral-PC (Fig. 37) shows that the motions of phenyl rings occur above $-80\,°$C. Furthermore, the doublet shape observed for the ^{1}H spectrum above room temperature presents a splitting constant of 25 ± 0.2 G, which corresponds to the static interaction between the 2,3 phenyl protons, indicating that the phenyl motions do not affect the dipole–dipole interaction parallel to the 1,4 phenyl axis. A quantitative analysis of the intra- and intermolecular contributions to $\langle \Delta H^2 \rangle$ leads to the conclusion that the phenyl motions correspond to either isolated or concerted rotations around the 1,4 axis, with little (if any) reorientation of this axis. In addition, it excludes other motions as crankshaft motions, or motion of the phenyl-ethylenic unit as a group. The decrease of $\langle \Delta H^2 \rangle$ above $-40\,°$C could be intermolecular in nature.

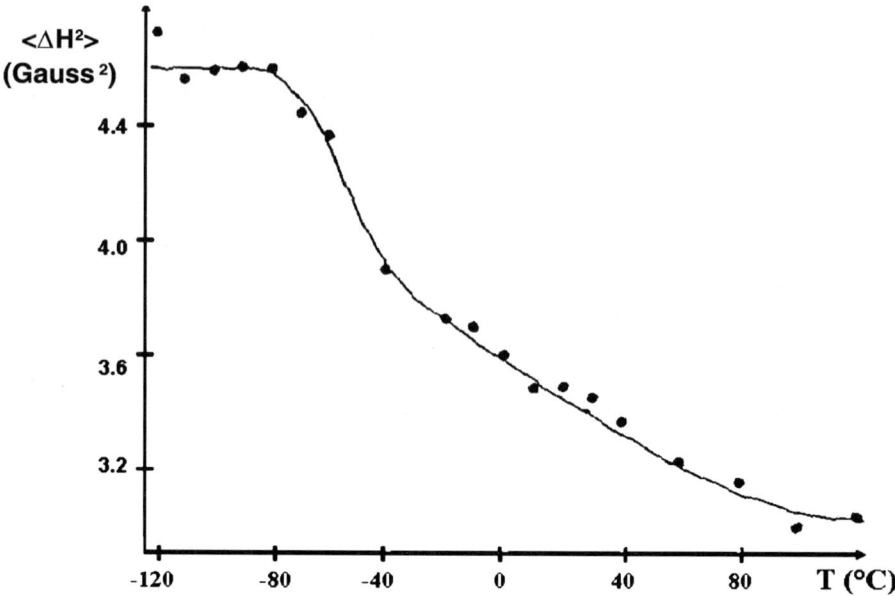

Fig. 37 Temperature dependence of the ^1H second moment for chloral-PC (from [30])

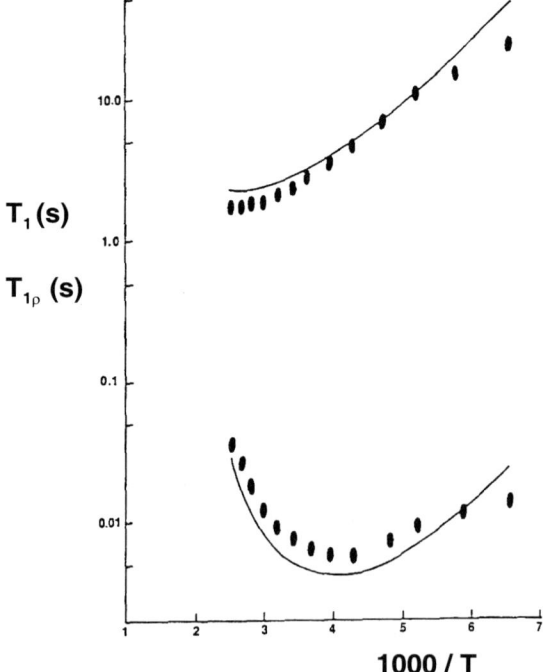

Fig. 38 ^1H T_1 and $T_{1\rho}$ versus reciprocal temperature for BPA-d_6-PC. The *solid line* corresponds to a fit with a Williams–Watts correlation function (from [32])

Another way of using ^{1}H NMR to study the dynamics of phenyl protons in BPA-PC consists in selective deuteration of the methyl groups (BPA-d_6-PC) [32]. Thus, the temperature dependence of the ^{1}H spin-lattice relaxation time, T_1, and spin-lattice relaxation time in the rotating frame, $T_{1\rho}$, has been determined, and is shown in Fig. 38.

Without using any motional model, the temperature positions of T_1 and $T_{1\rho}$ minima can be assigned an appropriate frequency: 90 MHz at 120 °C from T_1 and 43 kHz at − 34 °C from $T_{1\rho}$. These two results fit quite well on the relaxation map of BPA-PC obtained from dynamic mechanical and dielectric relaxation. They support the fact that phenyl ring motions are involved in the β relaxation of BPA-PC. Furthermore, the T_1 and $T_{1\rho}$ data can be simulated by considering the Williams–Watts fractional correlation function [33]:

$$\phi(t) = \exp(-t/\tau_p)^\alpha , \tag{5}$$

where τ_p is the centre correlation time and α the width parameter of the distribution reflecting the spatial inhomogeneity of the phenyl ring motions [34]. A good fit is obtained with $\alpha = 0.18$, as shown in Fig. 38.

5.3.2
Deuteron NMR

By using a selective deuteration, ^{2}H NMR allows investigation of motions of specific groups within a polymer chain.

5.3.2.1
Methyl Motion

In the case of BPA-PC, selective deuteration of methyl groups has been performed and the ^{2}H NMR obtained from − 160 °C [35]. First, when decreasing temperature, changes in the line shape appear at about − 80 °C (as shown in Fig. 39) in the transition region, indicating a methyl motion at a frequency lower than 10^6 Hz. The experimental data are perfectly simulated by considering a methyl motion consisting of threefold jumps about the C_3-axis and a log-Gaussian distribution of the jump rate, Ω, 2.3 decades in width, with the centre of the distribution, Ω_0, indicated at each temperature. Furthermore, the observed non-exponential spin-lattice relaxation demonstrates that the distribution of correlation times for the methyl rotation in glassy BPA-PC is heterogeneous in nature and probably due to differences in local packing in the glassy state.

At temperatures higher than − 70 °C, the methyl deuteron line shape corresponds to rapid rotation of the methyl group. This rotation in the rapid exchange limit leads to an averaged field gradient tensor, whose principal axis is along the spatially fixed C − CH$_3$ bond direction. As there is not any

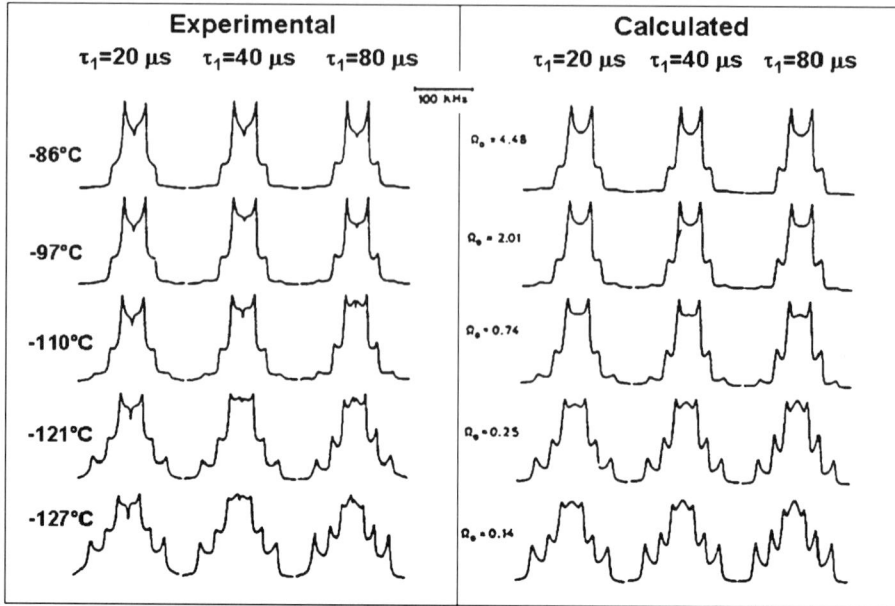

Fig. 39 Experimental and calculated ^{2}H spectra of methyl deuterated BPA-PC at various temperatures and different evolution times, τ_1 (from [35])

change in the ^{2}H spectra from -70 to $100\,^\circ$C [36], it means that conformational changes of the main chain do not occur until $100\,^\circ$C. However, small angle fluctuations up to $\pm 15^\circ$ do not affect the line shape significantly and cannot be ruled out (such fluctuations have been reported from ^{13}C NMR, as described in Sect. 5.3.3).

5.3.2.2
Phenyl Ring Motion

In order to investigate the phenyl ring motions by ^{2}H NMR, deuterated phenyl BPA-PC (BPA-d_4-PC) has been prepared by deuteron substitution in *ortho* position to the carbonate link, and studied in the temperature range from -110 to $120\,^\circ$C [36].

In the high-temperature range above room temperature [37], the ^{2}H NMR spectra shown in Fig. 40 strongly deviate from the Pake doublet characteristic of a rigid system, indicating the presence of rapid molecular motions. Satisfactory simulated spectra (Fig. 41) can be obtained by considering π-flip motions of the phenyl ring, augmented by small angle fluctuations about the same axis, the rms amplitude increasing from $\pm 15^\circ$ at room temperature to $\pm 35^\circ$ at $107\,^\circ$C. In addition, a Gaussian distribution of the rms amplitudes of the small angle fluctuations, with 10° variance for both temperatures, has to

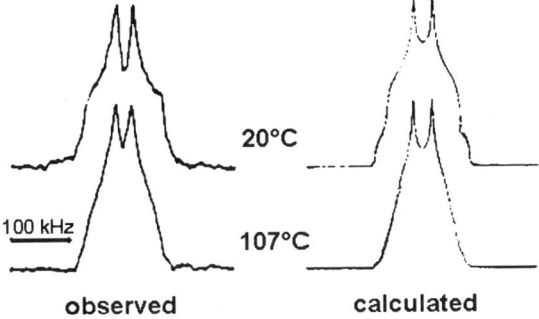

Fig. 40 Experimental and calculated ^{2}H spectra of phenyl deuterated BPA-PC at 20 and 107 °C (from [36])

be introduced, reflecting the heterogeneity between spatially different phenyl groups in the glassy state.

In the temperature range below room temperature, the fully relaxed and partially relaxed ^{2}H NMR spectra [36] are shown in Fig. 41. Whereas at – 113 °C the fully relaxed spectrum only shows a rigid Pake doublet, at higher temperatures the line shapes can be considered as a superposition of spectra in the rigid and rapid exchange limit with a weighting dependent on temperature. The simulated spectra (Fig. 41) have been calculated by consid-

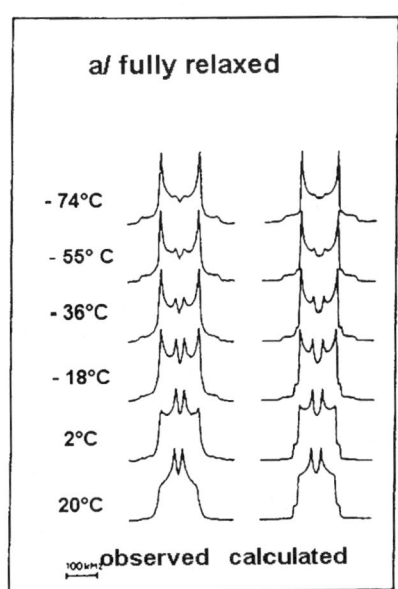

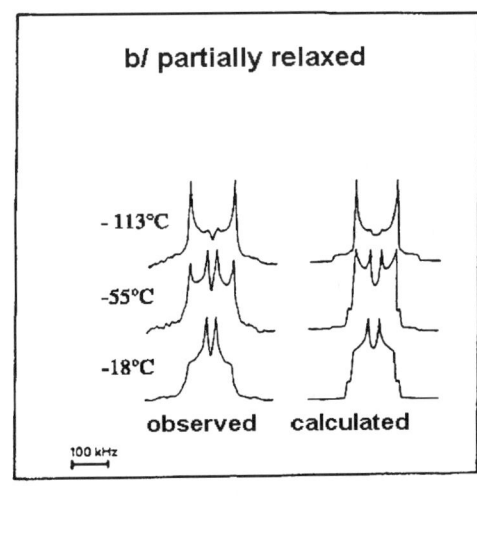

Fig. 41 Experimental and calculated ^{2}H spectra of phenyl deuterated BPA-PC at various temperatures: **a** fully relaxed spectra and **b** partially relaxed spectra (from [36])

ering only π-flip motions with a log-Gaussian distribution of jump rates and a temperature-dependent width varying from 2.6 decades at room temperature to 4.3 decades at $-113\,^{\circ}$C. However, neither partially relaxed nor even fully relaxed observed spectra can be exactly described by a simple π-flip motion. As for the high-temperature range, the motion of the phenyl groups consists of a π-flip augmented by additional small angle fluctuations, different for different phenyl groups. This latter feature is clearly demonstrated by the comparison of the fully and partially relaxed spectra in Fig. 41, as well as through the non-exponential spin-lattice relaxation. Indeed, in the partially relaxed spectra the phenyl groups detected are those with the highest mobility. As the partially relaxed spectra differ strongly from the fully relaxed ones at low temperatures (Fig. 41), the distribution of jump rates must be heterogeneous in nature. Thus, at $-113\,^{\circ}$C, the fully relaxed spectrum does not show any motional narrowed contribution, whereas the partially relaxed one in Fig. 41 proves that even at this low temperature phenyl groups with jump rates greater than 10^5 Hz are present.

For describing this heterogeneous distribution of jump rates, a symmetric log-Gaussian distribution is not satisfactory, as mentioned. An asymmetric distribution with a more rapid decay at low rates, similar to the Williams–Watts distribution, yields an excellent agreement with the experimental data.

As regards the activation energy associated with the phenyl ring motions, the two types of distributions lead to the same mean value of 37 kJ mol^{-1}.

Finally, 2D solid-state ^{2}H NMR has been applied [38] to BPA-d_4-PC in order to investigate the phenyl ring motions in a much lower frequency range (10–1 Hz). Two-dimensional exchange NMR detects the reorientations that occur during a mixing time, t_m, from 100 ms to 1 s, by measuring the angular-dependent NMR frequencies before and after t_m. The 2D frequency spectrum $S(\omega_1, \omega_2, t_m)$, shown in Fig. 42, represents the probability of finding a unit with a frequency ω_1 before t_m and a frequency ω_2 afterwards. If no reorientation takes place during t_m, $\omega_1 = \omega_2$ and the spectrum is confined to the diagonal of the ω_1, ω_2 frequency plane. If molecular reorientation occurs during t_m, then off-diagonal intensity will appear in the ω_1, ω_2 frequency plane. For molecular reorientation occurring by discrete jumps of given amplitude, elliptical ridges are observed in the 2D spectrum, from which the reorientational angle can be obtained directly and model-free. This latter feature is quite important, compared to what is required for line-shape analysis. Furthermore, the distribution of reorientational angles can be directly determined for the considered state. In the particular case of BPA-d_4-PC, with $t_m = 500$ ms, elliptical ridge patterns caused by jumps around the 1,4 axis of the phenyl rings appear below 0 $^{\circ}$C and their intensity increases with decreasing temperature. The 2D plot shown in Fig. 42, obtained at $-40\,^{\circ}$C, yields 120° for the reorientation of the C–^{2}H bonds, which correspond to a 180° flip of the phenyl rings. In addition, the measurements could be fitted, nearly independently of temperature, by considering Gaussian reorientational angle

Fig. 42 Contour plot of a 2D exchange spectrum of phenyl deuterated BPA-PC at – 40 °C, with t_m = 500 ms (from [38])

distributions centred at 0° and 120° and having widths of about 20°. This result is quite important for it clearly demonstrates, without requiring any model, the precise nature of the motions of the phenyl rings in BPA-PC.

5.3.3
^{13}C NMR

The 15.1 MHz MAS-CP ^{13}C NMR spectrum of BPA-PC at room temperature [39] is shown in Fig. 43. It consists of five lines identified, from left to right in order of increasing magnetic field, as: a combination line arising from the carbonyl and non-protonated aromatic carbons, two lines due to the protonated aromatic carbons (the first one corresponding to the carbon *ortho* to

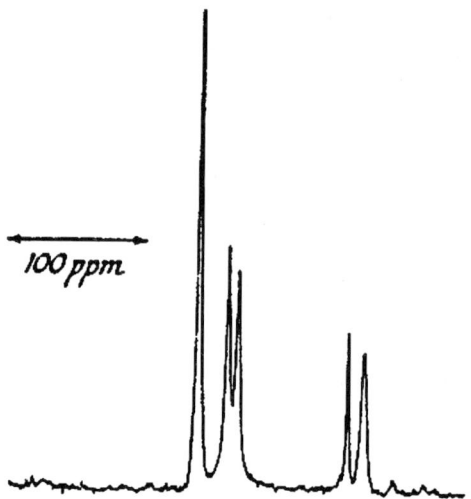

Fig. 43   15.1 MHz MAS-CP ^{13}C NMR spectrum of BPA-PC at room temperature (from [39])

the isopropylidene unit, the second one to the carbon *ortho* to the carbonate), the aliphatic quaternary carbon line and the methyl carbon line.

5.3.3.1
Phenyl Ring Motions

^{13}C NMR has been used to investigate the phenyl ring motions occurring at room temperature in BPA-PC, by using the dipolar rotational spin echo technique [40]. Indeed, the reduction in dipolar coupling between carbons and directly attached protons arising from molecular motion (of frequency

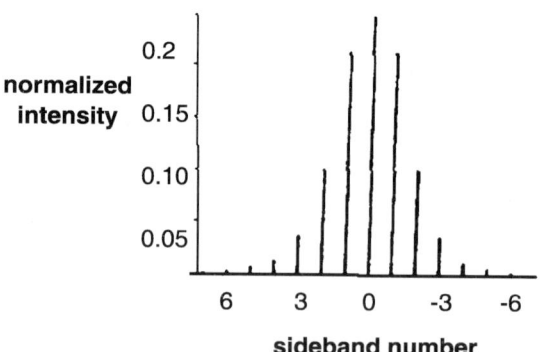

Fig. 44   Experimental dipolar patterns for an aromatic CH pair under MAS at 1.894 kHz for BPA-PC at room temperature (from [40])

comparable to or higher than the dipolar coupling of about 10^5 Hz) allows a measure of the nature and amplitude of the molecular motion. With a spinning frequency of 1.894 kHz, the dipolar line shapes are broken up into spinning sidebands separated by 1.894 kHz, as shown in Fig. 44. In addition to the shape, the ratio of intensity of the second to the first dipolar rotational sidebands constitutes a sensitive measure of the averaging of CH dipolar coupling by molecular motion. At room temperature, this ratio n_2/n_1 is equal to 0.47 for aromatic carbons (MAS at 1.894 kHz). Different motions lead to different dipolar patterns (as shown in Fig. 45) and different intensity values for the sidebands (Table 5).

Any small-amplitude motion is excluded for it could not lead to the large observed dipolar tensor reduction, as well as 90° flips about the 1,4 axis or

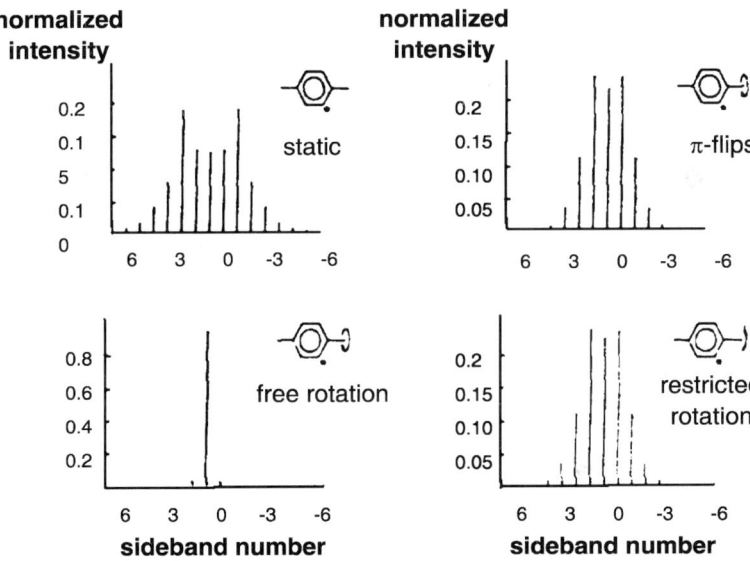

Fig. 45 Calculated dipolar patterns for an aromatic CH pair undergoing MAS at 1.894 kHz, based on four different molecular motions (from [40])

Table 5 Calculated dipolar rotational sideband intensities for an aromatic CH pair undergoing various molecular motions, under MAS at 1.894 kHz (from [40])

Motional model	n_2/n_1
Static	1.49
180° flips	0.47
90° flips	0.36
15° (rms) isotropic motion + 180° flips	0.40
9° (rms) isotropic motion + 11° (rms) C_2 rolls + 180° flips	0.37

free rotation. In contrast, 180° flips of the phenyl rings about the 1,4 axis or oscillations of ± 65° around an equilibrium position can agree with the experimental data. However, a detailed comparison of the intensities of the carbon *ortho* to the isopropylidene unit sidebands on each side of the centre line leads to rejection of the large oscillation motion alone. The best agreement with the dipolar pattern observed at room temperature is obtained by considering for the phenyl ring motion a π-flip mixed with a 14° rms oscillation motion about the 1,4 axis and a 9° rms isotropic main-chain reorientation motion (see Sect. 5.3.2).

The frequency heterogeneity of ring motions in BPA-PC is directly shown by the non-exponential decay of protonated aromatic ^{13}C spin-lattice relaxation, either T_1 or $T_{1\rho}$ [39], as illustrated in Fig. 46 for T_1 and $T_{1\rho}$. From T_1 measurements at 15 and 50 MHz, the occurrence of phenyl ring motions in BPA-PC, at room temperature, with a frequency around 15×10^6 Hz is deduced. From $T_{1\rho}$ measurements, ring motions around 3×10^5 Hz are also present, resulting in frequency heterogeneity over about 1.5 decades. Such

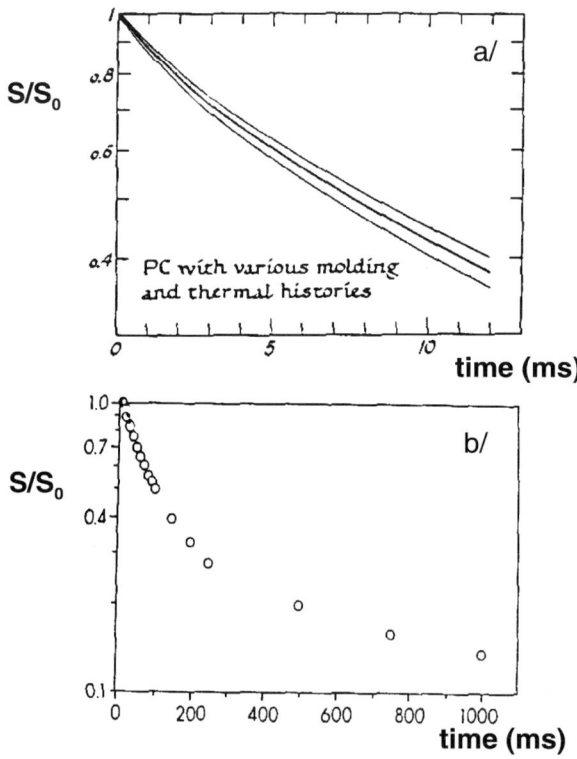

Fig. 46 Relaxation decays at room temperature of the protonated aromatic carbon region of BPA-PC: **a** 2.1 kHz MAS $T_{1\rho}$ decay and **b** 15.1 MHz T_1 decay (from [39, 44])

a heterogeneity quite likely originates from the difference in packing around the moving phenyl rings.

Confirmation of the type of phenyl ring motion has been obtained on a ^{13}C-labelled BPA-PC, where one of the two carbons in each phenyl group *ortho* to the carbonate is isotopically enriched [41]. Chemical shift anisotropy line shape analysis has been performed both in the low-temperature limit at − 160 °C and in the high-temperature limit at 120 °C. The spectra are shown in Fig. 47. The low-temperature data are well simulated with a powder pattern function corresponding to an absence of ring motions. For the high-temperature data, the best simulation is obtained by considering phenyl ring π-flips plus restricted rotation of ± 36° [32].

The temperature dependence of the phenyl ring motions has also been studied by ^{13}C NMR at two different frequencies, 22.6 MHz [32] and 62.9 MHz [34], through the chemical shift anisotropy of the aromatic protonated carbons.

First, it is worth noting that the activation energy value derived from the 22.6 MHz data strongly depends on the motional model considered. Thus, with a two-site π-flip jump model, leading to a single exponential correla-

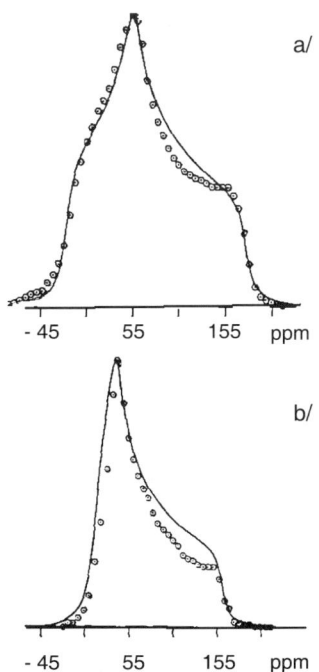

Fig. 47 ^{13}C chemical shift anisotropy line shapes for ^{13}C-enriched BPA-PC. Experimental (*points*) and simulation (*continuous line*) data at **a** low-temperatures and **b** high-temperatures (from [41])

tion function, a value of 11 kJ mol^{-1} is obtained for the phenyl ring motion. In contrast, by considering a simultaneous model in which π-flips are superimposed on low-amplitude vibrations, both about the 1,4 axis of the phenyl ring, an activation energy of 26 kJ mol^{-1} is obtained [42].

By considering the data obtained at the two frequencies, the motions are differently probed, allowing for a discrimination between a single exponential and a distribution, and in the latter between inhomogeneous or homogeneous. The analysis has been performed on data obtained below 20 °C, for between 120 and 20 °C the ring π-flips are in the limit of rapid motion and the line shapes are all of the same general shape as observed at + 120 °C (Fig. 47b). The best simulations [34] for data from 0 to – 80 °C are shown in Figs. 48 and 49 at 62.9 MHz and 22.6 MHz, respectively. The simulations are obtained by considering the simultaneous model with a temperature-dependent oscillation amplitude (from ± 32° at + 120 °C to ±12° at – 80 °C) and an inhomogeneous distribution of correlation times described by the Williams–Watts expression (with a value of 0.154 for α). It is worth noting that the same simulation conditions account quite well for the phenyl proton T_1 and $T_{1\rho}$ data shown in Fig. 38; a slightly better fit is achieved that the one shown in Fig. 38. This proves that the model can account for data corresponding to a much broader frequency range than the one involved in the chemical shift anisotropy line shape.

As regards the activation energy deduced from the temperature dependence of the centre correlation time, τ_p, obtained from line shape analysis, it is now equal to 50 kJ mol^{-1}.

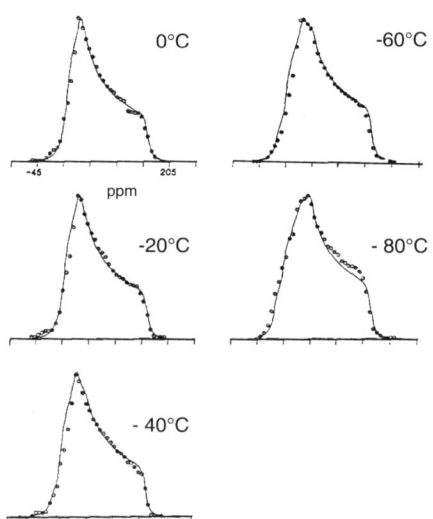

Fig. 48 ^{13}C chemical shift anisotropy line shapes at 62.9 MHz at several temperatures: experimental (*points*) and simulation (*continuous line*) (from [42])

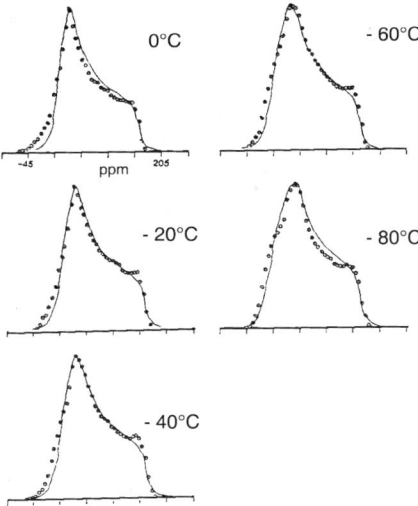

Fig. 49 ^{13}C chemical shift anisotropy line shapes at 22.6 MHz at several temperatures: experimental (*points*) and simulation (*continuous line*) (from [42])

5.3.3.2
Main-Chain Reorientation

In order to get a more precise description of the molecular motions occurring in BPA-PC, it is interesting to check whether, in addition to the ring motions about the 1,4 axis already described, main-chain reorientation takes place.

For this purpose, analysis of the methyl group behaviour is appropriate, as mentioned for ^{2}H NMR investigations (Sect. 5.3.2). Indeed, the ring motions about the 1,4 axis do not drag the methyl groups along, whereas main-chain motions do.

Methyl dipolar coupling has been applied [39] at room temperature to investigate the occurrence of main-chain reorientation in BPA-PC. Due to the

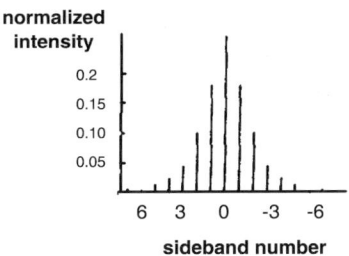

Fig. 50 Experimental dipolar pattern for methyl carbon under MAS at 947 Hz for BPA-PC (from [40])

rapid rotation at room temperature of the methyl group around its symmetry axis, the methyl has the same dipolar coupling as an ordinary CH pair. The experimental dipolar pattern under MAS at 947 Hz is shown in Fig. 50. It can be accounted for by considering a 10° rms isotropic main-chain motion [43], which corresponds to a 20–25° total angular main-chain displacement. These motions occur in a 10^5 Hz frequency range at room temperature and T_1 relaxation shows a non-exponential decay [44], indicative of a heterogeneity.

5.3.3.3
Carbonate Motions

The overlap of the carbonate carbon and the non-protonated aromatic carbon lines in MAS-CP ^{13}C NMR does not allow one to study the carbonate motions in BPA-PC.

To bypass this difficulty, BPA-PC isotopically labelled with ^{13}C at the carbonyl position has been synthesised and chemical shift anisotropy line shape has been studied from 100 °C down to – 255 °C [45].

The ^{13}C NMR spectrum obtained at 25 MHz and 12 °C is shown in Fig. 51b. Up to 100 °C, essentially the same spectra are observed. On decreasing temperature, there is a slow expansion of the band and at – 177 °C (Fig. 51a) the shoulder on the right-hand side is slightly more obvious. Furthermore, measurements performed at 15 MHz show that no substantial changes in the pattern occur between – 96 and – 255 °C.

In spite of the limited change of line shape with temperature, dielectric relaxation unambiguously evidences motions of the carbonate group of BPA-

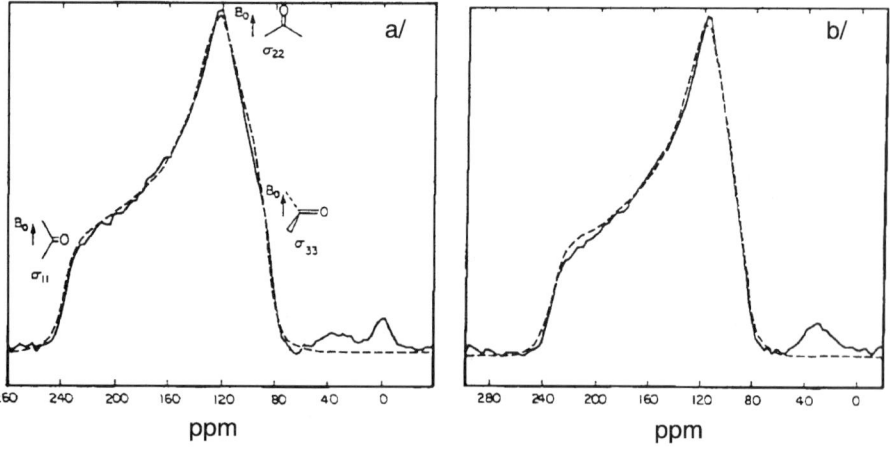

Fig. 51 Experimental (*solid line*) and simulated (*dotted line*) ^{13}C NMR spectrum at 25 MHz for BPA-PC isotopically labelled with ^{13}C at the carboxyl position: **a** $T = -177$ °C and **b** $T = 12$ °C (from [45])

PC in the considered temperature range. Thus, simulation of the patterns has been performed by using the chemical shift tensor elements shown in Fig. 51a. At – 177 °C, the simulation corresponds to a rigid system. At 12 °C, the motional model considers a jump of the carbonyl between two sites of equal energy separated by an energy barrier; at this temperature it is assumed that reorientation is very fast on the effective NMR time scale. A reasonable fit is obtained by considering that the carbonyl rotates around an axis perpendicular to the C = O bond in the plane formed by the three oxygen atoms, with an angle between the two sites of 40° (such a motion corresponds to a partial averaging of the σ_2 and σ_3 tensor elements). However, a better fit, shown in Fig. 51b, is achieved by superimposing to this motion a rotation of 15° about the C = O bond. Of course, this latter motion implies that it is accompanied by a reorientation of the phenyl ring 1,4 axis and, therefore, of the main chain.

5.3.4
Hydrostatic Pressure Effect

All the solid-state NMR investigations described provide a detailed description of the intramolecular motions performed in glassy BPA-PC. Unfortunately, they do not give any information on the intermolecular cooperativity that could be associated with some of them.

To address this question, an interesting study has been carried out on the effect of hydrostatic pressure on the proton NMR line width of BPA-PC [44]. Actually, the transverse relaxation time, ^1H T_2, inversely proportional to the line width, is a more convenient parameter.

The temperature dependence of ^1H T_2 at various hydrostatic pressures is shown in Fig. 52 for BPA-PC (Fig. 52a) and for a BPA-PC deuterated at all positions except in aromatic position *ortho* to the isopropylidene group, BPA-d_{10}-PC (Fig. 52b).

The increase of T_2 observed in BPA-PC between – 170 and – 120 °C does not exist in BPA-d_{10}-PC, it is assigned to the occurrence of the methyl rotation around its C_3 axis. Furthermore, the temperature dependence is unaffected by pressure, indicating a motion with zero or small activation volume, consistent with a methyl rotation.

At higher temperatures, a pressure-dependent process occurs from – 100 to 60 °C, which is reversible. Such a broad temperature range indicates a site heterogeneity and a broad distribution of activation energies. By considering a thermally activated motion, an estimate of the mean activation energy leads to 38 kJ mol^{-1} and for the activation volume to 25 cm^3 mol^{-1} for the two types of polycarbonates.

The shift of the temperature dependence of T_2 towards higher temperatures by about 18 °C under 1760 bars in the case of BPA-PC, and about 35 °C under 1450 bars for BPA-d_{10}-PC, clearly proves that there is an intermolecu-

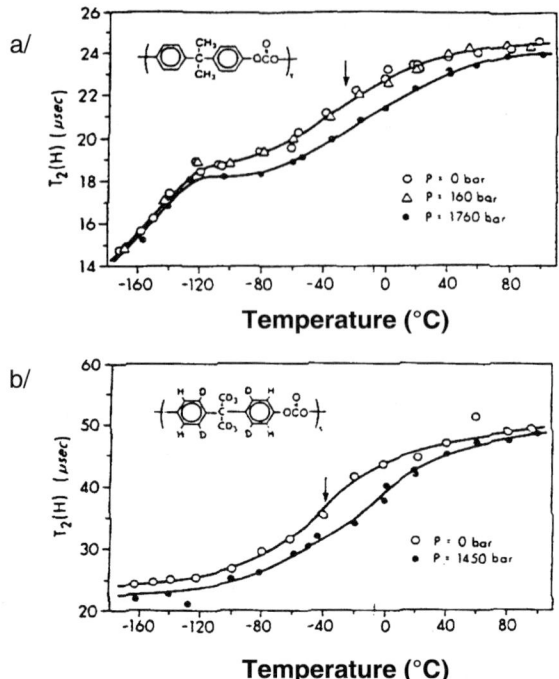

Fig. 52 Temperature dependence of ^{1}H T_2 at 21.3 MHz at different external hydrostatic pressures. The *arrow* indicates the approximate centre of the pressure-dependent line narrowing process: **a** BPA-PC and **b** BPA-d_{10}-PC (from [44])

lar contribution in the ring motions concerned in the ^{1}H T_2 change, which is modified by the increased packing generated by the external pressure. However, it is not possible to get more details about the type of motion (ring π-flip, ring oscillation or main-chain reorientation) that is most sensitive to local packing.

5.3.5
Conclusion on Solid-State NMR Investigations

As described, solid-state ^{1}H, ^{2}H and ^{13}C NMR provide quite a detailed picture of the motions occurring in glassy BPA-PC.

The methyl groups undergo rotational motions about the C_3 axis at a frequency of 10^5 Hz at – 170 °C and reach the rapid motion limit at – 100 °C. The associated activation energy is about 8 kJ mol^{-1}. In addition, a heterogeneity for spatially different methyl groups exists. However, this heterogeneity seems to be of intramolecular origin since the methyl motions are unaffected by an external hydrostatic pressure, which they would be if the heterogeneity were of intermolecular origin.

The main chain exhibits, at room temperature, an isotropic rotational motion with 10° rms at a frequency around 105 Hz, also showing a wide distribution of frequencies.

The phenyl ring motions can be considered a result of π-flips on which are superimposed: (i) oscillations whose amplitude (rms) increases from 12° at -80 °C to 32° at 120 °C, and (ii) isotropic main-chain reorientation of 10° rms. Quite a large frequency heterogeneity exists between spatially different phenyl groups: 3×10^5 Hz to 15×10^6 Hz at 20 °C, and 4 decades at -113 °C. This heterogeneous distribution is well accounted for by a Williams–Watts distribution with an exponent α equal to 0.154. These phenyl ring motions possess an intermolecular contribution for they slow down on applying an external hydrostatic pressure. The temperature shift is around 20 °C for 1.5 kbar, leading to an activation volume of about 25 cm^3 mol^{-1}. It is worth noting that the π-flip ring motions are directly proven by 2D ^{2}H NMR measurements, without any model requirement.

The carbonyl motions of the carbonate group can be accounted for by considering jumps between two sites, corresponding to a rotation around an axis perpendicular to the C $=$ O bond in the plane formed by the three oxygen atoms, with an angle of 40° between the two sites, over which is superimposed a rotation of 15° about the C $=$ O bond. This latter motion requires a main-chain reorientation motion.

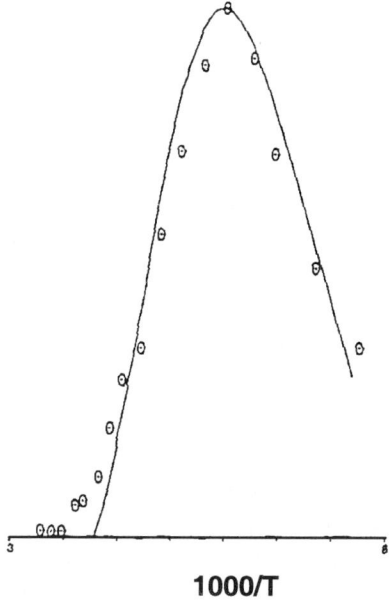

1000/T

Fig. 53 Dynamic mechanical loss spectrum of BPA-PC. The *solid line* is the result of simulation using the phenyl ring motion characteristics (from [34])

As regards the activation energy, it should be noted that the resulting value depends on both the width of the frequency range, and on the relevance of the motional model used. Thus, for phenyl ring motions, by considering the contributions to the motions and the Williams–Watts distribution with a equal to 0.154, an activation energy of 50 kJ mol^{-1} is derived. However, it is fair to point out that some NMR measurements, like spin-lattice relaxation times T_1 and $T_{1\rho}$ as a function of temperature, lead to the direct determination of a temperature-frequency pair, independently of any model. Thus, in the case of phenyl ring motions, ^{1}H T_1 and ^{1}H $T_{1\rho}$ lead to results in good agreement with the relaxation map obtained from mechanical and dielectric measurements on BPA-PC, corresponding to an activation energy of about 50 kJ mol^{-1}.

In conclusion, it is unambiguous from the solid-state NMR investigations that phenyl ring motions are involved in the mechanical β transition of BPA-PC. Additional support for this statement comes from the fact that the position and shape of the mechanical dynamic loss, G'', can be well simulated by using the activation parameters and the Williams–Watts exponent deduced from the analysis of the phenyl ring motions [34], as shown in Fig. 53.

5.4
Atomistic Modelling

From the various experimental investigations, it is clear that carbonate motions, as well as phenyl ring motions, are involved in the mechanical β transition of BPA-PC. The intermolecular contribution has been evidenced by several authors and the cooperative character of the motions has been pointed out. However, neither of the considered techniques can provide detailed information about the nature of intra- or inter-cooperativity occurring in the glassy state.

Atomistic modelling applied to a single BPA-PC chain provides information on the local chain conformations and allows determination of the energy barriers encountered for methyl, phenyl ring, or carbonate motions. More interestingly, atomistic modelling can be used to analyse chain dynamics in bulk BPA-PC and give insight on the cooperativity (intra- and intermolecular) associated with the phenyl ring or carbonate motions.

These two approaches will be considered.

5.4.1
Single BPA-PC Chain

Since the late 1960s, BPA-PC has been studied by various atomistic modelling approaches: molecular mechanics or quantum mechanics, either ab initio or semi-empirical. The references about these studies are given in [46].

The repeat unit of BPA-PC is shown in Fig. 54 in the conformation where all the torsion angles are zero, corresponding to a fully planar structure.

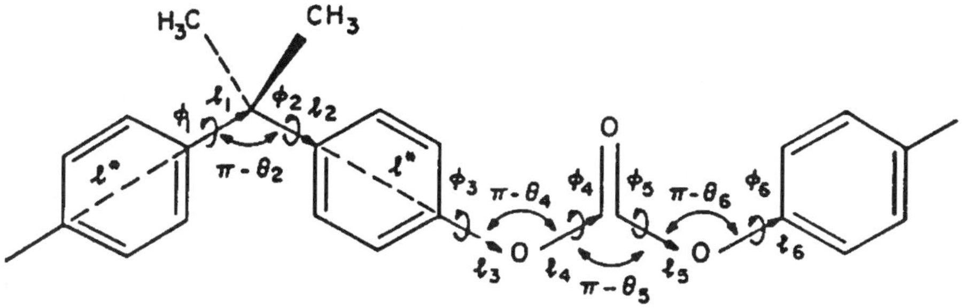

Fig. 54 Repeat unit of BPA-PC. In the given planar reference conformation, all the torsion angles assume a value of zero (from [46])

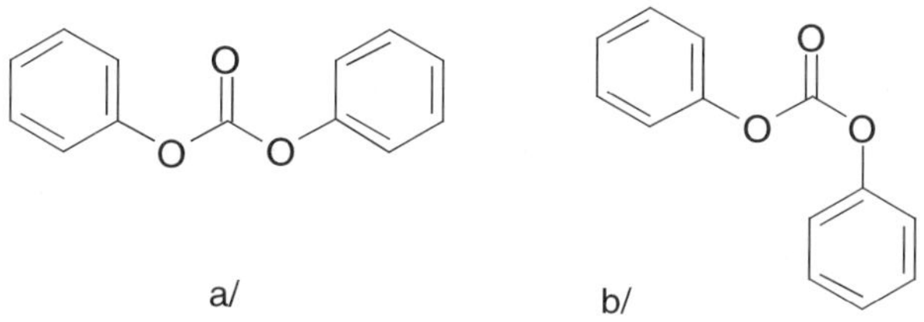

a/   b/

Fig. 55 a *trans-trans-* and **b** *trans-cis-*Diphenyl carbonate (from [50])

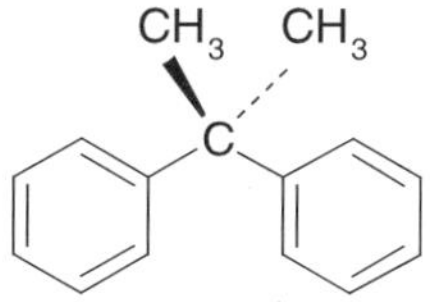

Fig. 56 2,2 Diphenyl propane

Such a repeat unit can be modelled by considering the conformational characteristics of the fragment molecules, diphenyl carbonate (DPC) and 2,2′ diphenyl propane (DPP), shown in Figs. 55 and 56, respectively.

In spite of ab initio calculations performed on these molecules [47], more detail can be found in the semi-empirical quantum mechanics approach developed in [48]. The most relevant results will be described.

The convention used for defining dihedral angles is: (i) to set them equal to 0° when two planes defining them are coplanar, and (ii) for positive values of the angle of rotation, atoms are being rotated about an axis in a counter clockwise direction when looking down from the first atom defining the rotation axis toward the second atom defining the rotation axis (since a 360° rotation about any axis is a full rotation, a negative rotation by − n degrees has the same effect as a positive rotation by (360 − n) degrees). However, in discussing motions, it is often more convenient to refer to the value of a given torsion angle in terms of how much it differs from the same angle in the minimum-energy geometry of the molecule.

5.4.1.1
2,2′ Diphenyl Propane

The optimised geometry of DPP has C_2 symmetry. The phenyl rings are both tilted by 48.1° relative to the $C_p - C_1 - C_p$ plane (where C_p denotes the phenyl carbon linked to C_1); X-ray data give 55°.

Concerning the methyl groups, there are three minima at 0°, 120° and 240° about the $C_1 - C_m$ axis. There are three maxima at methyl rotations of 60°, 180° and 300°. The barrier to methyl group rotation exhibits its minimum value if one methyl group is at its most favoured conformation while the other one is going through its maximum-energy conformation. This leads one to consider a simultaneous out-of-phase rotation of the two methyl groups in the same direction. When the barrier heights for the methyl are calculated, allowing re-optimisation of the phenyl ring torsional orientations about the $C_1 - C_p$ axes, a value of 28 kJ mol^{-1} is obtained. A small reorientation by about 15° of one of the two phenyl rings is required to remove a steric repulsion between a methyl hydrogen and a phenyl hydrogen in the *ortho* position.

As regards the rotations of the phenyl rings, it appears that torsional oscillations of $\pm 30°$ only require 28 kJ mol^{-1}. In contrast, for a π-flip of one phenyl ring with the other one being held fixed at any other torsion angle, quite a high energy barrier, around 100 kJ mol^{-1}, has to be overcome. Therefore, the two phenyl rings will have to be moving synchronously in the lowest barrier pathway for phenyl rotation, which corresponds to the simultaneous rotation of the two phenyl rings by the same magnitude but with opposite signs about their respective $C_1 - C_p$ axes. Two barriers of about 28 kJ mol^{-1} are obtained at rotations of 90° and 270° relative to the minimum-energy position.

5.4.1.2
Diphenyl Carbonate

At first, the carbonate group is planar due to electron delocalisation. The phenyl rings are rotated out of the plane of the carbonate group because of opposing forces. On the one hand, electron delocalisation between the carbonate group and the phenyl rings favours an all-planar conformation, on the other hand the steric hindrance between the *ortho* hydrogen on the phenyl ring and the carbonate oxygen favours a perpendicular arrangement of the phenyl rings and the carbonate group. As a result, the phenyl rings are tilted relative to the carbonate plane by a torsion angle of $- 44°$ around the $O'' - C_p$ (or the $O' - C_p$) axis; X-ray leads to 45°.

As regards the rotation of the phenyl rings about the $O - C_p$ bonds, there is a maximum at 176°, corresponding to a torsion angle of 220° relative to the minimum-energy conformation. The energy barrier is 7 kJ mol^{-1}, which is rather low.

In DPC, the carbonyl group can rotate around the $O' - O''$ axis. There are two energy maxima at torsion angles of $45°$ and $-26°$ about the $O' - O''$ axis, with the same maximum value of $13 \, kJ \, mol^{-1}$. The difference in torsion angles comes from the fact that the phenyl rings are tilted by the same amount and in the same direction relative to the plane of the carbonate linkage. Consequently, positive and negative $C = O$ rotations are non-equivalent. So, rotations of $C = O$ over an amplitude of $70°$ can occur. Larger amplitudes would lead to a rapid increase of energy.

Another feature deals with the *trans, trans* and *trans, cis* conformations of DPC, as shown in Fig. 55. The calculations indicate that the *trans, trans* conformation of DPC is preferred by $4.7 \, kJ \, mol^{-1}$ over the *trans, cis* conformation, with a barrier of $4.7 \, kJ \, mol^{-1}$ between these two conformations. The *trans, cis* energy minimum is at a torsion angle of $150°$ about the $C_c - O''$ axis relative to the *trans, trans* geometry.

5.4.1.3
BPA-PC Repeat Unit

Actually, the methyl and carbonyl dynamics within the BPA-PC repeat unit (Fig. 54) can be conveniently analysed by considering the methyl motion of DPP and the carbonyl motions of DPC. However, in the case of phenyl ring motions, neither DPP nor DPC correspond to the situation within BPA-PC. Indeed, in the latter the isopropylidene and the carbonate groups bonded to the phenyl rings are *para* to one another, with the $C_1 - C_p$ bonds lying almost on the same axis as the $O - C_p$ bonds.

This question has been addressed [49] by performing the same semi-empirical quantum calculations applied to DPP and DPC to the model molecule shown in Fig. 57 and denoted "large carbonate".

The lowest barrier rotation pathways for the phenyl ring rotation occurs for the synchronous motion of the two phenyl rings. Two energy barriers are encountered at rotations of $90°$ and $270°$ relative to the minimum-energy conformation, corresponding to an energy value of $42 \, kJ \, mol^{-1}$. It is worth noting that this value is very close to that found in DPP, $42 \, kJ \, mol^{-1}$.

Another interesting feature deals with the probability of occurrence of *trans, cis* (or *cis, trans*) conformation of the carbonate group in BPA-PC. The

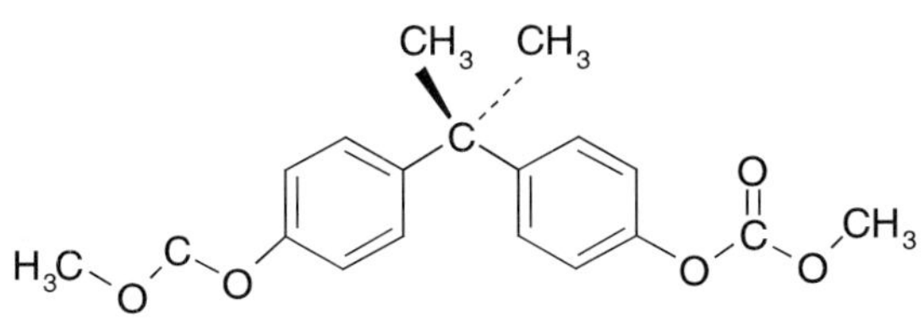

Fig. 57 Representation of the "large carbonate" model molecule

energy difference relative to the *trans, trans* conformation calculated in DPC leads, at the glass–rubber transition temperature (150 °C), to about 34% of *trans, cis* or *cis, trans* conformation in a BPA-PC chain.

5.4.2
Chain Dynamics in Bulk BPA-PC

In order to investigate the intra- and intermolecular cooperativity associated with phenyl and carbonyl motions in glassy BPA-PC, it is interesting to perform dynamic molecular modelling in bulk. Two different modelling approaches have been used to yield information on the nature of cooperativity.

5.4.2.1
Quasi-static Modelling

A method for the detailed atomistic modelling of amorphous glassy polymers has been developed [50] and applied to atactic polypropylene.

The model system is a cube of glassy polymer with 3D periodic boundaries, filled with chain segments at a density corresponding to the experimental value for the considered polymer. The entire contents of the cube are formed from a single "parent chain" with the chemical structure of the polymer. The cube can thus be considered as part of an infinite medium, consisting of displaced images of the same chain, as shown on Fig. 58.

The model rests on the following assumptions:

– Thermal motions are not incorporated, i.e. it is a static model. Temperature enters only indirectly, through specification of the density and statistical weights.
– Bond lengths and bond angles are fixed. Molecular movements occur exclusively through rotation around skeletal bonds.

An initial guess structure is constructed by generating the chain in a bond-by-bond fashion in the cube, observing periodic boundary conditions wherever a bond–border intersection occurs. At each step the probability of the internal rotation angle defined by the next bond takes into account both the statistical weights of the various conformations, according to the rotational isomeric state model (RIS) specific to the considered polymer, and the long-range interactions. In this way, the generated polymer conformation does not depart much from an unperturbed random coil.

The initial guess structure is subjected to a total energy minimisation by slightly modifying the internal rotation angles in order to reach a microscopic structure in a mechanical equilibrium, i.e. corresponding to a true potential energy minimum.

The quasi-static modelling of chain dynamics [51] involves starting with an energy-minimised structure in which one degree of freedom is selected

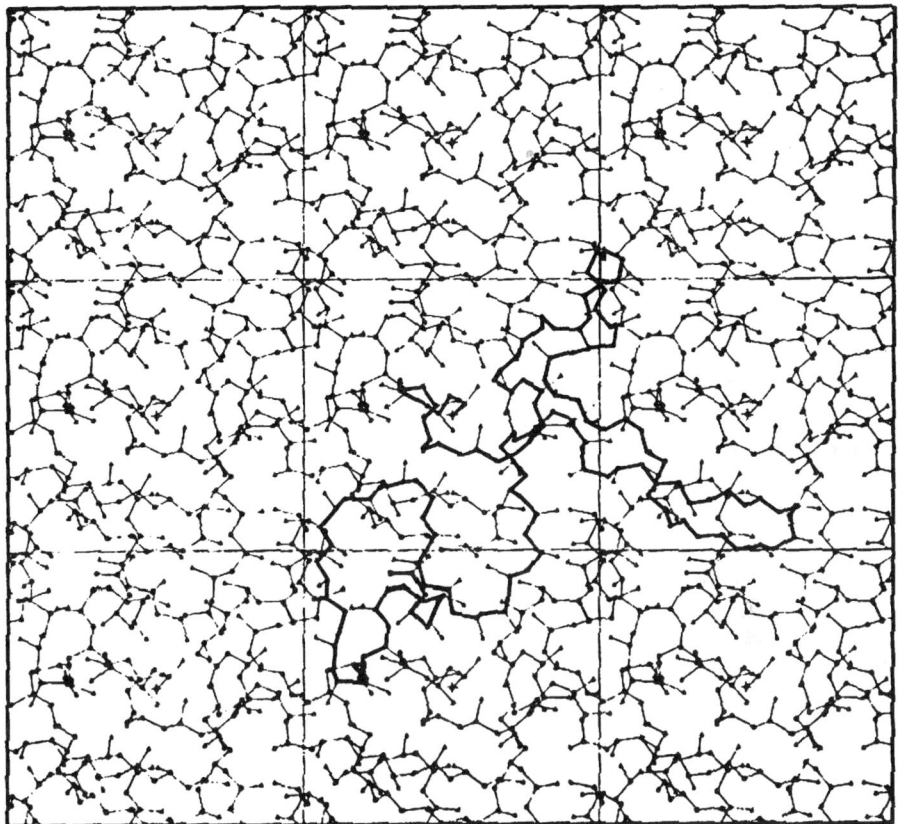

Fig. 58 Model structure in the cube (*centre*) and height of its neighbouring images, projected on the *xy* plane. The parent chain is traced with a *bold line*, the image chains are drawn as *thin lines*. Only carbons are indicated (from [50])

(for example, the internal rotation angle of a specific main-chain C – C bond or of a bond between a side chain and the main chain). This degree of freedom is changed by a small amount, and then held fixed at the new value while all the other degrees of freedom in the system are systematically adjusted to minimise the potential energy of the microstructure. This new state is generally of higher energy than the starting structure; it is said to represent a "constrained minimum". By repeating this process of small imposed microstructural changes and constrained minimisation, a path in the overall potential energy is traced, going from the initial value of the chosen degree of freedom to the final value.

In the case of BPA-PC [51], the chosen microstructures had cube-edge lengths of 18.44 Å and a degree of polymerisation of 35 ($Mw = 4532$). There are 485 atoms or atom groups in each cube (methyl groups are represented as spherical pseudo-atoms). The force field [46] has fixed bond lengths and fixed bond angles. The energy minimisation performed after each step in changing the degree of freedom only involves torsion angles. Rotations around $C_1 - C_p$ (between the phenyl ring and the isopropylidene group) and $O - C_p$ bonds were examined, and incremental rotation steps of $20°$ were applied. All bonds of the same type ($C_1 - C_p$ or $O \neg C_p$) from one structure were driven until the saddle point conformations, corresponding to the first energy maximum, so that a total of 60 energy barrier "events" were probed.

Phenyl Ring Flips

The quasi-static simulation of ring flips led to the following results:

- When rotation was driven around a $C_1 - C_p$ bond, the predominant process was a flip of the adjoining phenyl ring.
- When rotation was driven around a $O - C_p$ bond, of the 30 simulations performed, in 8 the resulting process was a flip of the adjoining phenyl ring, and in the remainder it was a conformational rearrangement involving the adjoining carbonate group.
- The mean value of the energy barrier for the 38 resulting phenyl ring flips was 45 ± 28 kJ mol^{-1}. The peak energy barriers showed a broad distribution, which could be fitted to a Williams–Watts distribution function with a between 0.1 and 0.2. By applying the transition state theory, the distribution of ring flip frequencies could be derived, as shown in Fig. 59.

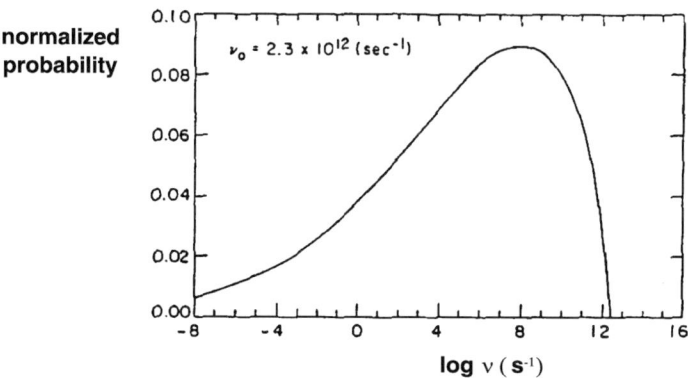

Fig. 59 Distribution of ring π-flip frequencies derived from simulations (from [51])

Carbonate Group

When the torsion angle around the $O - C_p$ bond is driven, the system responds differently depending on which conformational change is energetically more favourable: adjacent ring flip, conformational orientation change of the carbonate group, or a combination of the two. The intermolecular packing about the torsion angle is the deciding influence on whether the ring will flip or the carbonate group will change its conformation. The carbonate group changes its conformation through changes in the inner torsion angles (angles ϕ_4 and ϕ_5 in Fig. 54).

The mean value of the energy barrier for carbonate group rearrangements is 43 ± 28 kJ mol^{-1}. The energy barrier distribution is quite similar to that obtained for ring flips.

The wide distributions observed in both cases have their origin in the variety of local environments governed by intermolecular interactions and the strength of molecular packing influences. Thus, there is clearly an intermolecular contribution in the energy barriers.

Main-Chain Motion

Main-chain motion was analysed through the movements of the BPA unit as a whole.

In the case where rings flipped, the process was predominantly "rocking": the BPA group rotated about an axis parallel to the chain backbone (i.e. parallel to the $O' - O''$ axis) and the rms average of the main-chain motion was around $13°$ (67% of BPA changes were less than $15°$).

Where carbonate groups changed conformations, the rms average of the main-chain motion of the neighbouring BPA groups was around $11°$ (80% of changes were less than $15°$).

Intramolecular Cooperativity

For the phenyl ring flips, strong cooperativity across the isopropylidene unit between the two adjoining rings has been found. This is clearly illustrated in Fig. 60, which shows the potential energy contour map for the two torsion angles ϕ_1 and ϕ_2 (Fig. 54) in DPP. The minimum energy conformations are indicated by $\times$ and the lowest energy pathways by dotted lines. This figure also shows the actual paths found during several different ring flip simulations, where the end of the path indicated by the empty symbol is the starting position, and the position marked with a filled symbol is the last simulated point (corresponding to the first energy maximum of the whole amorphous cell).

For the carbonate conformation change, it is observed in some cases, that phenyl rings on both sides rotate, indicating an intramolecular cooperativity between carbonate and phenyl groups.

Effect of Intermolecular Packing

When the energy barrier for phenyl ring π-flips found in bulk BPA-PC (45 ± 28 kJ mol^{-1}) is compared to the value calculated for the BPA-PC repeat unit (42 kJ mol^{-1}), there is not too much difference. However, this energy barrier

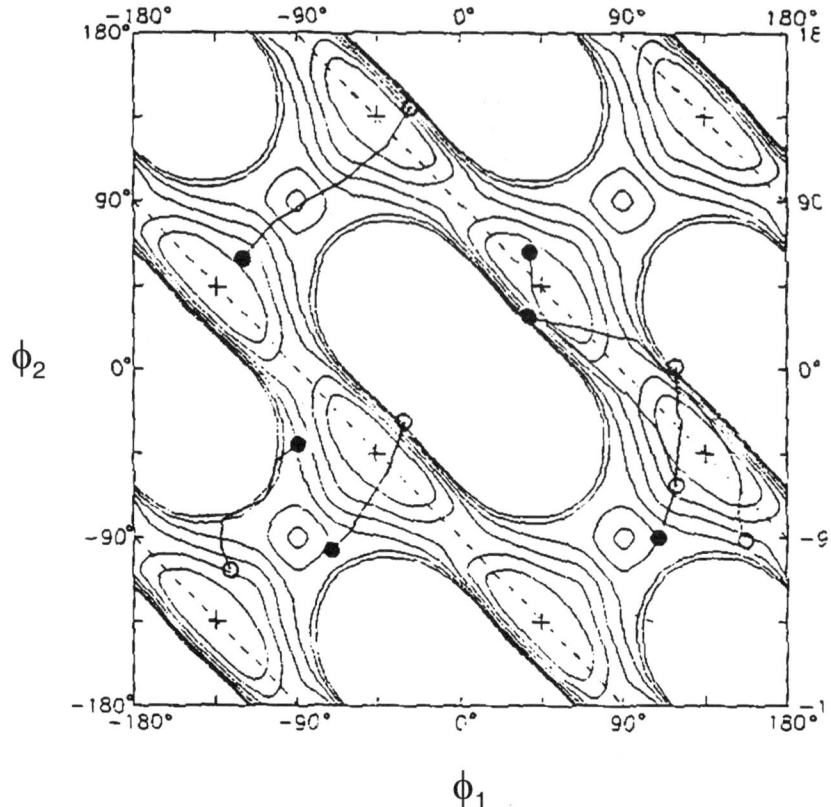

Fig. 60 Potential energy contour map of torsion angles ϕ_1, ϕ_2 for DPP. Some pathways observed in simulations are drawn (from [51])

refers to the mean value obtained in the bulk and the broad distribution originates from the large distribution of local density and intermolecular packing.

The effect of intermolecular packing is quite well illustrated in considering the potential energy contour map of the ϕ_1, ϕ_2 torsion angles (Fig. 60). Indeed, the starting positions of the torsion angles at the initial system energy minima are not close to the intramolecular minima calculated from DPP and represented by the × symbols. Such a behaviour comes from the strong influence of the intermolecular interactions.

The phenyl ring π-flip not only has the short range intramolecular influence described, but also a long range effect, as illustrated in Figs. 61 and 62.

Figure 61 shows the change of the various torsion angles along the chain, between the initial conformation and that corresponding to the energy peak when the driven torsion angle is ϕ_9, as indicated with an arrow. In addition to the large changes of the near neighbour torsion angles along the chain sequence, which have been discussed in the intramolecular cooperativity, large

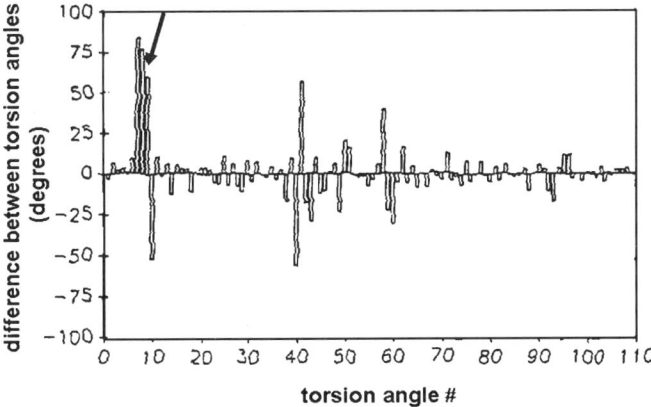

Fig. 61 Difference between the values of the torsion angles of the chain in the initial conformation and the conformation of the energy peak. The driven torsion angle ϕ_9 is indicated with an *arrow* (from [51])

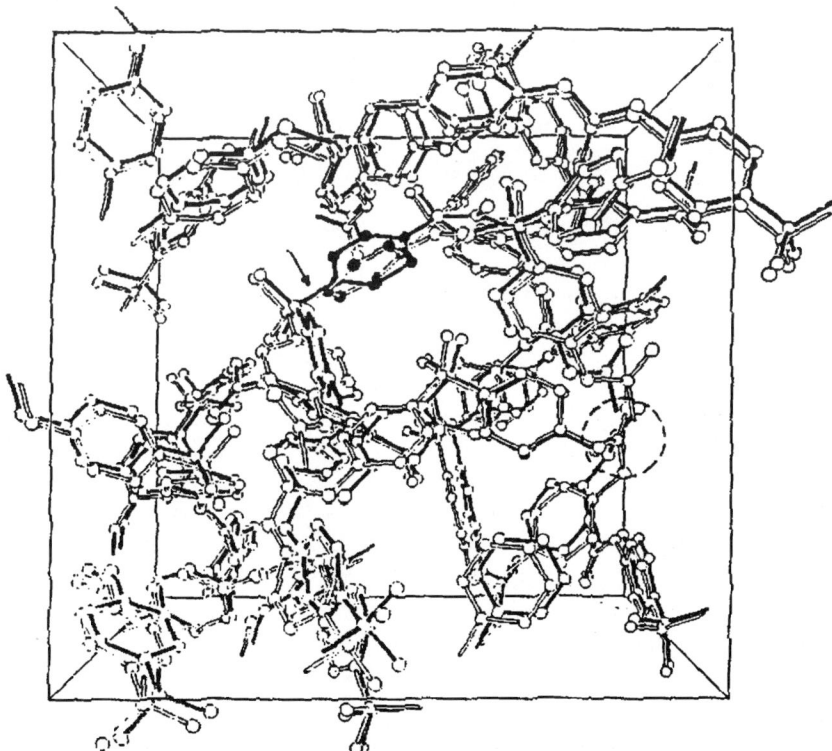

Fig. 62 Simulation cube, with hydrogens omitted and atomic radii decreased for clarity. The *faint pattern* shows the initial conformation of the system, while the *bold pattern* gives the conformation at the peak energy. The third phenyl ring that is rotating is *shaded black* and an *arrow* identifies the driven torsion angle (from [51])

changes are observed for torsion angles located around ϕ_{40} and ϕ_{60}, corresponding to bonds far-away along the chain backbone. The torsion angles around ϕ_{40} correspond to the carbonate group circled in Fig. 62. This particular carbonate group is far from the intramolecular *trans, trans* energy minimum and resides in a very broad local potential energy well. This same carbonate frequently changed conformations when other ring flip simulations were performed, independently of how spatially far or near it was to the flipping ring. Thus, the effects of a ring π-flip are generally very far-reaching in the structure. Furthermore, Fig. 62 evidences that all the groups within the cubic cell undergo displacements to accommodate the phenyl ring flip perturbation.

For the carbonate group, the activation energy obtained in the bulk cannot be compared to that determined for DPC as it does not concern the same rotation. In the latter case, calculations refer to a rotation around the $O'O''$ axis,

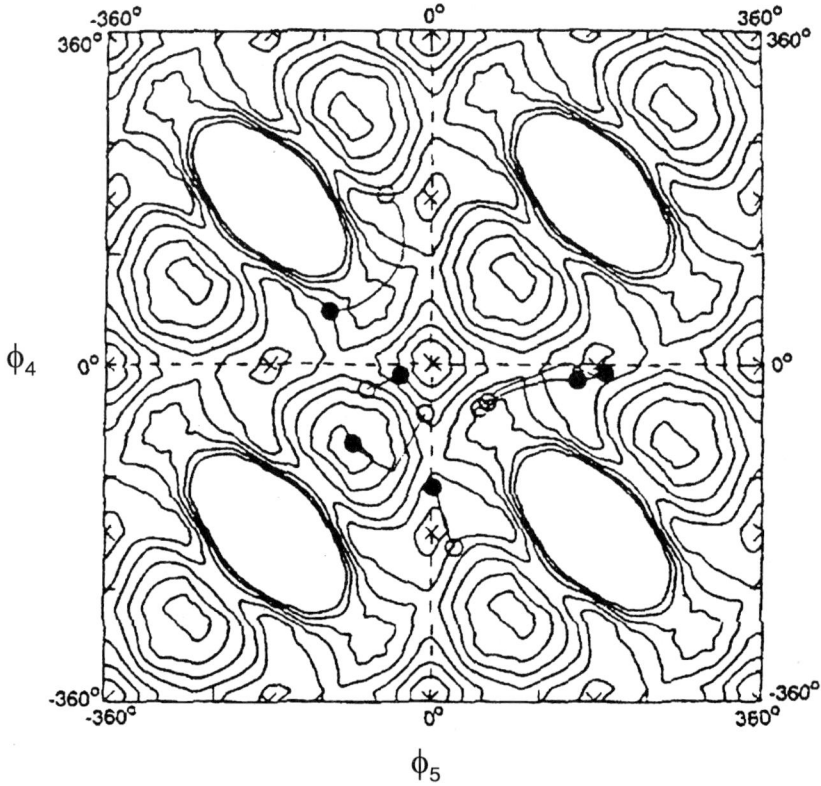

Fig. 63 Potential energy contour map of torsion angles ϕ_4, ϕ_5 for PDC. The intramolecular energy minima are represented by the symbol ×. Some pathways observed in simulations are drawn. The end of the path indicated by the *empty symbol* is the starting position, and the position marked with a *filled symbol* is the last simulated point (from [51])

whereas in the bulk it is a conformation change through changes in the inner torsion angles ϕ_4 and ϕ_5, indicated in Fig. 54.

Nevertheless, the intermolecular packing effect is reflected in the potential energy contour map of the torsion angles ϕ_4 and ϕ_5 (Fig. 63), where, as observed with phenyl rings, the starting positions at the initial system energy minima are located at positions corresponding to higher energy than the intramolecular minima calculated for DPC (represented by the $\times$ symbols).

Concerning the long range effects induced by conformation change of carbonate, they appear smaller than those due to phenyl ring π-flips.

5.4.2.2
Molecular Dynamics

Molecular dynamics (MD) simulations have been recently carried out on bulk BPA-PC [52].

The MD approach is one of the most elaborated techniques for simulating the dynamic behaviour of molecules. In this approach, spatial coordinates and velocity components of each atom are considered. At each time step the whole set of equations of motion, corresponding to all the atoms, is solved in order to define the new positions and velocity components of the atoms. Time steps are in the range of femtoseconds, the dynamics is usually performed (for computer time reason) over a rather short time, typically a few hundred picoseconds, in such a way that a limited number of events are picked up along the considered trajectories.

In the case of bulk BPA-PC, MD simulations were performed at 27 and 127 °C. A time step of 0.2 fs was used and trajectories of 80 ps were performed.

The BPA-PC chain contained 31 repeat units, with diphenyl isopropylidene units attached at each end. The chain contains 64 phenyl rings and 32 carbonate units in total. The solid amorphous system was prepared according to the following procedure. Initially, a single BPA-PC chain was put in a cubic box with the box dimensions much larger than the chain size of the polymer. MD simulation was performed to "equilibrate" until the kinetic energy of the system was fluctuating within 10% of the $3NkT/2$ value, where N is the total number of atoms in the system. The size of the box was then reduced slightly and the system was allowed to equilibrate. This process was repeated until the density of the system matched its experimental value (1200 kg m^{-3}). Three-dimensional periodic boundary conditions were used. Although only one BPA-PC chain was initially selected, due to the 3D boundary conditions used, the system is essentially multichain in nature and includes the interchain interactions.

Analysis of the BPA-PC resulting structure was in very good agreement with the experimental and calculated results obtained on DPP and DPC. In the latter case, it is found that there is no *trans* to *cis* or *cis* to *trans* conversions in the glassy state, the *trans, trans* conformation being more favoured than the *trans, cis* (or *cis, trans*) one.

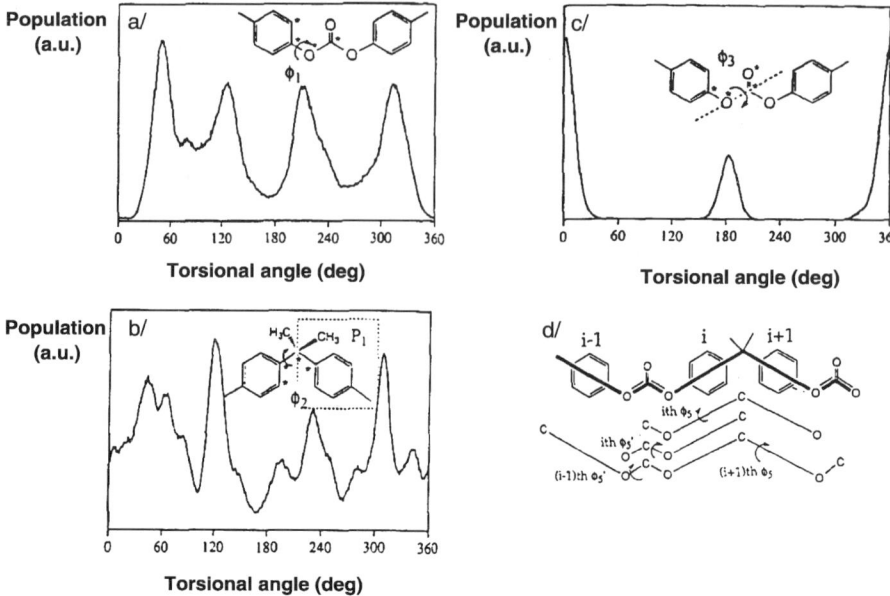

Fig. 64 Presentation of the various torsion angles in BPA-PC. In each case, the torsion angle ϕ is defined by the sequentially marked atoms: **a** ϕ_1, **b** ϕ_2, **c** ϕ_3 and **d** ϕ_5 (from [52])

As regards the dynamic analysis, in reality the characteristic times of phenyl ring π-flip and carbonate conformation change in bulk BPA-PC are of the order of microseconds. In a MD simulation of 80 ps, most of the phenyl rings and carbonate groups oscillated with small amplitude around their equilibrium position. In addition to this, a few conformational transitions of phenyl rings and carbonate groups are accidentally detected. Nevertheless, the analysis of these transitions gives insight into the motions occurring in bulk BPA-PC and into the short- and long-range consequences of the conformational transition.

The various torsion angles, ϕ_i, discussed in MD simulation of BPA-PC are represented in Fig. 64.

Short-Range Dynamics

The trajectories of the ϕ_1, ϕ_2 and ϕ_5 torsion angles of the 22nd and 23rd phenyl rings, located on each side of a carbonate group, are shown in Fig. 65. Figure 65a shows that the ϕ_1 values for these two phenyl rings change simultaneously around 28 and 59 ps, suggesting their cooperative rotation. However, Fig. 65b shows that there are no significant angle changes for ϕ_2 around 28 and 59 ps, as should be for a phenyl rotation. Actually, the changes observed in Fig. 65a are due to the wagging motion of the carbonate groups, confirmed by the changes of ϕ_5 (Fig. 65c) occurring at the same times. An-

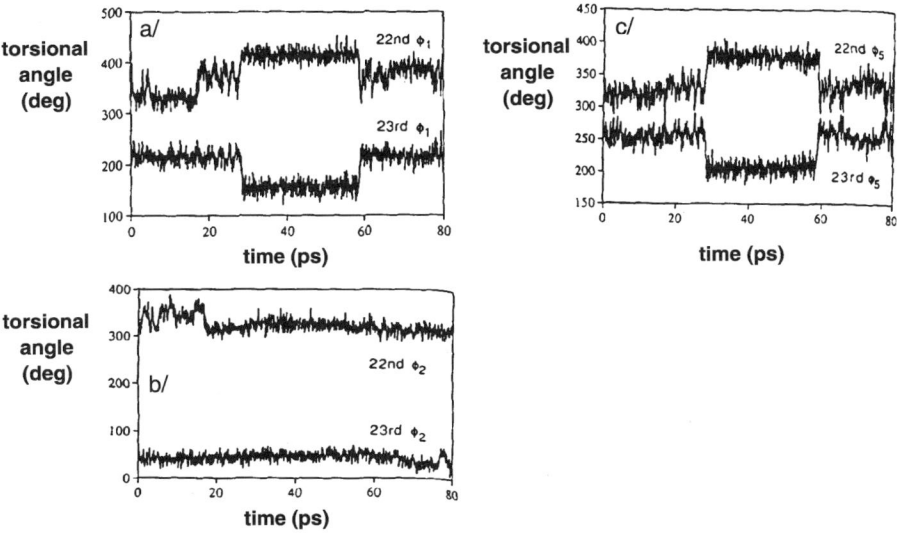

Fig. 65 Trajectories of ϕ_1, ϕ_2 and ϕ_5 for the 27th and the 23rd phenyl ring at 127 °C: **a** ϕ_1, **b** ϕ_2 and **c** ϕ_5 (from [52])

other interesting feature concerns the changes of both ϕ_1 and ϕ_2 occurring around 18 ps for the 22nd phenyl ring. These changes indicate independent rotation of the 22nd phenyl ring (about 80°). After this independent ring rotation, the neighbouring carbonate group rotates, showing a coupling of these rotations, as found in quasi-static simulation [54].

Another observed transition deals with the 1st and 2nd phenyl rings, which are attached to the 1st isopropylidene unit. Figure 66 shows that ϕ_2 of the 1st phenyl ring changes from 60° to – 50° between 55 and 80 ps, while ϕ_2 of the 2nd phenyl ring changes in the same direction exactly within the same duration. This is an in-phase cooperative rotation of the rings attached to the

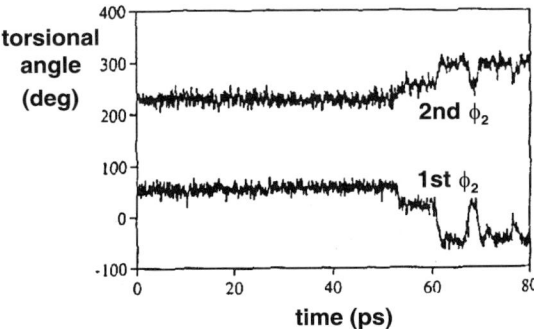

Fig. 66 Trajectories of ϕ_2 of the 1st and 2nd phenyl rings at 127 °C (from [52])

same isopropylidene unit. A similar transition is found at 65 ps between the 57th (from 120 to 40°) and the 58th (from 160 to 20°) phenyl rings, both attached to the same isopropylidene unit. Such observations agree with results found both by quasi-static simulation [51], and by atomistic calculation on DPP [48].

Long-Range Interactions

Figure 67 shows the angle changes of the segment composed of the 21st to 24th phenyl rings around 17–18 ps. It demonstrates that an intramolecular cooperative motion may extent to a BPA–carbonate–BPA sequence. In this motion, the phenyl rings on both sides of the carbonate unit rotate when the carbonate rotates, which is in agreement with the quasi-static simulation results. It is worth mentioning that no larger intramolecular cooperative motion has been found in MD simulations.

Quite interestingly, MD simulation provides evidence of intermolecular cooperativity, as illustrated in Figs. 68 and 69, analysing the effect of the rotation of the 42nd phenyl ring on two other phenyl rings, the 58th and the 62nd, initially at centre-to-centre distances of the 42nd equal to about 6 Å and 7 Å, respectively (Fig. 68). Figure 69 shows that the 42nd phenyl ring undergoes a rotation around 17 ps, which induces a smaller rotation of the 58th

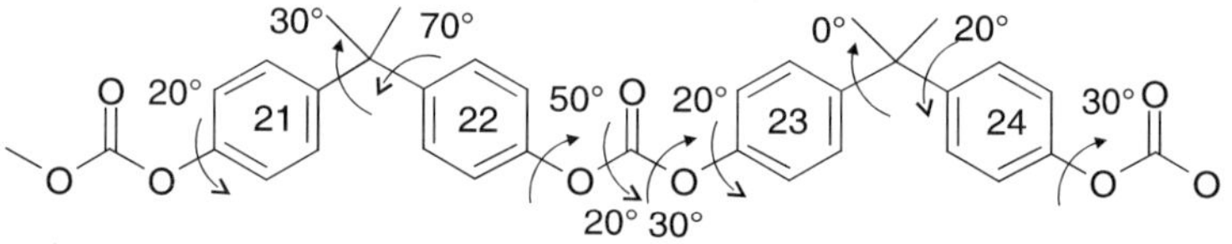

Fig. 67 Approximate angle changes of the 21st–24th phenyl rings with 17–18 ps at 127 °C. The *double arrows* indicate the directions of rotation (from [52])

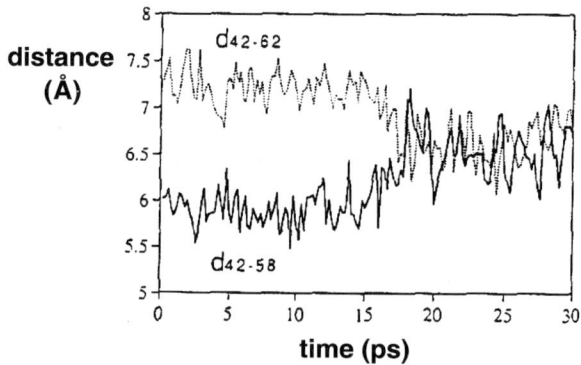

Fig. 68 Evolution of the centre-centre distances of the 42nd–58th and 42nd–62nd phenyl rings at 127 °C (from [52])

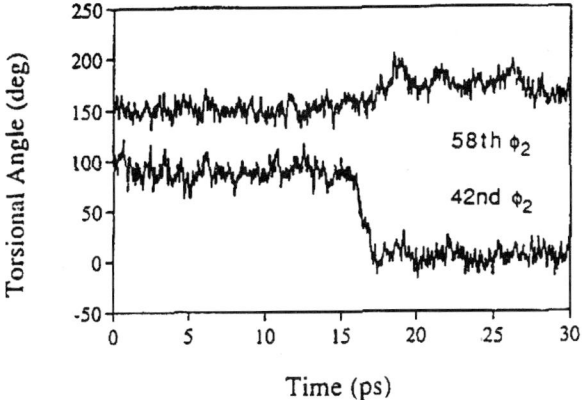

Fig. 69 Trajectories of ϕ_1 of the 42nd and the 58th phenyl rings at 127 °C (from [52])

phenyl ring. As a consequence of the rotation of the 42nd phenyl ring, the 58th phenyl ring moves away, whereas the 62nd comes closer. This provides quite detailed information on the intermolecular cooperativity: the 42nd ring repels the neighbouring rings to generate free volume for rotation. After the rotation, other rings move closer to occupy the vacancy produced by the rotation. It shows that the rotation of a phenyl ring would affect other rings within 7 Å, which again agrees with the extent of perturbation induced by a phenyl ring rotation obtained from quasi-static simulation [51].

5.4.3
Conclusion on the Atomistic Modelling

The results reported clearly show how powerful the atomistic modelling can be when applied to model molecules (DPP, DPC, BPA-PC repeat unit) and also to bulk BPA-PC.

The most relevant features are:

1. In-phase motions of the phenyl rings attached to the same isopropylidene unit, observed in both isolated molecule (DPP) and bulk BPA-PC.
2. Cooperativity of phenyl ring rotation with carbonate conformation change.
3. Intramolecular cooperativity limited, in the bulk, to the BPA–carbonate–BPA sequence.
4. Intermolecular cooperativity associated with the phenyl ring rotation, which concerns groups as far as 7 Å from the moving phenyl ring, as well as a rearrangement of all the units within a volume of about 1 nm³.
5. Inhomogeneous local packing, which leads to a broad distribution of the activation energy of both phenyl ring and carbonate group motions. This

distribution can be well fitted with a Williams–Watts stretched exponential distribution, with an exponent α between 0.1 and 0.2.

6. Rocking main-chain motions, which occur in bulk BPA-PC when phenyl rings flip and carbonate groups change conformation, with rms averages around 13° and 11°, respectively.

7. Favoured *trans* conformation of carbonate group, the *trans,cis* conformation representing only about 34% in the glassy BPA-PC. Furthermore, no *trans* to *cis* (or *cis* to *trans*) transition occurs in the glassy state.

5.5
Effect of Small Molecule Antiplasticisers

Some small molecule additives, called antiplasticisers, miscible with BPA-PC, lead to an increase of both modulus and yield stress at room temperature [14, 15]. In contrast, the toughness and the elongation at break decrease.

It has been recognised for a long time that the antiplasticisers induce a decrease of the β transition peak intensity [16].

Such an effect of small molecule antiplasticiser is not specific to BPA-PC. It seems to occur with most polymers undergoing a β transition originating from motions in the main chain. In the present paper, the effects of antiplasticisers on the β transition of poly(ethylene *tere*-phthalate) and epoxy networks are analysed in Sects. 4 and 7, respectively.

One of the most efficient antiplasticiser for BPA-PC is Aroclor 1254, which consists of a polychlorinated bi-phenyl with five chlorine substituents; it will be denoted AP. The results reported here deal with this additive.

Dependence of the dynamic mechanical $\tan \delta$ on temperature [17] is shown in Fig. 70 for pure BPA-PC and for mixtures with 10 and 30 wt % AP.

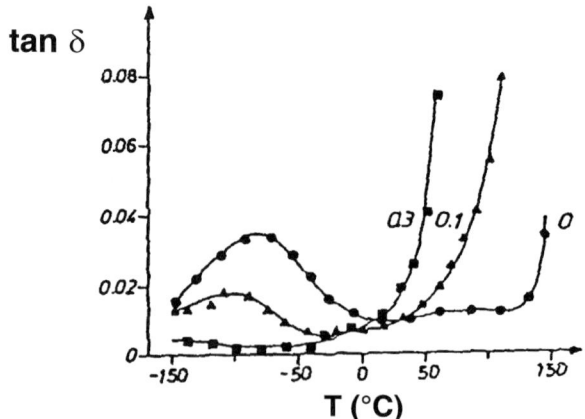

Fig. 70 Dynamic mechanical $\tan \delta$, at 10 Hz, as a function of temperature for BPA-PC and various concentrations (wt/wt) of antiplasticiser AP (from [17])

It appears that it is mostly the high-temperature part of the β transition which is suppressed, leading to a downward shift of the β peak maximum. As presented in Fig. 71, the strength of the β relaxation, expressed by the area under the dynamic mechanical loss, E'', peak, R_{sec}, decreases much more rapidly with increasing the AP concentration that it should according to a dilution effect. It is worth pointing out that a similar behaviour has been reported, a long time ago, on the same system [53] and is encountered in BPA-PC-antiplasticiser mixtures, irrespective of the specific nature of the antiplasticiser small molecule [16, 17, 53, 54].

Analysis of the motions involved in the β transition of BPA-PC clearly shows that the high-temperature part of the mechanical β transition peak is associated with phenyl ring π-flips.

In order to get a more direct proof of the effect of AP additive on the phenyl ring motions, ^2H NMR studies have been performed as a function of temperature on BPA-PC whose aromatic hydrogens in *ortho* position to the carbonate group have been substituted by deuterons (BPA-d_4-PC) [17, 36]. It is worth pointing out that, in ^2H NMR, the line shape changes require jump frequencies of 10^5 to 10^6 Hz, i.e. at much higher frequency and temperature than the jumps observed in 10 Hz dynamic mechanical measurements. The fully relaxed ^2H NMR spectra of the phenyl rings in BPA-d_4-PC and in a mixture with 25% AP, are shown in Fig. 72 on the left and right sides, respectively. Whereas below $-55\,^\circ C$ the spectra are identical, at higher

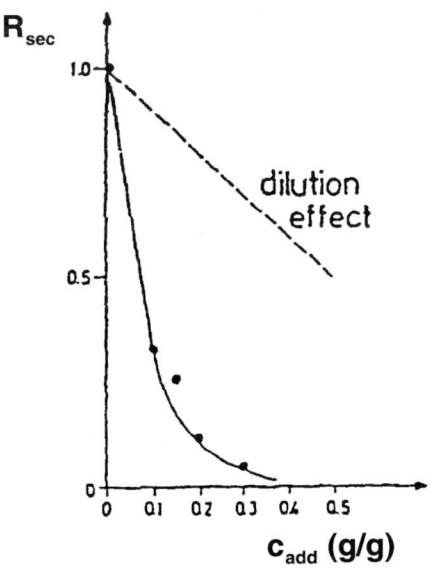

Fig. 71 Relaxation strength of the β transition as a function of AP concentration (from [17])

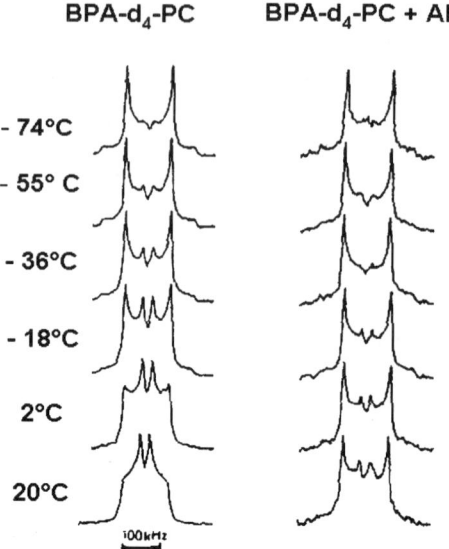

Fig. 72 Fully relaxed ^{2}H NMR spectra of the phenyl rings in pure BPA-d_4-PC (*left side*) and in a mixture with 25% AP (*right side*), at the indicated temperatures (from [35])

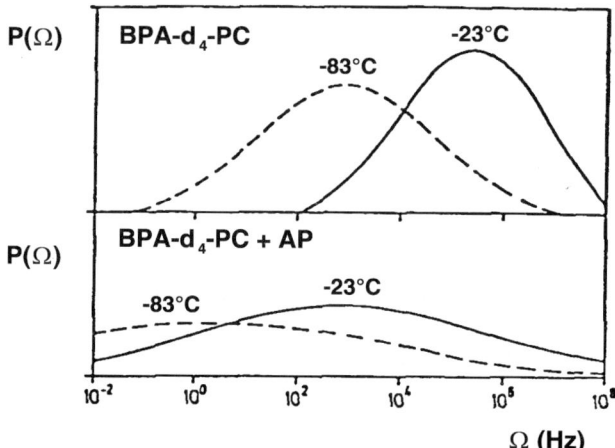

Fig. 73 Distribution of correlation frequencies for the phenyl ring π-flips in BPA-d_4-PC and in a mixture with 25% AP, at the indicated temperatures, as obtained from ^{2}H NMR spectra (from [17])

temperatures the inner lines are considerably lower in the mixture than in pure BPA-d_4-PC. Thus, a substantial fraction of deuterons that are mobile in the pure polymer become rigid in the mixture, enhancing the weight of the Pake spectrum. However, the effect of the antiplasticiser is not only to shift the distribution of correlation frequencies, Ω, to lower values, but also to considerably broaden the distribution, as deduced from partially relaxed spectra with variable waiting times. It turns out that, by assuming a symmetric log-Gaussian distribution, the width varies from 2.6 decades at room temperature to 4.3 decades at $-110\,°C$, instead of 5.2 to 9.1 decades at the same temperatures for the mixture with AP. This is illustrated in Fig. 73. So, it is clear that the antiplasticiser molecule hinders the phenyl ring π-flips.

5.6
Conclusion on the Molecular Motions Involved in the Secondary Transitions of BPA-PC

Before describing the molecular motions involved in the β transition of BPA-PC, as presently understood, it is worth pointing out that some models were developed in the middle of the 1980s. Among them is the defect diffusion model [55], based on diffusion of carbonate *cis* conformation up and down the long axis of the chain, with rings flipping as defects pass; it is a single chain model. In contrast, another model [56] emphasises the intermolecular cooperativity between ring flips of locally parallel BPA-PC chain segments (bundles). Neither of these models being satisfactory, they will not be discussed further.

Various experimental techniques (dielectric relaxation, dynamic mechanical analysis, ^1H, ^2H and ^{13}C solid-state NMR) have been used for investigating the secondary transitions of BPA-PC, and the block copolymers of BPA and TMBPA carbonates as well as compatible blends of BPA-PC and TMBPA-PC. They have provided lots of information on the motions of methyl, phenyl ring and carbonate units in bulk BPA-PC. The effect of intermolecular packing has also been clearly evidenced.

Atomistic modelling leads to a more detailed description of the motions and yields a clear insight on the extent of intra- and intermolecular cooperativity.

By combining all the results, a comprehensive picture of the motions occurring in bulk BPA-PC is reached. At very low temperatures, the first motions to occur are the rotations of the methyl groups of the isopropylidene unit around the C_3 axis. The activation energy associated with this γ transition is around $8\,kJ\,mol^{-1}$.

At higher temperatures, around $-100\,°C$ at 1 Hz, the β transition involves motions of both carbonates and phenyl rings, accompanied by main-chain reorientation, which can be described in more details as follows.

In the low-temperature part of the dielectric β relaxation, independent motions of limited amplitude of carbonate groups occur. In the high-temperature part, the conformation change of the carbonate group is coupled to motions of the adjacent phenyl rings. This intramolecular cooperativity is also associated with the perturbation of the surroundings induced by carbonate conformation change and phenyl ring motions. As a consequence, this high-temperature part is affected by the compatible blending of BPA-PC with TMBPA-PC. In the same way, the coupling carbonate–phenyl ring accounts for the large shift towards high temperatures observed between BPA-PC and TMBPA-PC. The activation energy of the dielectric β transition of BPA-PC is 54 ± 2 kJ mol^{-1}.

The dynamic mechanical β transition of BPA-PC has an activation energy of 60 kJ mol^{-1} and an activation entropy of 110 J K^{-1} mol^{-1}, indicative of quite cooperative motions. The dynamic mechanical response mostly originates from the phenyl ring motions and the perturbation of the surroundings (at a scale as large as 10 Å). The conformation change of the carbonate group leads to a lower contribution due to the smaller size of the moving units, the smaller amplitude of the motion and a weaker perturbation of the surrounding, as shown by the quasi-static simulations. The phenyl rings undergo both oscillations, with amplitude increasing with temperature from 12° to 32°, and π-flips. The latter motions are more coupled with the surroundings and sensitive to local packing. Consequently, due to fluctuations of local packing, the frequencies of the phenyl ring π-flips are broadly distributed (Williams–Watts distribution with an exponent α equal to 0.154).

As regards the intramolecular cooperativity between phenyl ring motions, whereas there is some cooperativity between the two phenyl rings on each side of the carbonate group, there is a strong motional cooperativity between the two phenyl rings attached to the same isopropylidene unit, since they undergo in-phase motions. However, there is no intramolecular cooperativity along several (five to seven) repeat units, as previously suggested [26, 27].

The phenyl oscillations constitute most of the motions involved in the low-temperature part of the dynamic mechanical β transition, whereas the phenyl ring π-flips mostly contribute to the high-temperature part of the β transition. That is the reason why the high-temperature part is sensitive to the surrounding medium, as shown in introducing rigid TMBPA-PC in BPA-PC-compatible blends. Furthermore, this assignment is supported by the fact that the β transition of TMBPA-PC does not correspond to a shift towards a higher temperature of the whole β peak of BPA-PC, but it presents a very broad low-temperature part (from -120 to 50 °C at 1 Hz). Indeed, phenyl ring oscillations happen in TMBPA-PC, even if they have a smaller amplitude than in BPA-PC, but most hindered are the phenyl ring π-flips.

Quite an important feature of the phenyl ring π-flip mechanism of the BPA-PC β transition is the intermolecular cooperativity that can exist between the flip of a phenyl ring and the motion of another phenyl ring as far away as 7 Å (centre-to-centre distance), or that of a carbonate group in an out-of-equilibrium conformation.

Low-amplitude (rms 10° at room temperature) main-chain motions are also present, and are mostly driven by conformation changes of carbonate groups.

Finally, it is interesting to notice that small molecule antiplasticisers added to BPA-PC decrease the strength of the mechanical β transition, mostly the high-temperature part where phenyl ring π-flips occur. ^{2}H NMR analysis confirms the hindering of these π-flips, but in addition it shows quite a substantial increase of their frequency distribution.

6
Aryl-aliphatic Copolyamides

Aryl-aliphatic copolyamides (Ar-Al-PA) are industrial technical polymers whose uses as transparent plastics require performances in terms of temperature behaviour, or resistance to solvents and stress cracking. The transparency of materials necessitates a completely amorphous structure, which is obtained by using not only *para*-substituted phenyl rings, but also *meta*-substituted phenyl rings that decrease the chain regularity.

In the glassy state, these Ar-Al-PA exhibit local chain dynamics which are largely controlled by the chemical structure. Recently, the local motions that may occur in the glassy state and might take part in secondary transitions, have been investigated on a series of Ar-Al-PA of various chemical structures by using dielectric relaxation, ^{13}C and ^{2}H solid-state NMR and dynamic mechanical experiments [57–60].

The polymers considered in the study correspond to two different groups. In the first one, the xT_yI_{1-y} copolyamides, whose repeat unit contains x lactam-12 sequences, y and $1 - y$ *tere-* and *iso*-phthalic moieties, respectively, and a 3,3'-dimethyldicyclohexyl methane unit in the regular order:

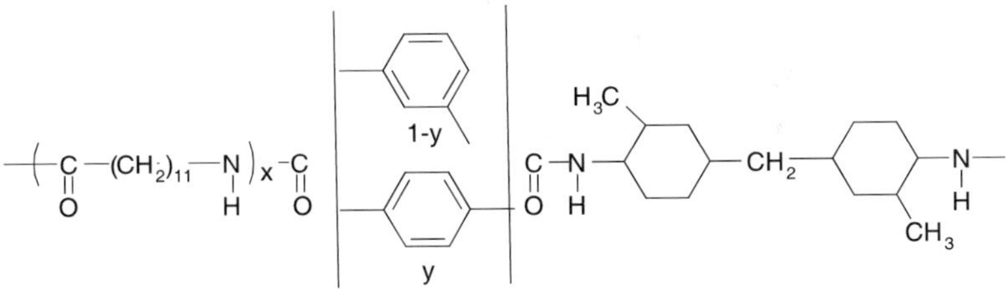

The second group is represented by the following copolyamide, MT, obtained from the condensation of *tere*-phthalic acid onto 1,5 diamino-2-methylpentane:

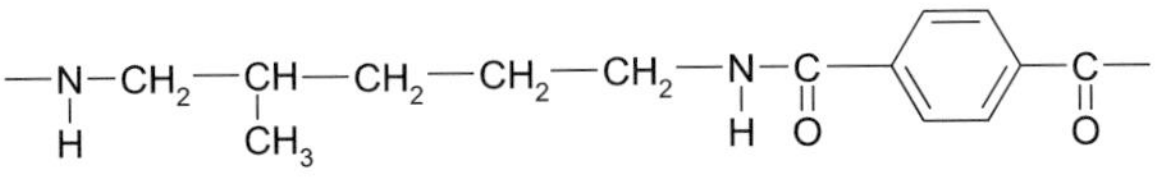

The chemical formulae and code names of the investigated polymers are given in Fig. 74.

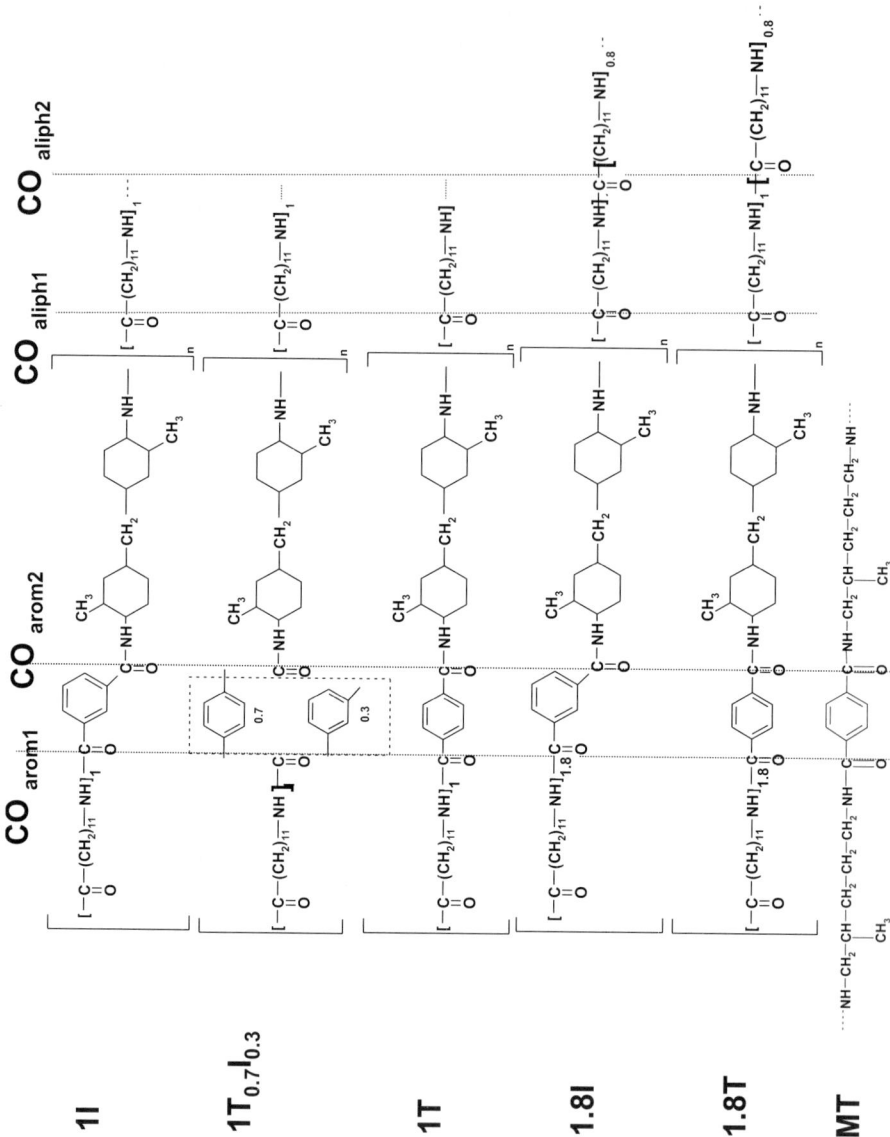

Fig. 74 Chemical formulae and code names of the aryl-aliphatic copolyamides studied

6.1
Dielectric Analysis

Dielectric measurements in the glassy state, over a range of five different frequencies between 1 Hz and 10 kHz, at temperatures from – 140 to 150 °C, have been performed [57]. As examples, the loss tangent at different frequen-

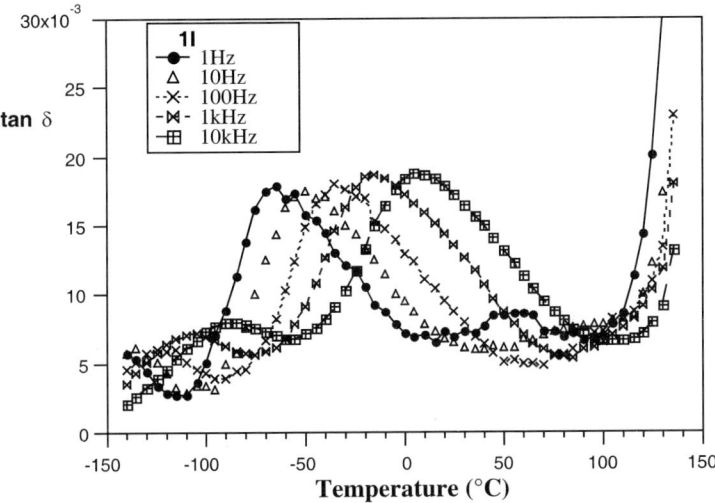

Fig. 75 Dielectric loss tangent vs temperature for 1 I at different frequencies (from [57])

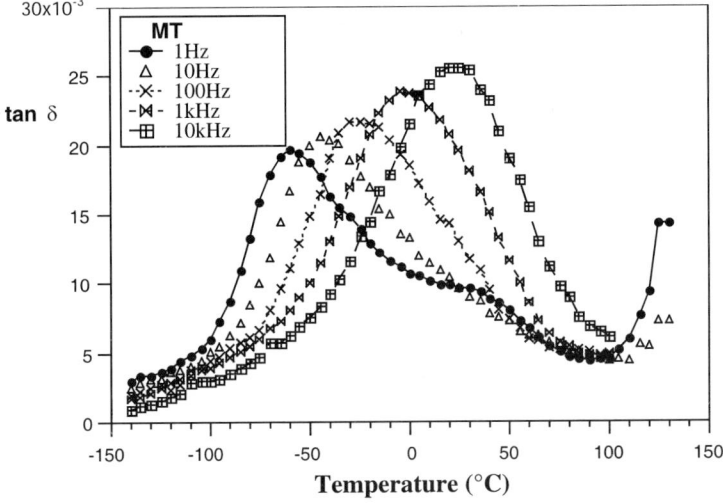

Fig. 76 Dielectric loss tangent vs temperature for MT at different frequencies (from [57])

cies as a function of temperature is shown in Figs. 75 and 76 for 1I and MT, respectively. For comparison, the loss tangent at 1 Hz and 1 kHz as a function of temperature is shown for the whole series of polymers in Figs. 77 and 78, respectively.

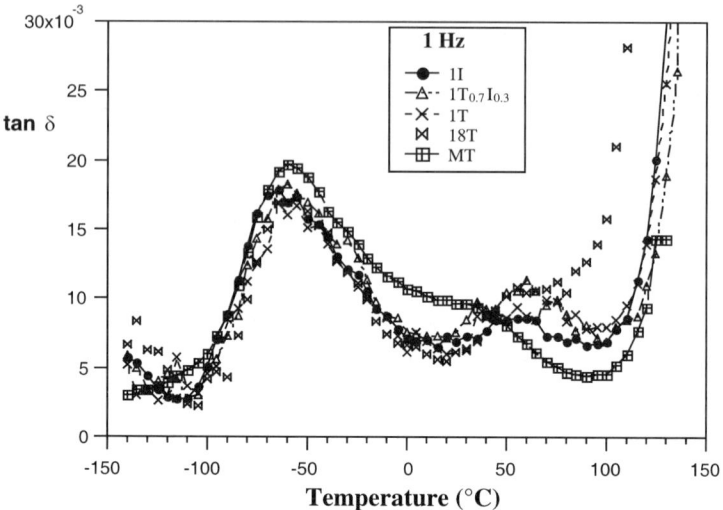

Fig. 77 Dielectric loss tangent vs temperature determined at 1 Hz for the different polymers (from [57])

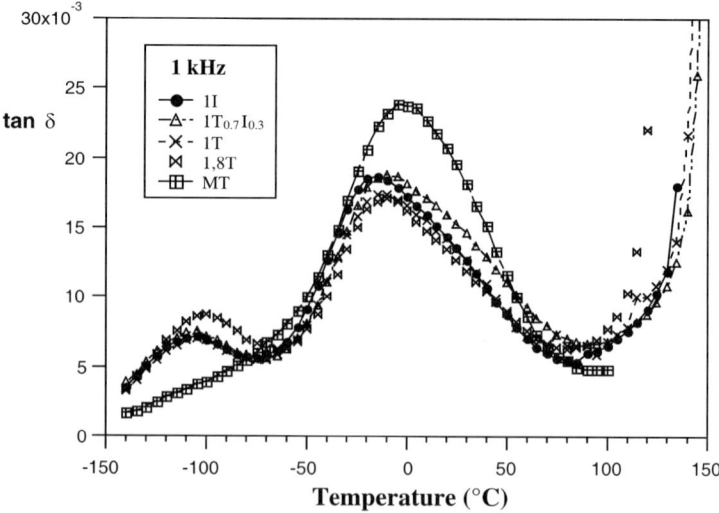

Fig. 78 Dielectric loss tangent vs temperature determined at 1 kHz for the different polymers (from [57])

6.1.1
Experimental Results

As regards the xT_yI_{1-y} copolyamides, the dielectric experiments point out the existence of two main secondary transitions, observed in the whole range of frequencies: a β transition around $-70\,°C$ at 1 Hz and a γ transition, with a smaller intensity, which can be seen around $-110\,°C$ at 1 kHz. At 1 Hz the γ relaxation peak maximum occurs below the lowest temperature investigated, i.e. $-150\,°C$. In addition to these main secondary transitions, the four samples exhibit a low-intensity broad peak, noted as ω, occurring around 60 °C at 1 Hz, which progressively merges with the glass transition phenomenon at higher frequencies.

The activation energies corresponding to the γ and β processes, $E_{\alpha\gamma}$ and $E_{\alpha\beta}$, are listed in Table 6, together with the activation entropies, $\Delta S_{\alpha\gamma}$ and $\Delta S_{\alpha\beta}$, calculated according to the Starkweather analysis described in Sect. 2.3. The low values of the activation entropy ($10-20\,\mathrm{J\,K^{-1}\,mol^{-1}}$) obtained for the γ transition means that this transition deals with localised motions. The β transition, which has a higher activation entropy ($50-80\,\mathrm{J\,K^{-1}\,mol^{-1}}$), should correspond to motions with some cooperative character.

The MT copolyamide exhibits several differences with respect to the results observed in xT_yI_{1-y} copolyamides. The first difference comes from the absence of a transition in the low-temperature region, where the γ transition of the xT_yI_{1-y} polymers is observed. The second point concerns the characteristics of the broad secondary transition lying in the temperature range from -100 to 80 °C. On increasing the frequency, the maximum of the peak is shifted to a higher temperature, the peak intensity increases and the line shape is modified. At 1 Hz, two distinct contributions can be distinguished in the broad line. The low-temperature part corresponds to a peak centred around $-60\,°C$, comparable to the β peak observed in the xT_yI_{1-y} samples; this contribution will be denoted as β. The high-temperature part exhibits a broad shoulder in the range from 0 to 50 °C. A more detailed investiga-

Table 6 Activation energies and entropies determined for γ and β relaxations of different copolyamides (from [57])

Polyamide	Ea_γ [kJ mol^{-1}]	ΔSa_γ [J K^{-1} mol^{-1}]	Ea_β [kJ mol^{-1}]	ΔSa_β [J K^{-1} mol^{-1}]
1 I	31 ± 5	12 ± 35	61 ± 5	56 ± 25
1 T$_{0.7}$I$_{0.3}$	32 ± 5	14 ± 35	63 ± 5	62 ± 25
17	32 ± 5	15 ± 35	64 ± 5	67 ± 25
1.8 T	33 ± 5	18 ± 35	66 ± 5	74 ± 25
MT			57 ± 5	34 ± 25

tion [57] shows that this latter contribution does not shift with frequency, and that DSC measurements allow one to assign it to a quasi-melting phenomenon. Consequently, it is not related to the molecular motions under concern here.

The activation energy and entropy of the β transition in MT are listed in Table 6. The activation energy, $E_{a\beta}$, is very similar to the activation energies determined in xT_yI_{1-y} polymers for the β transition. In contrast, the activation entropy is quite low, indicating that the β transition of MT is less cooperative than the β transition of xT_yI_{1-y} polymers.

6.1.2
Assignment of the Dielectric Relaxations

In the copolyamides under consideration, the dipoles that are responsible for the dielectric relaxations are associated with the $C = O$ groups of the amide functions. Due to the quasi-conjugated character of the $CO - NH$ bond, the amide group takes on a rigid plane conformation in such a way that the dielectric relaxations of copolyamides should correspond to motional modes that involve amide groups and not only the carbonyls, in contrast to what happens with the ester groups encountered in polyethylene *tere*-phthalate (Sect. 4.1.2).

As shown in Fig. 74, depending on their degree of conjugation with the neighbouring groups and on the rather rigid or flexible character of these neighbours, three different CO groups can be distinguished:

1. CO_{aliph} groups, which are non-conjugated carbonyl groups situated between two aliphatic units, a flexible lactam-12 on one side and a rigid cycloaliphatic unit on the other side. In addition, the 1.8T copolyamide also possesses a $C = O$ group, noted $CO_{aliph\ 2}$, located between two lactam-12 segments.
2. $CO_{arom\ 1}$ groups, which are located between a flexible lactam-12 or methylpentane unit and a phenyl ring and share some conjugation with the phenyl ring.
3. $CO_{arom\ 2}$ groups, which are located between a rigid cycloaliphatic unit and a phenyl ring and share some conjugation with the phenyl ring.

Regarding the assignment of the transitions, the simplest compound is the MT polymer for it contains $CO_{arom\ 1}$ groups only. As its dielectric trace exhibits only one transition, this β transition can be unambiguously assigned to the motion of these $CO_{arom\ 1}$ dipoles.

The dielectric β transition of the xT_yI_{1-y} polymers (Figs. 77 and 78), whose characteristics in terms of temperature and activation energy are very close to those of MT, can also be assigned to the $CO_{arom\ 1}$ motions.

In addition to the β transition, the dielectric traces of the xT_yI_{1-y} copolyamides (Figs. 77 and 78) exhibit two other transitions, γ and ω, which occur in very different temperature ranges. However, in the case of 1.8T, there are some differences in the region of the γ transition, relative to the other polymers of the series. As a consequence, discussion of the γ transition of 1.8T will be considered later.

To assign the γ and ω transitions of 1I, $1T_{0.7}I_{0.3}$ and 1T, two kinds of carbonyl groups can be involved: the $CO_{arom\ 2}$ group, which is conjugated with the phenyl ring, and the CO_{aliph}, which is not. The existence of some conjugation implies that the correlation between the $CO_{arom\ 2}$ group and the phenyl ring is much stronger than the correlation between the CO_{aliph} and its aliphatic environment. In addition, due to the proximity of the flexible lactam-12 unit, the CO_{aliph} groups should exhibit a higher mobility and their motions should occur at lower temperatures than the $CO_{arom\ 2}$ motions. The ω transition in all the xT_yI_{1-y} copolyamides (including 1.8T) can thus be assigned to the motions of $CO_{arom\ 2}$ groups.

Concerning the 1.8T copolyamide, in the γ transition region, its dielectric trace at 1 Hz (Fig. 77) is similar to those of the $1T_yI_{1-y}$ polymers, but with a higher intensity. However, at 1 kHz (Fig. 78), its dielectric trace is not only more intense, but is also more spread out towards the high-temperature side whereas the onset of the transition, in the low-temperature part, is unchanged. This result shows the existence of an additional contribution, denoted γ', occurring at a higher temperature than the CO_{aliph} process of the $1T_yI_{1-y}$ polymers. As shown in Fig. 79, this γ' transition is centred around $-92\ ^\circ C$ at 1 kHz. The associated activation energy is 42 kJ mol^{-1} and the activation entropy is 59 J K^{-1} mol^{-1}. The latter value is significantly higher than the 12–15 J K^{-1} mol^{-1} $\Delta S_{a\gamma}$ values derived for the $1T_yI_{1-y}$ copolyamides, indicating a more cooperative character for motions responsible for the γ'

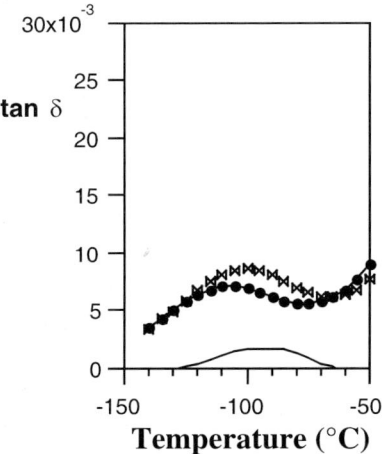

Fig. 79 Dielectric loss tangent vs temperature determined at 1 kHz for 1.8 T ($\times$) and 1 I ($\bullet$); difference trace (–)(from [57])

contribution. In 1.8T, two kinds of aliphatic carbonyl groups are present: in addition to the CO_{aliph} groups, situated between a lactam-12 sequence and cycloaliphatic ring and similar to those found in CO_{aliph} polymers, there are also $CO_{aliph\ 2}$ groups located between two lactam-12 sequences. Comparison with the characteristics of the γ transition in aliphatic copolyamides [57] leads to assignment of the low-temperature part of the γ peak, observed in all the considered xT_yI_{1-y} copolyamides to CO_{aliph} modes. This is possibly enhanced by the methylene motions of the flexible sequence, whereas the high-temperature γ' contribution in 1.8T originates from $CO_{aliph\ 2}$ motional processes with a spatial scale larger than the CO_{aliph} ones, involving not only the aliphatic carbonyl groups located between two lactam-12 units, but also the adjacent methylene units on both sides.

6.2
Solid-State NMR Analysis

Dielectric relaxation experiments applied to Ar-Al-PA have evidenced γ (and γ'), β and ω transitions and led to an assignment of the amide group motions involved in each of them. However, dielectric relaxation exclusively focuses on the motions of electric dipoles, i.e. amide groups in the considered polymers. In contrast, ^{13}C NMR provides specific information on the motional behaviour of each magnetically inequivalent carbon contained in the repeat unit and should permit characterisation of the participation of each part of the constitutive unit of the molecule – cycloaliphatic diamine, lactam-12 unit and phenyl rings – to the secondary transitions of the considered Ar-Al-PA [58]. Furthermore, a $1T_{0.7}I_{0.3}$ copolyamide, specifically perdeuterated on the *tere*-phthalamide phenyl rings, has been investigated by ^{2}H NMR in order to obtain more information about the phenyl ring motions [59].

6.2.1
^{13}C Solid-State NMR

As already mentioned in the NMR investigation of molecular motions in poly(ethylene *tere*-phthalate) (Sect. 4.1.3), the most relevant experiments are:

- ^{13}C chemical shift anisotropies of the unsaturated carbons. The chemical shift parameters used in the Ar-Al-PA for the unprotonated and protonated aromatic carbons, as well as for the carbonyl groups, are shown in Fig. 80.
- $t_{1/2}$ cross-polarisation times for the CH and CH_2 groups.

A motional model is required in order to interpret, in terms of motions, the change as a function of temperature of either the chemical shift anisotropy or $t_{1/2}$. An approach similar to the one applied to PET has been adopted for Ar-Al-PA [58], leading to the following model for the phenyl rings:

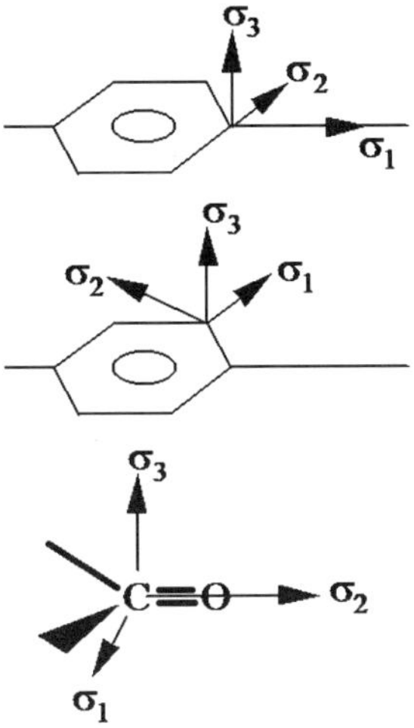

Fig. 80 Directions of the chemical shift parameters used in Ar-Al-PA (from [58])

- For the *para*-substituted phenyl rings (i) rapid oscillations about the *para*-axis of the ring between angles α and $-\alpha$, corresponding to a 2α amplitude of motion, and (ii) fast π-flips of the phenyl rings
- For the *meta*-substituted phenyl rings, only oscillations occurring about an axis joining the two N atoms

As a consequence, the motional behaviour of phenyl rings is described at a given temperature in terms of (i) an average oscillation amplitude and (ii) a fraction of phenyl rings undergoing fast π-flips.

6.2.1.1
MT Copolyamide

As observed in dielectric measurements, the MT copolyamide is the simplest Ar-Al-PA to analyse for it presents only one secondary transition, the β one.

The CP MAS DD ^{13}C NMR spectrum at 23 °C is shown in Fig. 81. The 166.9, 136.8, 127.3 and 17.4 ppm lines correspond to the C = O carbons, unprotonated and protonated aromatic carbons, and methyl carbons, respectively. The set of lines at 46.3, 40.7, 32.3 and 27.4 ppm correspond to aliphatic CH and CH_2 carbons. The temperature dependence of the $t_{1/2}$ values of these aliphatic carbons is shown in Fig. 82. The carbons associated with the 40.7 and 43.6 ppm lines, whose onset of motions occurs above 100 °C, should correspond to the CH_2 carbons in a position with respect to the amide group. Consequently, the 27.4 and 32.3 ppm lines, associated with carbon atoms which undergo motions at temperature equal or higher than 20 °C, should correspond to the central CH_2 carbons of the aliphatic sequence.

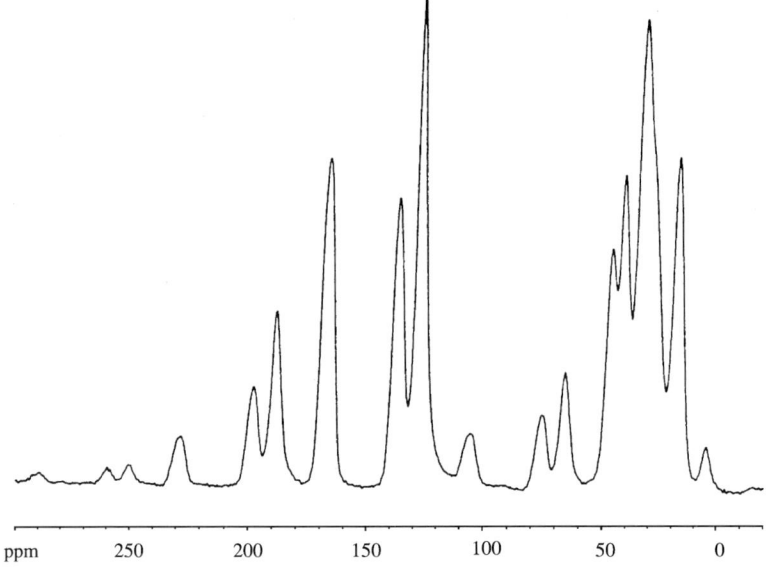

Fig. 81 CP MAS DD ^{13}C NMR spectrum of MT at 23 °C (from [58])

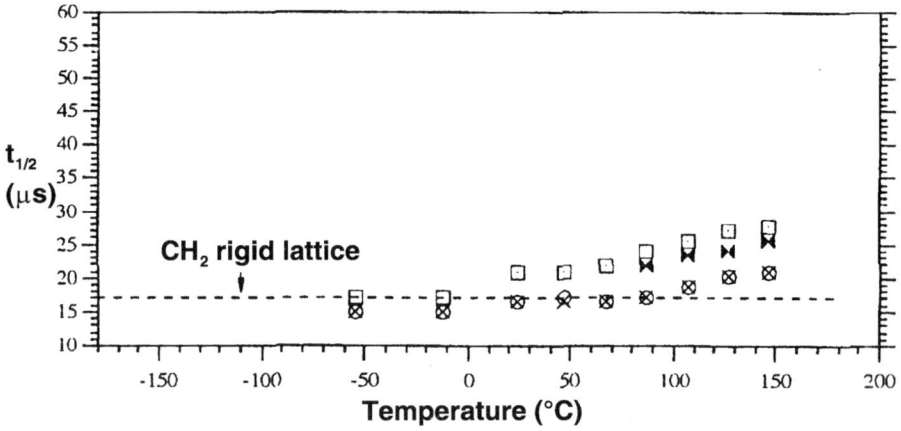

Fig. 82 Temperature dependence of the $t_{1/2}$ values, measured for all the lines of the aliphatic carbons in the MT copolyamide: × 46.3 ppm, ○ 40.7 ppm, □ 32.3 ppm and ▶◀ 27.4 ppm (from [58])

As regards the aromatic rings in MT, the chemical shift anisotropy of the unprotonated aromatic *para* carbons are only sensitive to oscillations, whereas the protonated aromatic *ortho* and *meta* carbons reflect both oscillations and phenyl ring π-flips. The oscillation amplitudes of these two types of carbons are shown in Fig. 83. In contrast, the $t_{1/2}$ values of the protonated aromatic carbons reflect the ring flips. The fraction of phenyl rings undergo-

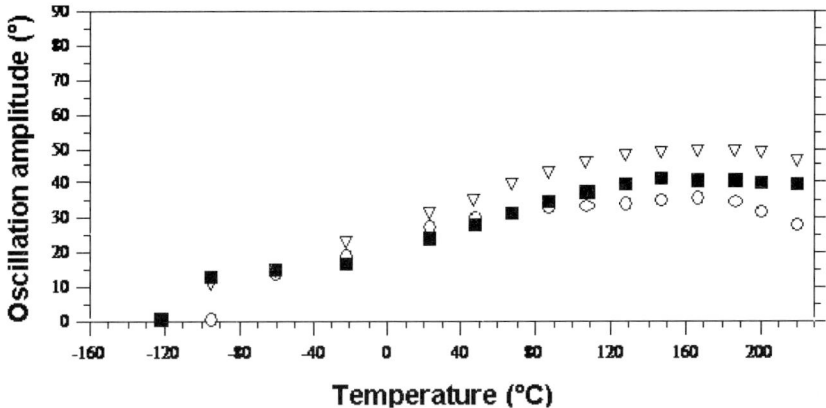

Fig. 83 Oscillation amplitudes (°) determined for the protonated aromatic carbons (■), unprotonated aromatic carbons (▽) and carbonyl carbons (○) of the MT copolyamide (from [58])

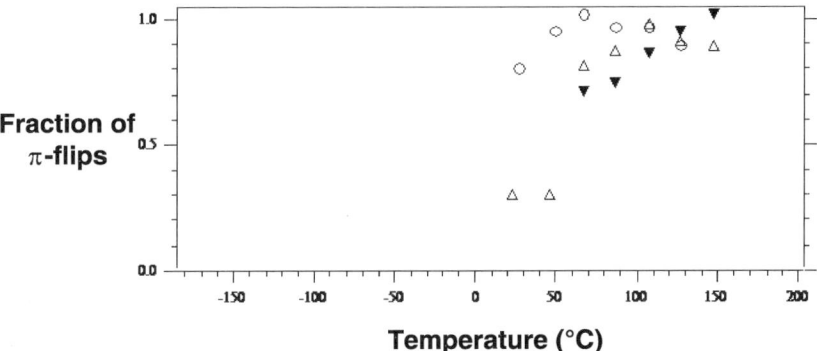

Fig. 84 Fraction of phenyl rings undergoing rapid π-flips in the MT (▼), 1 T (△) and 1.8 T (○) copolyamides (from [58])

ing π-flips as a function of temperature is reported in Fig. 84. At temperatures higher than 60 °C, an increasingly large part of the phenyl rings takes part in π-flips. It is worth noting that ^{2}H NMR investigations [59] support this analysis in terms of oscillations and π-flips.

Concerning the carbonyl groups in MT, the corresponding chemical shift anisotropy analysis leads to the consideration of fast oscillations of amplitude $\pm \alpha$ about the *para* axis of the adjacent phenyl ring.

The temperature dependence of the oscillation amplitudes associated with the carbonyl groups, shown in Fig. 83, is very similar to that of the aromatic carbons, indicating that the C = O groups and adjacent phenyl rings are involved in correlated motions. Such a conclusion is consistent with the quasi-conjugated character exhibited by the phenyl–amide bond.

6.2.1.2
xT_yI_{1-y} Copolyamides

As an example, the solid-state NMR spectrum of 1T at 32 °C is shown in Fig. 85.

The 137.1 and 127.3 ppm lines correspond to the unprotonated and proto-nated aromatic carbons, the 54.6, 34.0 and 19.4 lines are assigned to CH, CH$_2$ and CH$_3$ carbons of the cycloaliphatic units, respectively. The lactam-12 carbon line is observed at 29.7 ppm. Concerning the carbonyl groups, the C = O groups in an entirely aliphatic environment resonate at 172.9 ppm whereas the two C = O adjacent to the phenyl rings share the same resonance line at 166.1 ppm, independently of their aliphatic environment (i.e. lactam-12 or 3,3'-dimethyldicyclohexyl methane unit), in contrast to what happens in dielectric relaxation (Sect. 6.1.2).

Dealing first with the dynamics of the lactam-12 units, the C = O groups included in these units undergo fast oscillations about an axis joining the N atom to the cycloaliphatic ring.

These oscillation amplitudes are plotted as a function of temperature in Fig. 86 for the 1T, 1.8T and 1I polymers.

The onset of oscillations is detected at very low temperatures, around – 80 °C. The amplitude increases on increasing temperature and does not depend on the exact chemical nature of the xT_yI_{1-y} polymer, within experimental accuracy. The temperature dependence of the $t_{1/2}$ values for the methylene carbons of the lactam-12 units (29.7 ppm line) is shown in Fig. 87 for 1.8T, 1T and 1I polymers.

These low temperature motions clearly occur in the very temperature range where small oscillations of the C = O groups in an aliphatic environment are observed. This result indicates that the oscillations of the C = O groups are accompanied by simultaneous motions of the lactam-12 sequence.

Concerning the *iso-* and *tere*-phthalamide units, the amplitudes of oscillations of the *iso-* and *tere*-phthalamide C = O groups and phenyl rings, about an axis parallel to the local symmetry axis of the phenyl ring, are plotted as a function of temperature in Fig. 88a,b for the 1I and 1T polymers.

In the 1I copolyamide, data obtained for *iso*-phthalamide C = O groups and phenyl rings indicate the existence of small oscillations, with similar amplitude for both groups, over the whole temperature range, indicating that

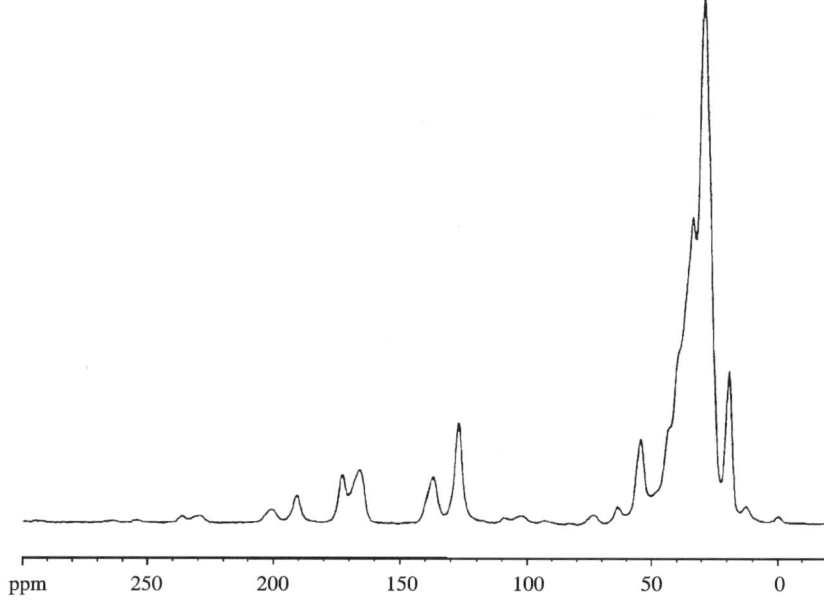

Fig. 85 CP MAS DD ^{13}C NMR spectrum of 1 T at 23 °C (from [58])

these two groups are involved in correlated oscillation motions. It is consistent with the expected restricted mobility of *meta*-substituted phenyl rings. Furthermore, $t_{1/2}$ values of protonated aromatic carbons of the 1I polymer indicate that *meta*-substituted rings do not undergo π-flips.

In the 1T copolyamide, below room temperature, C $=$ O groups do not exhibit oscillations of noticeable amplitude and for the phenyl groups only low

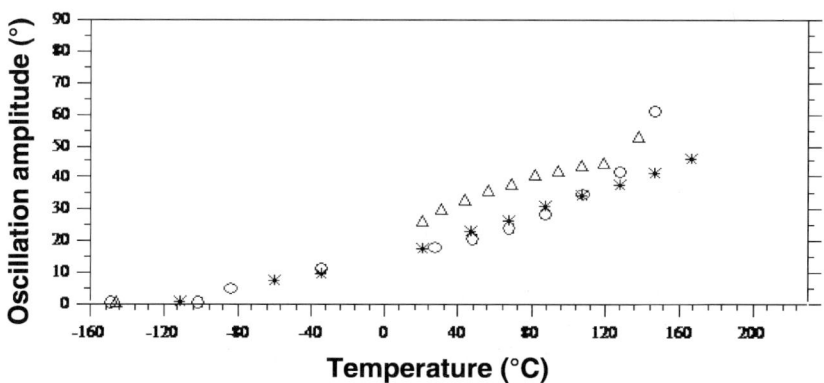

Fig. 86 Temperature dependence of the oscillation amplitudes (°) of the aliphatic C $=$ O groups in the 1 T ($\triangle$), 1.8 T ($\bigcirc$) and 1 I (*) polymers (from [58])

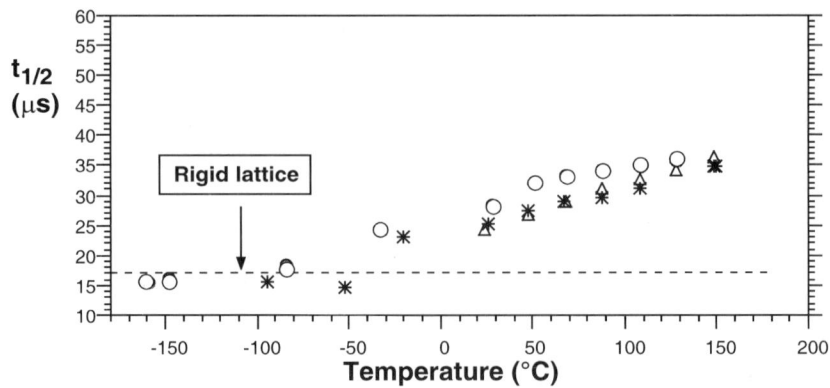

Fig. 87 $t_{1/2}$ values for the lactam-12 methylene carbons in the 1 T ($\triangle$), 1.8 T ($\bigcirc$) and 1 I ($*$) polymers (from [58])

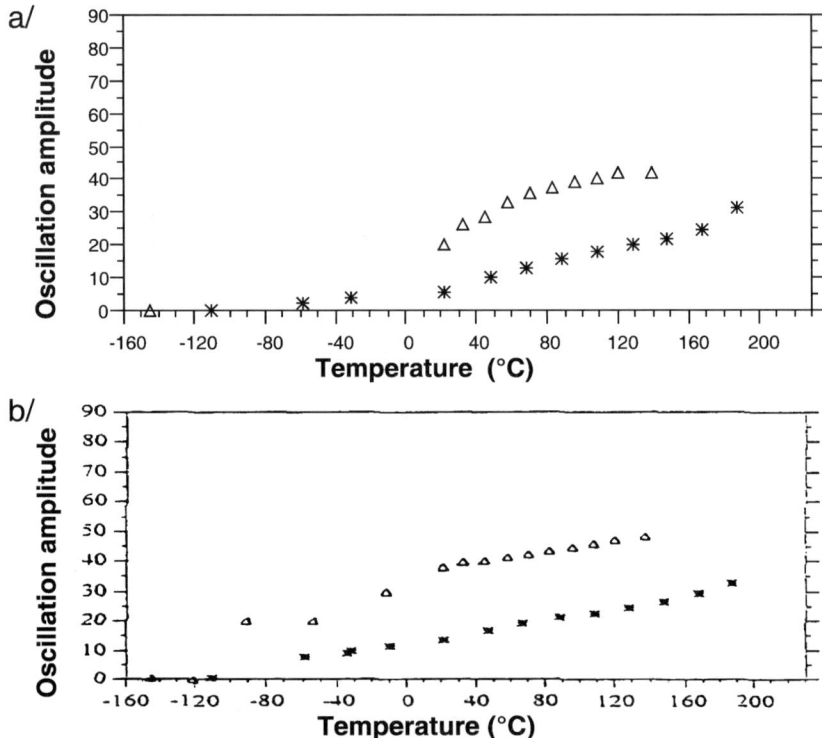

Fig. 88 a Amplitudes of oscillations (°) of the *iso-* and *tere-*phthalamide $C = O$ groups vs temperature in the 1 I ($*$) and 1 T ($\triangle$) polymers. **b** Amplitudes of oscillations (°) of the phenyl rings vs temperature in the 1 I ($*$) and 1 T ($\triangle$) polymers (from [58])

amplitude motions are observed. Above room temperature, large oscillations occur with the same amplitude for $C = O$ groups and phenyl rings, indicating that the motions of these two groups are correlated in this temperature range. In addition, $t_{1/2}$ measurements for the protonated aromatic carbons of 1T and 1.8T indicate that a fraction of the *para*-substituted phenyl rings are undergoing π-flips, as shown in Fig. 84.

Finally, as regards the 3,3'-dimethyldicyclohexylmethane units, $t_{1/2}$ values determined for the CH and CH_2 carbons do not reveal any local motions of these units at the frequency corresponding to NMR experiments (10^5 Hz range) in the temperature domain considered.

6.2.2
Comparison with Dielectric Relaxation

As mentioned, the frequency range for the $t_{1/2}$ measurements is around 10^5 Hz. In order to check whether the molecular motions observed by NMR for the aliphatic units, aromatic rings and carbonyl groups can be associated with the transitions determined from dielectric relaxation, it is necessary to extrapolate the dielectric results to estimate the temperature range where each transition should occur at 10^5 Hz and compare this with the temperature range at which the motions are detected by NMR.

6.2.2.1
The γ Transition

The dielectric γ transition is observed for the xT_yI_{1-y} copolyamides only. The transition temperature, determined at 10^5 Hz by extrapolating the dielectric data, is around $-60\,°C$. Such a temperature agrees with NMR observations of the motions of the $C = O$ groups in an aliphatic environment (Fig. 86), as well as the motions of the lactam-12 methylene groups (Fig. 87).

Thus, the γ transition involves oscillations of the $C = O$ groups accompanied by simultaneous motions of the lactam-12 sequence.

6.2.2.2
The β Transition

In MT copolyamides, this is the only secondary transition. Extrapolation of the relaxation map of the β dielectric transition to a frequency of 10^5 Hz leads to a temperature of $58\,°C$ for the maximum of the loss peak. The ^{13}C NMR results (Fig. 83) indicate that, at $58\,°C$, the $C = O$ groups and the adjacent phenyl rings undergo oscillations with amplitude, $\pm \alpha$, of $30°$, which is a satisfying qualitative agreement. Furthermore, ^{13}C NMR indicates that the $C = O$ groups and the adjacent phenyl rings are involved in correlated motions and that phenyl ring π-flips occur (Fig. 82).

In the xT_yI_{1-y} copolyamides, the ^2H NMR data [59] obtained on $1T_{0.7}I_{0.3}$ copolyamide, specifically perdeuterated on the *tere*-phthalamide phenyl rings, lead to an activation energy of the phenyl ring motions of 63.5 kJ mol^{-1}, comparable to the 63 kJ mol^{-1} obtained for the β dielectric relaxation. Furthermore, in the 1T copolyamide, the extrapolation at 10^5 Hz of the β dielectric relaxation leads to 35 °C. The ^{13}C NMR results (Fig. 88) indicate that, at this temperature, the *tere*-phthalamide C = O groups exhibit a significant mobility, with an amplitude of $\pm 30°$. The adjacent phenyl rings undergo oscillations of similar amplitude, indicating the strong correlation between the motions of these two groups. In addition, π-flips of the phenyl rings occur around this temperature (Fig. 84).

Despite experimental evidence, it is very likely that, in both MT and xT_yI_{1-y} copolyamides, the π-flips of the *tere*-phthalamide phenyl rings are associated with correlating flips of the adjacent C = O groups.

6.3
Dynamic Mechanical Analysis

Once the molecular motions occurring in the glassy state have been characterised and assigned through dielectric relaxation, ^{13}C and ^2H NMR, it is interesting to investigate their effect on the dynamic mechanical response of Ar-Al-PA [60, 61].

6.3.1
*x*T*y*I*₁₋y* Copolyamides

The temperature dependence of tan δ at 1 Hz in the glassy state for the various xT_yI_{1-y} copolyamides [60] is shown in Fig. 89. Three secondary transitions γ, β and ω, in the order of increasing temperature, are clearly observed.

6.3.1.1
The γ Transition

The γ transition observed from dynamic mechanical analysis at 1 Hz, is centred around – 150 °C. However, the temperature range available experimentally does not permit observation of the whole γ relaxation. The temperature position is independent of the chemical composition of the xT_yI_{1-y} copolyamides.

Furthermore, a quantitative analysis [60] performed on the loss compliance, J'', shows that the intensity of the γ transition is mostly determined by the amount of lactam-12 units and does not depend on the nature, *iso*- or *tere*-, of the phthalic acid.

As the γ transition observed by dielectric relaxation at 1 Hz occurs at the same temperature and the ^{13}C NMR experiments show that the lactam-12 unit motions are also involved, it is very likely that the molecular motions mechanically active in the γ transition are motions localised in the $(CH_2)_n$ sequences of the lactam-12 units and adjacent amide groups.

6.3.1.2
The β Transition

In contrast to the γ transition, the β relaxation is observed in its entirety in the temperature range available. As shown in Fig. 89, the β transition occurs in the temperature range from -100 to $0\,°C$ at 1 Hz. It exhibits very similar characteristics in the 1I, $1T_{0.7}I_{0.3}$ and 1.8T copolyamides. In contrast, the position of the β peak in 1.8I copolyamide is shifted to lower temperature and the intensity of the transition is lower.

In order to investigate the influence of the nature of the acid, the data obtained for the loss modulus, E'', as a function of temperature for the two polymers, 1.8I and 1.8T, are shown in Fig. 90, as well as the experimental reproducibility. First, the effect of the nature of the acid is much larger than the experimental uncertainty. Figure 90 shows that the low temperature part of the β transition is identical in the 1.8I and 1.8T samples. In the high-temperature region of this transition, an additional contribution is observed for the 1.8T copolyamide. Furthermore, the activation energies, $E_{a\,max}$ and $E_{a\,end}$, and entropies, ΔS_{max} and ΔS_{end}, determined at the maximum and end of the β transition, respectively, are reported in Table 7.

The value of ΔS_{max} for 1.8T is twice that for 1.8I, but this difference increases with increasing temperature and ΔS_{end} for 1.8T it is four times that

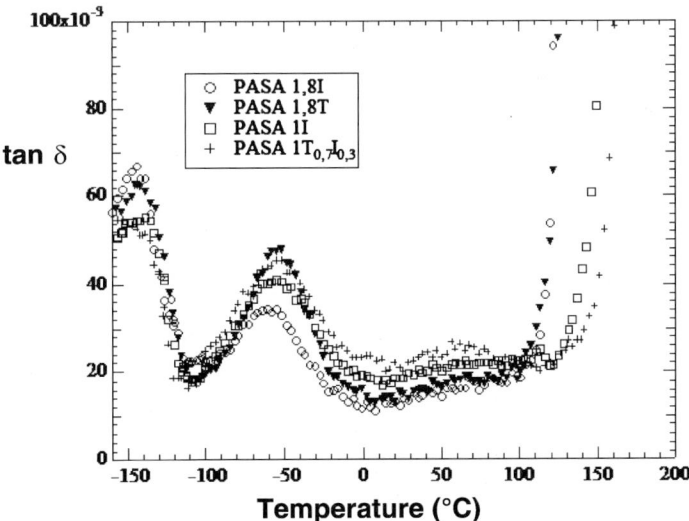

Fig. 89 Temperature dependence of $\tan\delta$ at 1 Hz for the different xT_yI_{1-y} copolyamides: ◯ 1.8 I, ▼ 1.8 T, □ 1 I and + $1T_{0.7}I_{0.3}$ (from [60])

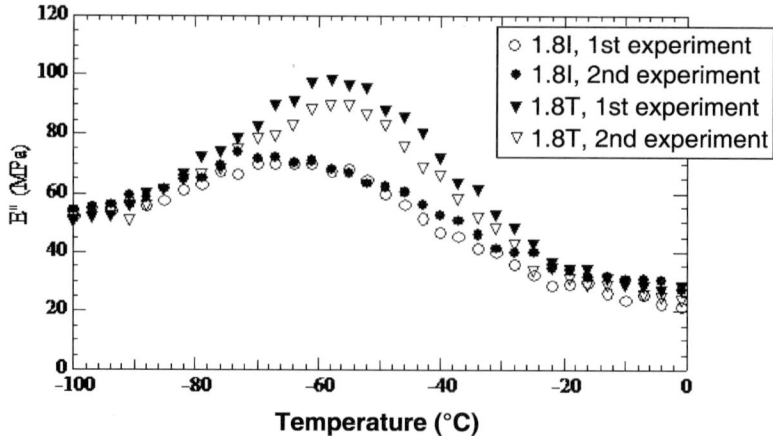

Fig. 90 Temperature dependence and reproducibility of the loss modulus, E'', for the 1.8 I and 1.8 T copolyamides (from [60])

Table 7 Activation energies and entropies determined at the maximum and end of the β transition for the different copolyamides (from [60])

Polyamide	Ea_{max} [kJ mol^{-1}]	ΔSa_{max} [J K^{-1} mol^{-1}]	Ea_{end} [kJ mol^{-1}]	ΔSa_{end} [J K^{-1} mol^{-1}]
1.8 I	55	30	70	35
1.8 T	65	60	95	120
1 I	55	30	70	25
1 T$_{0.7}$I$_{0.3}$	60	40	70	30
1 T	60	40	85	85

for 1.8I. It means that in the high temperature part of the β transition the motions are considerably more cooperative in 1.8T than in 1.8I.

It is worth noting that the β relaxation observed from dielectric measurements (Fig. 77) occurs at the same temperature as the one from dynamic mechanical measurements. In addition, the activation energies derived from the maximum of the mechanical β peak (Table 7) are close to those obtained from dielectric relaxation (Table 6).

By using the assignments of the motions involved in the β relaxation of the xT_yI_{1-y} copolyamides obtained from dielectric, ^{13}C and ^{2}H NMR experiments, the high-temperature contribution, present in the 1.8T copolyamide and absent in the 1.8I polymer, is due to the π-flips of the phenyl rings. In contrast, the low-temperature region of the β transition, which is identical in the two polymers, is likely to arise from oscillations of the phenyl rings and adjacent amide groups. Since these oscillations are mechanically active, they should be accompanied by a local reorganisation of the environment of the mobile unit.

It is worth noting that the investigation of motions in the glassy state of the poly(ethylene *tere*-phthalate) (Sect. 4) leads to a similar conclusion about the low- and high-temperature sides of the β transition.

Finally, let us mention that the lower intensity of the β transition in 1.8I, compared to 1I, can be quantitatively accounted [60] for by the dilution of the mechanically active groups which results from the increase in the lactam-12 content.

6.3.1.3
The ω Transition

The ω transition is a weak and wide transition centred around 60 °C, in the temperature range of the ω secondary relaxation observed by dielectric relaxation. In both cases, the position and width of the ω peak are independent of the copolyamide composition.

The common features shared by the ω mechanical and dielectric transitions tend to indicate that these two phenomena have the same origin, i.e. they arise from motions that involve the conjugated amide groups situated between the aromatic ring and aliphatic cyclodiamine.

6.3.2
MT$_y$ Copolyamide

Since the dynamic mechanical measurements require samples 2 mm thick, it was not possible to avoid a rather high degree of crystallinity of the MT samples. In order to reduce the extent of crystallinity a copolymer containing 0.9T and 0.1I, i.e. $MT_{0.9}I_{0.1}$, was used [61].

The temperature dependence of the loss modulus, E'', at 1 Hz of $MT_{0.9}I_{0.1}$, plotted in Fig. 91, shows quite a broad peak with a maximum around – 110 °C. Such a behaviour is very surprising, compared to the results obtained by dielectric relaxation (Fig. 76) where a broad peak is also observed from – 120° to about 50 °C, centred around – 60 °C.

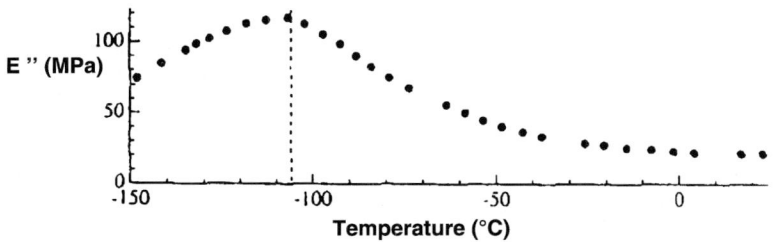

Fig. 91 Loss modulus, E'', vs temperature for the $MT_{0.9}I_{0.1}$ copolyamide (from [61])

The dielectric relaxation exclusively reflects the amide group motions, starting at $-120\,°C$. So, the mechanical loss, E'', occurring at a lower temperature originates from motions of the CH_2 units, as evidenced by ^{13}C NMR measurements (Sect. 6.2). Furthermore, these latter show that, at higher temperatures, the phenyl ring motions (oscillations then π-flips) are coupled to the amide group motions, in a way similar to that observed for poly(*tere*-phthalate) (Sect. 4).

Consequently, the broad mechanical β transition of $MT_{0.9}I_{0.1}$ copolyamide is a combination of a γ-like and a β transition.

6.4
Conclusion

The combined investigations of a series of aryl-aliphatic copolyamides (xT_yI_{1-y} and MT) by dielectric relaxation, solid-state ^{13}C and ^2H NMR, and dynamic mechanical experiments demonstrate the existence of three secondary transitions: γ, β and ω, in order of increasing temperature.

The lactam-12 units, whose non-conjugated amide groups and methylene carbons undergo motions in a low-temperature range are responsible for the γ transition. Furthermore, the intensity of the transition is directly related to the amount of lactam-12 units; this transition does not exist in the MT copolyamide.

The β transition involves the conjugated Φ-CONH sequence, where the relevant amide group is situated between the aromatic ring and the lactam-12 moiety. However, the mechanical response is closely related to the ring motions. Whereas the lower temperature part of the β transition can be assigned to localised motions (oscillations) of amide groups and *iso*- or *tere*-phthalic rings, the higher temperature range of the β transition is associated with intra- or intermolecular cooperative motions involving π-flips of the *tere*-phthalic rings. It is worth pointing out that such a behaviour is similar to that in poly(ethylene *tere*-phthalate) (Sect. 4).

Finally, the ω transition, present in xT_yI_{1-y} copolyamides, is due to motions that involve the amide groups located between the phenyl rings and the cycloaliphatic moiety.

7
Aryl-aliphatic Epoxy Resins

The polymers considered in Sects. 4 to 6, i.e. poly(ethylene *tere*-phthalate), bisphenol A polycarbonate and aryl-aliphatic polyamides, are linear polymers or copolymers without side chains, in which molecular motions occur in the glassy state. Detailed analysis of the transitions, mostly the β transition, reveals that several processes are involved in this β transition and that the cooperativity (as shown through ΔS_a) of the processes develops with increasing temperature.

In order to complete such a picture, it is interesting to investigate the effect of crosslinks on the secondary transitions, in particular the effect on the development with temperature of the cooperativity of the molecular motions involved.

A very convenient system for performing this investigation consists of epoxy networks formed by reacting aromatic epoxy on aliphatic diamine. Indeed, the different nature (aromatic or aliphatic) of the units allows one to apply ^{13}C NMR for identifying the groups involved in the transitions and their motions. Furthermore, in the same way as for poly(ethylene *tere*-phthalate), specific small molecule additives act as antiplasticisers, which can offer an additional possibility for investigation of the molecular motions and their cooperativity.

7.1
Pure Epoxy Networks

The epoxy resins [62, 63] are formed from the chemicals indicated in Table 8. The different types of networks that can be obtained are outlined in Fig. 92.

All the systems contain the digycidyl ether of bisphenol A (DGEBA) and the changes come from the type of amine with which it is reacted:

- With neat hexamethylene diamine (HMDA), each amine group leads to a crosslink and, consequently, DGEBA/HMDA networks have the highest crosslink density.
- With a mixture of HMDA with various amounts of hexylamine (HA), the crosslink density of the network is determined by the relative amount of diamine with respect to monoamine. Furthermore, pending hexamethylene units are present.
- With a mixture of HMDA with various amounts of dimethylhexamethylene diamine (DMHMDA), the crosslink density is determined by the relative amount of diamine with respect to the secondary diamine, but now the role of secondary diamine is to extend the mesh of the network by inserting flexible units into the chains between crosslinks.

It should be noticed that linear polymers DGEBA/HA or DGEBA/DMHMDA lead to very brittle samples; for this reason lightly crosslinked systems were used instead.

The code names, compositions and glass transition temperatures of the various networks are reported in Table 9.

It is worth pointing out that the stoichiometry between epoxy group and NH function has always been controlled; furthermore, the curing conditions have been defined in order to have an epoxide–amine reaction as complete as possible: 40 °C for 12 h, then at $T_g + 30$ °C for 24 h. ^{13}C and ^{15}N NMR [64]

Table 8 Characteristics of the chemicals used for the epoxy resins and antiplasticiser

Chemical	Formula
Diglycidylether of Bisphenol A (DGEBA)	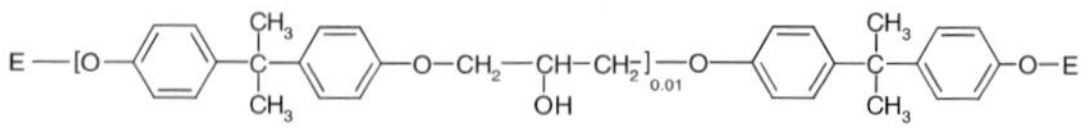 $E-[O-\bigcirc-\overset{CH_3}{\underset{CH_3}{C}}-\bigcirc-O-CH_2-\underset{OH}{CH}-CH_2]_{0.01}-O-\bigcirc-\overset{CH_3}{\underset{CH_3}{C}}-\bigcirc-O-E$ Width E = $-CH_2-CH-CH_2$ (epoxide)
Hexamethylene diamine (HMDA)	$NH_2 - (CH_2)_6 - NH_2$
Hexylamine (HA)	$CH_3 - (CH_2)_5 - NH_2$
N,N-dimethyl hexamethylene diamine (DMHMDA)	$CH_3 - NH - (CH_2)_6 - NH - CH_3$
Epoxyphenoxypropane	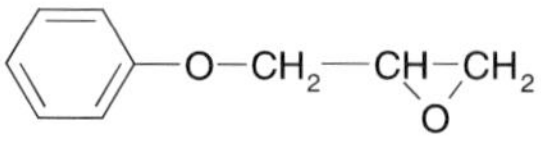 $\bigcirc-O-CH_2-CH-CH_2$ (epoxide)
Hydroxyacetanilide	$HO-\bigcirc-NH-\overset{O}{\underset{}{C}}-CH_3$
EPPHAA	$\bigcirc-O-CH_2-\underset{OH}{CH}-CH_2-O-\bigcirc-NH-\overset{O}{\underset{}{C}}-CH_3$

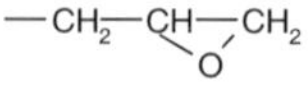

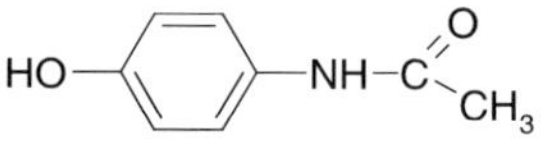

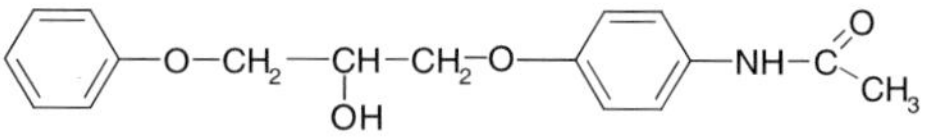

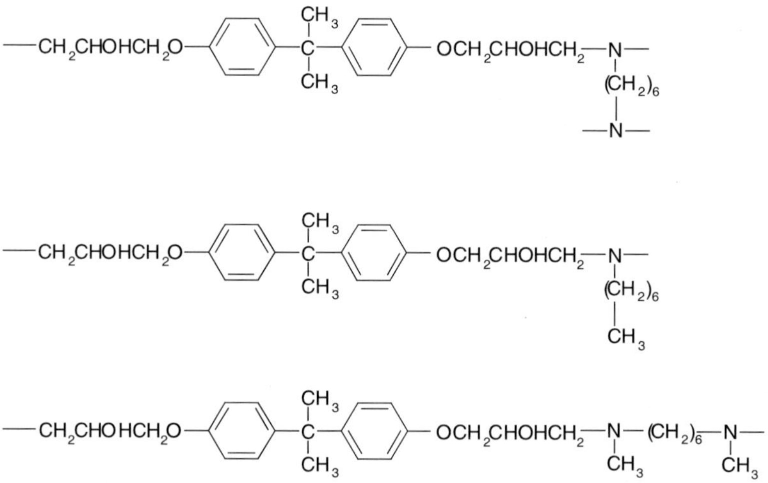

Fig. 92 Representation of the different types of networks: **a** DGEBA/HMDA, **b** DGEBA/HA and **c** DGEBA/DMHMDA

Table 9 Code names, compositions and glass transition temperatures of pure and anti-plasticised epoxy resins

Code name	Primary diamine [Mole %]	Primary monoamine [Mole %]	Secondary diamine [Mole %]	EPPHAA [wt %]	T_g [°C]
HMDA	100	0	0	0	121
HMDA/AP10	100	0	0	10	80
HMDA/AP19	100	0	0	19	68
HA60	25	75	0	0	70
HA60/AP19	25	75	0	19	45
HA95	2.5	97.5	0	0	56
HA95/AP19	2.5	97.5	0	19	39
DMHMDA60	25	0	75	0	60
DMHMDA60/AP19	25	0	75	19	33
DMHMDA95	2.5	0	97.5	0	32
DMHMDA95/AP19	2.5	0	97.5	19	28

confirm that the efficiency of crosslinking is $90 \pm 5\%$, i.e. all epoxy rings are opened and virtually all amine nitrogens are substituted. In addition, under these curing conditions the DGEBA does not produce side reactions with amines.

Due to the approach chosen for controlling the crosslink density and the curing conditions adopted, the considered systems can be viewed as "model" epoxy networks.

To illustrate the transitions occurring in the glassy state, the temperature dependencies of the dynamic mechanical loss tangent at 1 Hz of the DGEBA/HMDA, DGEBA/tetramethylene diamine and DEBA/dodecamethylene diamine systems [65] are shown in Fig. 93. They clearly exhibit two transitions:

1. At low temperatures, the γ transition is centred around – 150 °C. It has been assigned to motions of the central methylene groups of the amine moiety. Indeed, it appears with diamines containing at least four methylene units and increases in intensity with longer methylene sequences.

2. At higher temperatures, a broad β transition, present in all epoxy resins independently of the reacting species (amines or anhydrides), is much more complex than the γ transition. Its position and shape vary strongly with the chemical structure of the epoxy resin. The dynamic mechanical response and the solid-state ^{13}C NMR of the various model networks referred to in Table 9 focus precisely on the analysis of this β transition [62, 63].

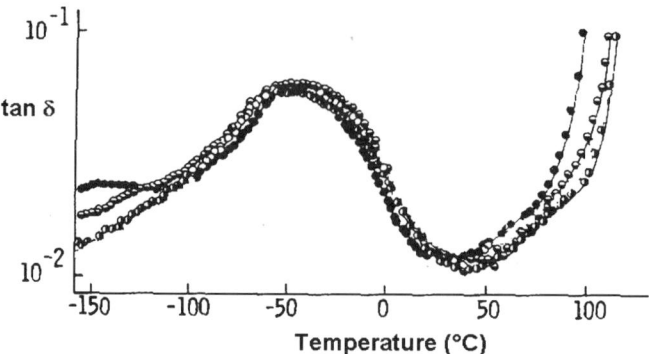

Fig. 93 Loss tangent versus temperature for DGEBA/HMDA (*circles with bottom side black*), DGEBA/tetramethylene diamine (*circles with right side black*) and DGEBA/dodecamethylene diamine (●) systems (from [65])

7.1.1
Dynamic Mechanical Analysis

The dynamic mechanical response of a material can be characterised through the loss modulus, E'', the loss tangent, tan δ, or the loss compliance, J''. However, as already mentioned for Ar-Al-PA (Sect. 6), the loss compliance can be considered the most relevant parameter for quantitatively comparing different materials, at least for additive purposes. For this reason, the semi-quantitative analysis and the comparison of viscoelastic data determined for different systems have been performed [63] in terms of J'', whereas the determination of activation energies and entropies are based on loss modulus data.

7.1.1.1
DGEBA/HMDA Densely Crosslinked Network

The dependence of the β transition on the test frequency is shown in Fig. 94a, where the loss modulus, E'', is plotted versus the reciprocal temperature, $1/T$. Besides the expected shift between the different isochrones, a surprising feature is the increase of the height of the maximum with increasing frequency. Such a result clearly indicates that the β transition does not correspond to a single relaxation process, but suggests the existence of several processes having very different activation energies and gradually merging when increasing the frequency. Indeed, at high frequency, the processes that are observed in the low-temperature side of the transition and that have a relative low activation energy tend to overlap the higher activation energy processes. Thus, that the observed β peak is narrower and the intensity of its maximum increases with frequency. This is confirmed by the impossibility of building a master curve over the entire β transition, as shown in Fig. 94b. Therefore,

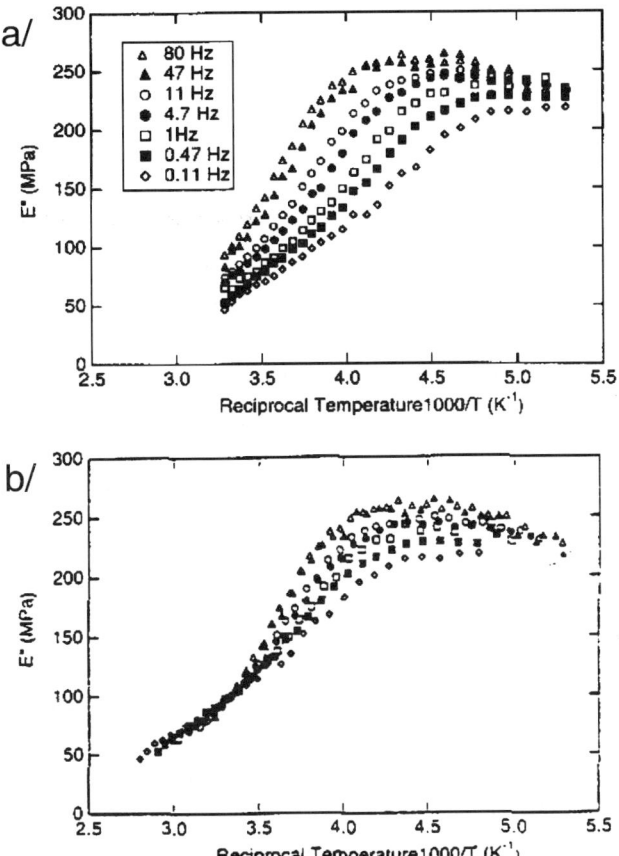

Fig. 94 a Dependence of E'' as a function of frequency in the HMDA system. **b** Result of the high temperature superposition (from [63])

the activation energies, E_a (determined from the frequency shift of the maximum of the β peak) and the activation entropies, ΔS_a (derived according to the Starkweather expression, Eq. 3 in Sect. 2.3) in the densely crosslinked network are apparent activation energies and entropies. For qualitative purposes, the activation energies and entropies have been determined in three different regions of the β transition:

1. High-temperature region corresponding to the last third of the transition (i.e. $E'' < E''_{max}/3$, $T > T_{max}$)
2. Low-temperature region corresponding to the first third of the transition (i.e. $E'' < E''_{max}/3$, $T < T_{max}$)
3. Middle of the transition by following the frequency dependence of the maximum of the β peak.

Table 10 Characteristic temperatures (T) and activation energies (E_a), enthalpies (ΔH_a) and entropies (ΔS_a), in the β relaxation region for the different systems, pure and antiplasticised (from [70])

Code name	$T_{beginning}$ [°C]	$Ea_{beginning}$ [kJ mol^{-1}]	$T_{maximum}$ [°C]	$Ea_{beginning}$ [kJ mol^{-1}]	$T_{end\ of\ peak}$ [°C]	$Ea_{end\ of\ peak}$ [kJ mol^{-1}]
HMDA	– 123	36	– 61	70	4	111
HMDA/AP10	– 123	37	– 64	60	– 38	71
HMDA/AP19	– 123	34	– 77	51	– 47	60
HA60	– 128	34	– 74	53	– 39	80
HA60/AP19	– 123	37	– 77	51	– 48	61
HA95	– 133	32	– 83	50	– 55	64
HA95/AP19	– 133	35	– 83	45	– 57	58
DMHMDA60	– 128	34	– 68	72	– 33	85
DMHMDA60/AP19	– 128	39	– 74	65	– 39	82
DMHMDA95	– 128	35	– 74	64	– 38	85
DMHMDA95/AP19	– 128	38	– 77	59	– 44	82

Table 10 Continued.

Code name	$\Delta H a_{beginning}$ [kJ mol^{-1}]	$\Delta Sa_{beginning}$ [J mol^{-1} K^{-1}]	$\Delta Ha_{maximum}$ [kJ mol^{-1}]	$\Delta Sa_{maximum}$ [J mol^{-1} K^{-1}]	$\Delta Ha_{end\ of\ peak}$ [kJ mol^{-1}]	$\Delta Sa_{end\ of\ peak}$ [J mol^{-1} K^{-1}]
HMDA	35	7	68	95	108	163
HMDA/AP10	36	14	58	52	69	66
HMDA/AP19	33	0	49	26	58	30
HA60	33	3	51	31	78	105
HA60/AP19	35	7	40	26	59	36
HA95	31	0	48	28	62	58
HA95/AP19	33	11	45	13	59	33
DMHMDA60	33	9	70	116	83	118
DMHMDA60/AP19	38	37	63	92	82	114
DMHMDA95	34	3	62	87	83	125
DMHMDA95/AP19	36	30	57	66	80	121

The results are gathered in Table 10.

The weak entropy determined at the onset of the β transition indicates that the low-temperature motions are localised. As the temperature increases along the β transition, the entropy increases as well and reaches very high values for the high-temperature part of the transition, indicating that more and more cooperative motions are taking place with increasing temperature.

7.1.1.2
Quasi-linear DGEBA/HA95 and Loosely Crosslinked DGEBA/HA60 Systems

The dependence of the loss compliances, J'', on temperature is plotted in Fig. 95 for the HA60 and HA95 systems. It is striking how much the width and height of the β transition decrease when the crosslink density decreases. The high-temperature part of the transition, which corresponds to the more cooperative motions, progressively disappears on decreasing the crosslink density. The changes observed in the series HMDA, HA60 and HA95 indicate that the distribution of motional modes decreases with decreasing the crosslink density. For the quasi-linear HA95 system a master curve can be built over the whole β transition.

Consequently, the results are a clear indication that the high-temperature part of the β transition in densely crosslinked networks originates from cooperative motions involving the crosslinks. However, there is some complexity in the cooperativity involving the crosslinks because the β transition of the HA60 is not a linear combination of the HA95 and HMDA β transitions. Indeed, rather than adding a single contribution, the increase of the crosslink density permits the development of more and more cooperativity. A crude

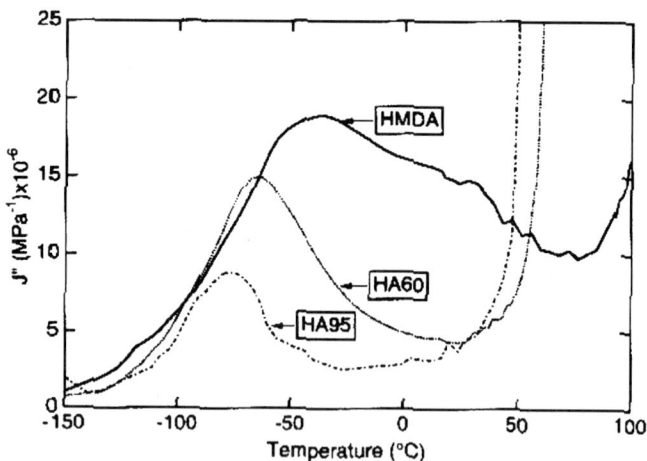

Fig. 95 Temperature dependencies of the loss compliances, J'', determined at 1 Hz for the HMDA, HA60 and HA95 systems (from [63])

estimate of the extent of the cooperative mobile segments can be achieved through the probability of occurrence of consecutive crosslinks. It turns out that the high-temperature region of the β transition, which is observed in the HMDA sample and not in the HA60 sample, can be assigned to cooperative motions involving more than six crosslinks.

7.1.1.3
Quasi-linear DGEBA/DMHMDA95
and Loosely Crosslinked DGEBA/DMHMDA60 Systems

The loss compliances, J'', versus temperature for DMHMDA95 and 60 are plotted in Fig. 96. For comparison, the results for HMDA are also shown. The corresponding activation energies and entropies are gathered in Table 10.

The low-temperature part of the β transition is independent of the crosslink density and of the amines, again indicating that the onset of the transition is due to very localised motions, occurring at the scale of an epoxy-amine repeat unit. Such a conclusion is supported by the low values of the activation entropies at the onset of the transition (Table 10).

It is interesting to compare the behaviours of DMHMDA- and HA-based systems at similar crosslink density. First, for the quasi-linear systems DMH-MDA95 and HA95, the amplitude of the β peak is higher by a factor of more than two in the first system and the activation entropy associated with the maximum is also higher. Similar observations have been made for DMH-MDA60 and HA60.

All these results can be understood by comparing the chemical architecture of the DMHMDA- and HA-based systems (Fig. 92). The DMHMDA sec-

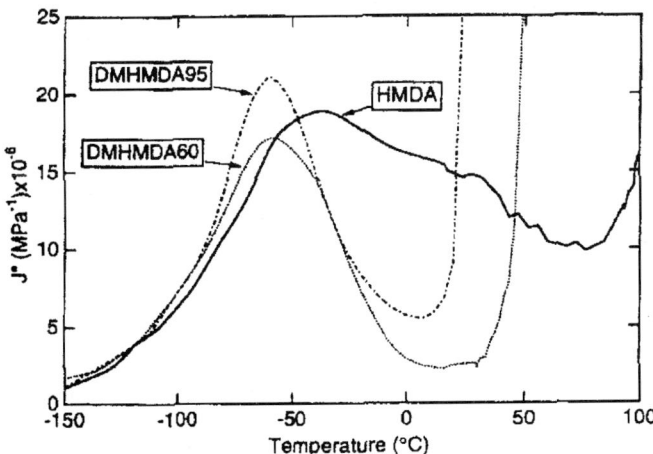

Fig. 96 Temperature dependencies of the loss compliance, J'', determined at 1 Hz for the HMDA, DMHMDA60 and DMHMDA95 systems (from [63])

ondary diamine plays the role of a mesh extender in the network, whereas the HA primary diamine is incorporated in the network as a pending chain. The results on the height of the β peak, the activation entropies, tend to indicate that the presence of the DMHMDA unit located between two epoxy groups allows the development of motions with a larger spatial extent and also induces an intramolecular cooperativity along the chain. It differs from the cooperativity induced by the crosslink points because the chain constraints are much weaker and, therefore, it manifests itself at lower temperature. Furthermore, as shown by the data reported in Table 10, the cooperativity associated with the chain extender does not seem to be affected significantly by the crosslink density.

At the end of the β peak, the entropies (Table 10) are smaller in the DMH-MDA95 and 60 systems than in the HMDA network, indicating that cooperativity is likely to have a larger spatial extent when it proceeds from the crosslink points. It thus appears that the secondary diamines lead to an enhancement of the mobility and a development of a cooperativity that is more limited than the one observed in densely crosslinked networks.

7.1.2
Solid-State ^{13}C NMR Analysis

As with the previously considered polymers (Sects. 3 to 6), solid-state ^{13}C NMR experiments can provide more information about the chemical sequences and the types of motions involved in the β transition of epoxy resins.

Due to the higher frequency involved in NMR, the broad distribution involved in the β transition observed by dynamic mechanical measurements will be reduced, the localised or cooperative processes tending to merge.

Furthermore, the comparison between dynamic mechanical results and NMR mobility observations will be more delicate since it requires extrapolation of the mechanical response over about 5 decades, and must take into account the experimental uncertainty of the activation energy values. In the other reported polymers, the use of dielectric relaxation techniques, which cover a frequency range up to 10^5 Hz, overcame the extrapolation difficulty. Consequently, for the epoxy resins the comparison will remain more qualitative [63].

7.1.2.1
DGEBA/HMDA Densely Crosslinked Network

The high resolution solid-state ^{13}C NMR spectrum of the HMDA network observed at 80 °C leads to separate lines for the various aliphatic carbons: DGEBA methyl carbons, HMDA methylene carbon in the β or γ position with respect to the nitrogen atom, CH_2 (of the hydroxypropyl ether) directly bonded to the nitrogen, and the $CHOH - CH_2 - O$ unit. For the aromatic part of the DGEBA, lines from protonated and unprotonated carbons are resolved.

Local Motions of Aliphatic Units

The temperature dependencies of the $(t_{1/2})_0/t_{1/2}$ ratio, where $(t_{1/2})_0$ is the $t_{1/2}$ value measured at room temperature, determined for the $CHOH - CH_2 - O$ and $CH_2 - N$ units of the hydroxylpropyl ether (HPE) sequence (Fig. 92) in the HMDA network [63] are shown in Fig. 97. It is worth noticing that the $t_{1/2}$ values of these two types of carbons have the same temperature dependence. Up to 60 °C, the $t_{1/2}$ values are constant and equal to the rigid-lattice values, indicating that the HPE sequence does not undergo any local motion at a frequency equal to or higher than 10^5 Hz in this temperature range. Above 60 °C, mobility develops, which leads at 100 °C to motions in the tens of kilohertz for the whole HPE sequence. These results are qualitatively confirmed by data on ^{13}C spin-lattice relaxation time in the rotating frame, $T_{1\rho}(^{13}C)$.

Local Motions of the Aromatic Units

At very low temperature, the DGEBA aromatic rings are immobile whereas, at room temperature, there are π-flips occurring at a rate of around 10^3 Hz [66].

Results of $t_{1/2}$ measurements performed as a function of temperature for the protonated aromatic carbons [63] are shown in Fig. 97. At room temperature, no motion at 10^5 Hz is observed, in agreement with that mentioned

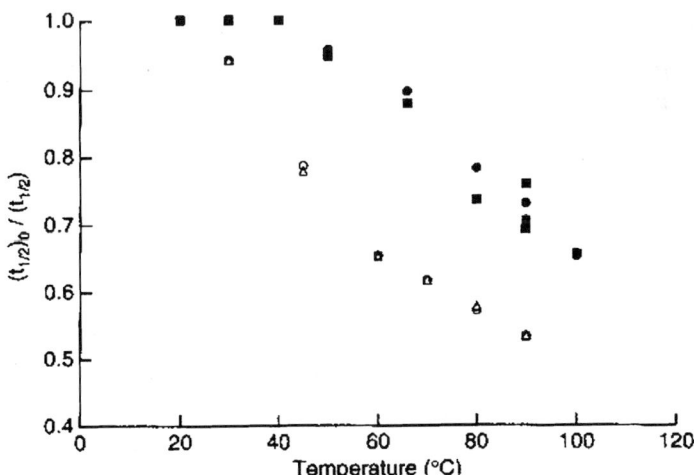

Fig. 97   Temperature dependence of the $(t_{1/2})_0/t_{1/2}$ ratio, measured for the $CHOH - CH_2 - O$ (●) and $CH_2 - N$ carbons (■) of the hydroxypropyl ether sequence and for the protonated aromatic carbons (△, ○) in the HMDA sample

above. At higher temperatures, an increase in mobility happens in the same temperature range as the HPE motion. $T_{1\rho}$ (^{13}C) experiments indicate the occurrence at 70 °C of π-flips at 64 kHz.

The unprotonated aromatic carbons do not reveal any reorientation of the *para* axis of the phenyl ring below the glass transition temperature.

Comparison with Results from Dynamic Mechanical Measurements

In addition to the difficulty in extrapolating the dynamic mechanical results over 5 decades in frequency, the complexity of the β relaxation in epoxy resins induces quite a broad range for the temperatures extrapolated at 10^5 Hz. The calculated values are equal to:

- 23 °C For the lower temperature processes
+ 26 °C For the maximum of the β peak
+ 90 °C For the high-temperature part of the transition

As a matter of fact, the onset of mobility, as observed from the $t_{1/2}$ and $T_{1\rho}$ (^{13}C) measurements performed on the CHOH – CH$_2$ – O and CH$_2$ – N groups, occurs in the upper part of this range. This result is consistent with previous assignment of the β transition to motional processes of the HPE sequence. Moreover, the parallel behaviour of the CHOH – CH$_2$ – O and CH$_2$ – N groups, observed in NMR, shows that the crosslink points are involved in the motional processes. However, the temperature of the NMR onset of mobility indicates that the sensitivity of the NMR experiments probes the cooperative motions rather than the isolated ones. This conclusion is in agreement with results obtained by ^{2}H NMR [67].

As regards the role of the DGEBA phenyl groups in the β transition of epoxy networks, Fig. 97 shows that the ring motions are detected at a lower temperature than the reorientation of the HPE sequences, in agreement with the ^{2}H NMR experiments [67]. Despite this observation, it is not possible to assign π-flip motions to the low-temperature side of the β transition. Indeed, the system based on HMDA and diglycidyl ether of resorcinol (DGERO) [68], whose *meta*-substituted rings do not flip, has exactly the same mechanical spectrum as the DGEBA/HMDA system in the low-temperature part of the β transition. However, it is worth pointing out that ^{13}C and ^{2}H NMR measurements indicate that, although the flip process is faster than the reorientation of the HPE sequences, the motions of the ring and the HPE units share very similar temperature dependencies. A likely conclusion to this observation is that the motions of these two parts of the molecules occur in a partly correlated way. It could be due to the fact that in DGEBA phenyl rings, as in phenyl rings of bisphenol A units of polycarbonate (see [28–30, 51, 63]), the steric hindrance is so high that some ring flips cannot occur without a slight increase of the bond angle between the two rings. Such a process could be associated with the motions of the oxygen atoms next to the phenyl ring and, therefore, with those of the HPE sequences. Indeed, it induces a correlation between the ring flips and the reorientation of the HPE groups. In such a view, ring flips would be indirectly coupled to the mechanical losses.

7.1.2.2
Loosely Crosslinked DGEBA/HA60 System

In addition to the carbons observed in the high resolution solid-state ^{13}C NMR spectrum of the HMDA part, the ^{13}C NMR spectrum of the HA60 system contains two more lines arising from the hexylamine carbons [63].

Local Motions of Aliphatic Units

The temperature dependencies of the $(t_{1/2})_0/t_{1/2}$ ratio, where $(t_{1/2})_0$ is the $t_{1/2}$ value measured at room temperature, determined for the CHOH – CH_2 – O and CH_2 – N units are shown in Fig. 98. An onset of mobility occurs around 70 °C for the CHOH – CH_2 – O unit and 80 °C for the CH_2 – N, in contrast to what happens in the densely crosslinked network where the two groups behave identically. This means that the methylene carbons directly bonded to the crosslinks are now slightly less mobile than the CHOH – CH_2 – O units and much less than they are in densely crosslinked HMDA networks.

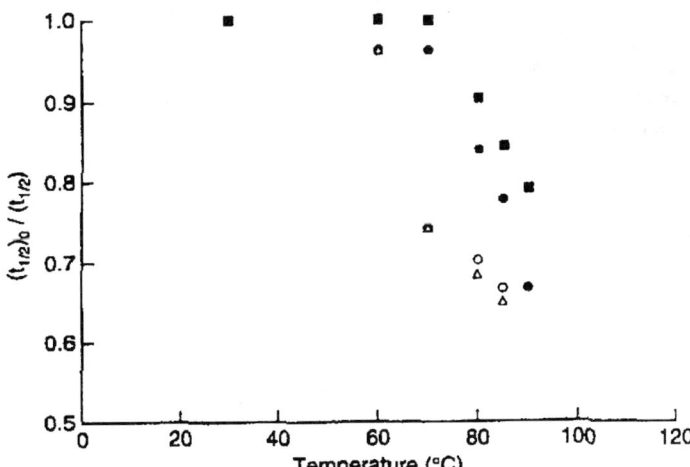

Fig. 98   Temperature dependence of the $(t_{1/2})_0/t_{1/2}$ ratio, measured for the CHOH – CH_2 – O (●) and CH_2 – N carbons (■) of the hydroxypropyl ether sequence and for the protonated aromatic carbons (△, ○) in the HA60 sample (from [63])

Local Motions of Aromatic Units

As shown in Fig. 98, the $t_{1/2}$ values for protonated aromatic carbons indicate an onset of mobility at 10^5 Hz around $70\,°C$, in the same temperature range as the HPE group, suggesting a correlation with the reorientations of the HPE units.

Comparison with Dynamic Mechanical Results

The extrapolation to 10^5 Hz of the different parts of the β transition identified in the mechanical analysis leads to the following temperatures:

- $28\,°C$ For the low-temperature processes
- $38\,°C$ For the maximum of the β peak
- $52\,°C$ For the high-temperature cooperative motions

These extrapolated temperatures are lower than the ones derived for the densely crosslinked HMDA network.

The onset of mobility observed around $70\,°C$ for the HPE units roughly corresponds to the most cooperative motions involved in the β transition of the loosely crosslinked system, supporting the conclusion, derived from HMDA results, that NMR is mainly sensitive to cooperative motions. Along this reasoning, the onset of mobility of the $CH_2 - N$ units, which appears at still higher temperature, tends to indicate that the nitrogen atoms of the diamine or monoamine moieties are less involved in the β process of the loosely crosslinked network than they are in the densely crosslinked network. This point is in good agreement with results derived from dynamic mechanical analysis, which point out a larger cooperativity of the β relaxation in the densely crosslinked network. Consequently, with the loosely crosslinked network, due to the less constrained chain structure, there are less cooperative motions and it is necessary to go to higher temperatures for getting cooperative motions that can be picked up by NMR measurements. This is the reason why, in spite of extrapolated temperatures lower for HA60 than for HMDA systems, the onset of mobility detected by NMR for HPE and $CH_2 - N$ units occurs nevertheless at higher temperatures for HA60 than for HMDA.

7.1.3
Conclusions on β Transition Motions in Pure Aryl-aliphatic Epoxy Resins

The dynamic mechanical results and the solid-state ^{13}C NMR measurements lead to a deeper insight of the motions occurring below the glass transition temperature in the considered pure aryl-aliphatic epoxy networks, in particular those involved in the β transition of these systems, and the nature of their cooperativity.

The β transition clearly originates from motions of the hydroxypropyl ether sequence, but the crosslink points are also involved. Whereas the motions involved in the low-temperature part of the β transition are quite isolated motions of HPE units, when temperature increases a cooperativity appears directly with the mobility of the crosslinks and, indirectly, with the π-flip motions of the DGEBA phenyl rings.

The overall crosslink density of the systems plays an important role in the high-temperature range of the β transition. For highly crosslinked HMDA networks, the extent of cooperativity of the HPE motions reaches more than six crosslinks. Decreasing the crosslink density reduces the cooperativity between the HPE moieties and the crosslink points, the latter being less involved, in such a way that the high-temperature part of the β transition disappears. However, it is worth noticing that such an effect of the crosslink density decrease depends on the architecture of the network. Thus, for the networks obtained by decreasing the crosslink density through monoamine and, consequently, pending hexamethylene units, the cooperativity disappears rapidly. For the corresponding quasi-linear system only isolated motions remain. In contrast, the networks with a chain extender (obtained from secondary diamine) have less cooperativity with the crosslink points and, by the way, are less sensitive to their density. However, they develop an intramolecular cooperativity which, nevertheless, is more limited than the spatial cooperativity with the crosslink points observed in densely crosslinked networks.

7.2
Epoxy Networks with Antiplasticiser Additives

In a way similar to that described for polyethylene *tere*-phthalate (Sect. 4.2), some antiplasticiser small molecules with a specific chemical structure are able to affect the β transition and the yield stress of epoxy resins, but they do not have any effect on the γ transition. In the case of HMDA networks, an efficient antiplasticiser, EPPHAA, whose chemical structure is shown in Table 8, has been reported [69]. The investigation of such antiplasticised epoxy networks by dynamic mechanical analysis as well as solid-state NMR experiments [70] can lead to a deeper understanding of the molecular processes involved in the β transition and of their cooperativity.

The antiplasticiser molecule is added to the reacting agents before curing. A single glass transition has been observed in the final samples, indicating that the antiplasticiser is fully miscible with the epoxy networks.

The code names of antiplasticised systems contain the reference APx, where x corresponds to the wt % of antiplasticiser in the system. They are gathered in Table 9.

7.2.1
Dynamic Mechanical Analysis

7.2.1.1
The DGEBA/HMDA/AP Densely Crosslinked Networks

Figure 99 shows the temperature dependence of the loss compliance, J'', at 1 Hz for the pure and antiplasticised HMDA networks.

It is clear that the three networks exhibit the same behaviour in the low-temperature part of the β relaxation. In contrast, the high-temperature component of the β peak is progressively suppressed in the presence of increasing amounts of antiplasticiser. More precisely, the antiplasticiser hinders all the high-temperature motions, whatever its amount (10 or 19 wt %), but in the intermediate temperature range the efficiency of the antiplasticiser increases with its concentration in the network. According to our conclusions on the degree of cooperativity of the β motions as a function of temperature in the case of pure epoxy networks, these results would imply that the capability of the antiplasticiser to hinder local motions is strong when dealing with cooperative processes. It decreases when the spatial scale of cooperativity decreases and finally vanishes when only isolated motions are occurring. The effect is such that for networks with a higher concentration of AP (i.e. 19 wt %) successful building of a master curve is achieved, giving evidence for a simple motional behaviour over the whole temperature range corresponding to the β transition.

These conclusions are fully corroborated by the E_a and ΔS_a values reported in Table 10. In particular, ΔS_a values as small as 30 J K^{-1} mol^{-1} or less are observed for the high-temperature part and the maximum of the β tran-

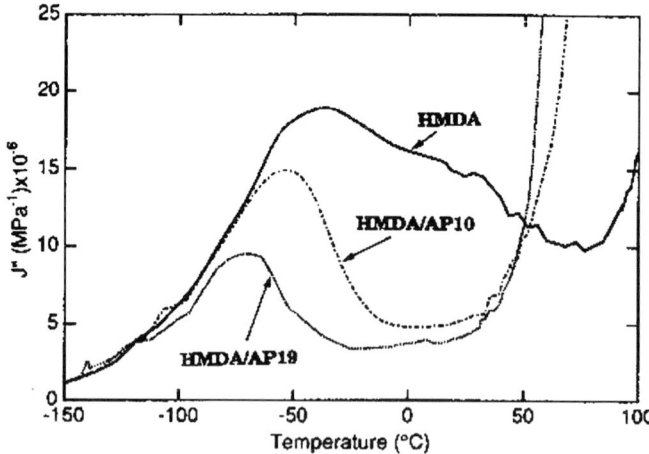

Fig. 99 Temperature dependence of the loss compliance, J'', at 1 Hz for the HMDA, HMDA/AP10 and HMDA/AP19 systems (from [70])

sition peak of the HMDA/AP19 system, confirming the isolated character of the motions, even in this temperature range.

7.2.1.2
Quasi-linear DGEBA/HA95/AP and Loosely Crosslinked DGEBA/HA60/AP Systems

As mentioned, the introduction of monoamine in the network induces a decrease of the crosslink density, associated with the presence of pending groups.

In the case of the HA95 quasi-linear network, J'' data plotted in Fig. 100a show that the shape of the J'' peak is nearly unchanged by the presence of the antiplasticiser and its amplitude is only slightly reduced.

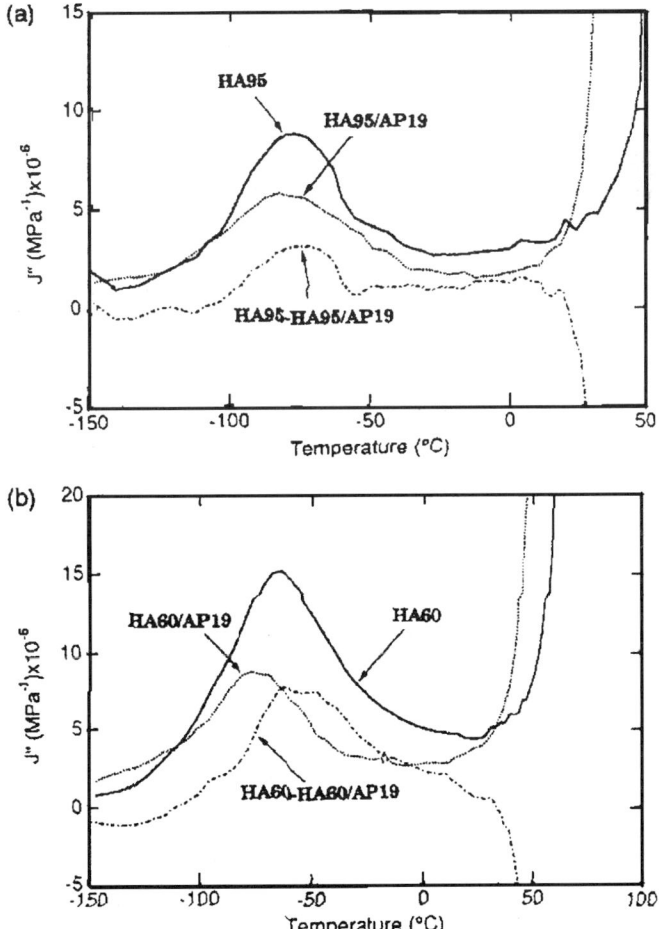

Fig. 100 Temperature dependence of J'' for **a** HA95 and HA95/AP19 systems and **b** HA60 and HA60/AP19 systems (from [70])

Furthermore, the already low ΔS_a values obtained for the pure system are slightly decreased. Such a behaviour is in agreement with assignment of the β transition in HA95 to isolated motions that are not affected by antiplasticisation.

In the HA60/AP19 system (Fig. 100b), the effect of antiplasticiser is intermediate between the behaviours observed in the densely crosslinked network and quasi-linear system, because some residual cooperative motions still occur in the pure network and can be hindered by the presence of the antiplasticiser.

7.2.1.3
Quasi-linear DGEBA/DMHMDA95/AP and Loosely Crosslinked DGEBA/DMHMDA60/AP Systems

Investigations on the pure networks (Sect. 7.1.1.3) led to the conclusion that the introduction of a mesh extender allows intramolecular cooperativity to develop. Consequently, it is interesting to determine whether the antiplasticiser prevents these intramolecular cooperative motions from occurring.

The J'' data of DMHMDA95 and DMHMDA95/AP19 systems are plotted in Fig. 101a. The β peak is asymmetrically reduced by the antiplasticiser, the high-temperature side being more affected. However, the effect is much less than for the HMDA network (Fig. 99). Furthermore, a master curve cannot be constructed in the high-temperature range with the data at various frequencies, similarly to that for pure DMHMDA95 systems. Finally, the ΔS_a values (Table 10) are unchanged by the addition of antiplasticiser. All these results clearly reveal the low efficiency of the EPPHAA antiplasticiser in this quasi-linear epoxy network.

As regards the DMHMDA60/AP69 system, in the J'' plot, shown in Fig. 101b, the influence of the antiplasticiser is seen in the high-temperature range. However, its role cannot be precisely described since both inter- and intramolecular cooperativities are likely to occur in the pure homologue.

7.2.2
^{13}C NMR Measurements in the DGEBA/HMDA/AP19 Network

In the case of antiplasticised networks, because of similarities in their chemical structure, in the solid-state ^{13}C NMR spectrum the lines arising from the $CHOH - CH_2 - O$ sequence of the polymer matrix and antiplasticiser overlap. It is the same for the lines corresponding to the protonated aromatic carbons.

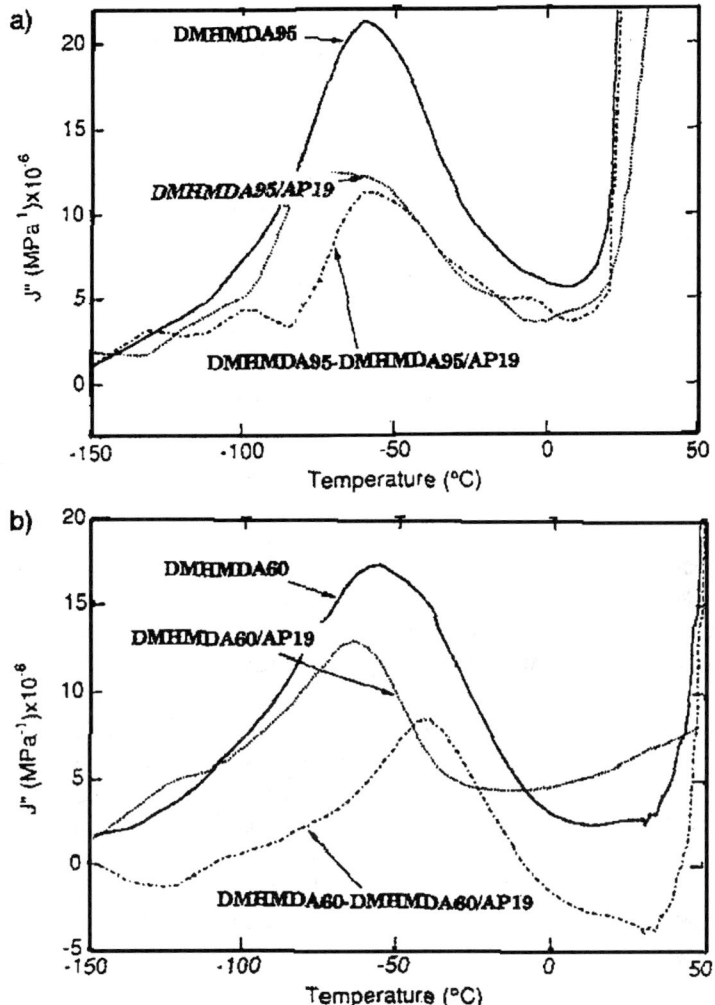

Fig. 101 Temperature dependence of J'' for **a** DMHMDA95 and DMHMDA95/AP19 and **b** DMHMDA60 and DMHMDA60/AP19 (from [70])

7.2.2.1
Local Motions of Aliphatic Units

As for the pure epoxy networks, the interesting measurements deal with $t_{1/2}$ and $T_{1\rho}$ (^{13}C).

The temperature dependence of the ratio $(t_{1/2})_0/(t_{1/2})$, where $(t_{1/2})_0$ refers to the room temperature value, measured for the CHOH $-$ CH$_2$ $-$ O and CH$_2$ $-$ N carbons in HMDA, HMDA/AP10 and HMDA/AP19 systems, is displayed in Fig. 102a and b, respectively. Over the temperature range investigated, the $(t_{1/2})_0/(t_{1/2})$ ratio of the $CH_2 - N$ units (Fig. 102b) is significantly different in the three systems. In contrast with results observed for the pure resin, in the HMDA/AP19 systems this ratio does not depend on temperature below the glass-transition temperature. It indicates the absence of motions with a frequency higher than 105 Hz at temperatures up to 90 °C. In the HMDA/AP10 network, the temperature dependence is very weak. Therefore,

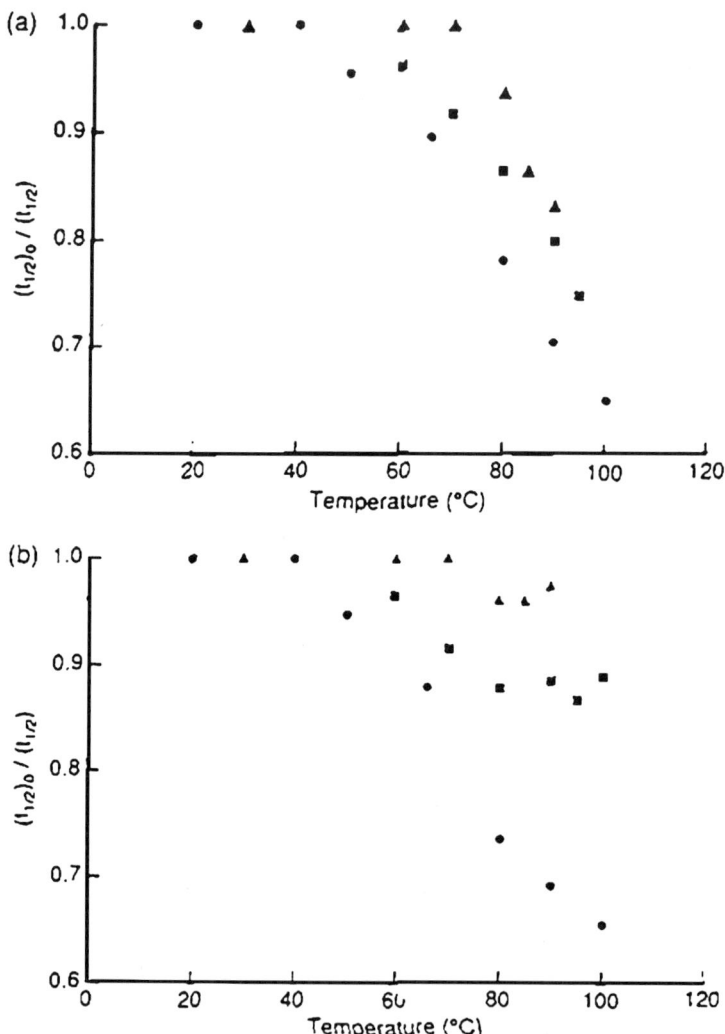

Fig. 102 Temperature dependence of the $(t_{1/2})_0/t_{1/2}$ ratio in HMDA (●), HMDA/AP10 (■) and HMDA/AP19 (▲) networks: **a** $CHOH - CH_2 - O$ carbons and **b** $CH_2 - N$ carbons (from [70])

the motions of the CH_2 group next to the crosslinks are strongly restricted in the HMDA/AP10 network and totally hindered in the HMDA/AP19 one.

As regards the HPE sequence (Fig. 102a), the difference in the behaviour in pure and antiplasticised networks is less pronounced, but an increase in the temperature at which the onset of mobility occurs is observed, in particular with 19 wt % of antiplasticiser. It means that the presence of antiplasticiser induces some slowing down of the HPE motions.

All these $t_{1/2}$ results are confirmed by the temperature dependence of $T_{1\rho}$ (^{13}C), leading to the same conclusion.

It is worth noting that in pure epoxy networks, the two groups, HPE and $CH_2 - N$, have the same mobility and behave similarly as a function of temperature.

7.2.2.2
Local Motions of Aromatic Units

Results of $(t_{1/2})_0/(t_{1/2})$ determinations as a function of temperature for the protonated aromatic carbons are shown in Fig. 103. The higher content of antiplasticiser (19 wt %) is required to see the occurrence of the onset of mobility of the aromatic carbons at a higher temperature than in pure matrix. Such a behaviour is quite similar to that observed for HPE units, in agreement with the conclusion, reached in pure epoxy networks, of a likely correlation between the motions of the aliphatic units and the ring flips.

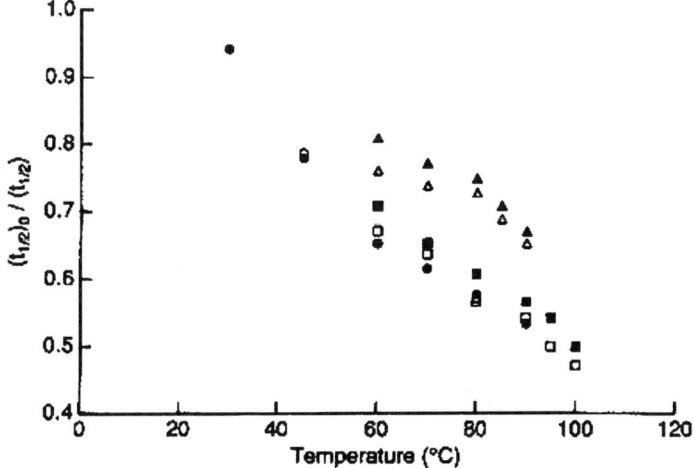

Fig. 103 Temperature dependence of the $(t_{1/2})_0/t_{1/2}$ ratio measured for the two protonated aromatic carbons of the HMDA ($\bullet$, $\bigcirc$), HMDA/AP10 ($\blacksquare$, $\square$) and HMDA/AP19 ($\blacktriangle$, $\triangle$) networks (from [70])

7.2.2.3
Local Dynamics of Antiplasticiser in the Network

In spite of the overlapping of the HPE lines of the polymer matrix and antiplasticiser, if the two different HPE sequences were to undergo different
dynamics, changing the cross-polarisation time would make it possible to
observe separately the lines corresponding to each type. For HMDA/AP19
networks at 80 °C, no line specific to the EPPHAA molecule can be distinguished, indicating that, within the sensitivity of the experiment, the HPE
sequences of the antiplasticiser and the matrix share similar local dynamics. Identical conclusions were derived from the behaviour of the protonated
aromatic carbons.

7.2.3
NMR Location of the Antiplasticiser in the Epoxy Network

^{2}H, ^{15}N and ^{13}C rotational-echo double-resonance (REDOR) NMR, applied to specifically labelled epoxy networks and the antiplasticiser moiety,
can provide information on the location of the antiplasticiser inside the
epoxy network. Such an investigation has been performed in the case of
HMDA/AP19 and HA60/AP19 networks [71] using [^{15}N$_2$] HMDA, [^{15}N] HA,
[^{2}H labelled on the bis-phenol residue] DGEBA and [^{13}C-carboxyl] EPPHAA.
The main conclusions are:

1. Homogeneous mixing and local ordering of epoxy, amine and antiplasticiser on a 10 Å scale.
2. The antiplasticiser has no preference in locating near free or crosslinked
   amino-nitrogen sites.
3. The distance between the ^{13}C-carboxyl of EPPHAA and the ^{15}N is $4.9 \pm$
   0.5 Å.
4. The distance between the ^{13}C-carboxyl of EPPHAA and the isopropylene
   carbon directly bonded to the two C(^{2}H)$_3$ groups of DGEBA is 6.7 ± 1 Å.
5. The intra-unit distance from an amine nitrogen to an isopropylidene quaternary carbon is approximately 10 Å. Thus, the antiplasticiser carbonyl
   carbon is about midway between one of the two amine nitrogens of the
   epoxy repeat unit and the quaternary carbon of the isopropylidene moiety; 4.7 Å to the former and 6.7 Å to the latter.
6. With just three distances, the antiplasticiser can be located approximately
   by triangulation relative to the nearest-neighbour network chain, but it is
   impossible to infer details of the orientation of the antiplasticiser within
   the network. However, the increase in density, observed for antiplasticised epoxy relative to pure epoxy, suggests, by similarity with the case of
   bisphenol A polycarbonate, that HPEs tend to align with one another in
   one direction and isopropylidene moieties in another.

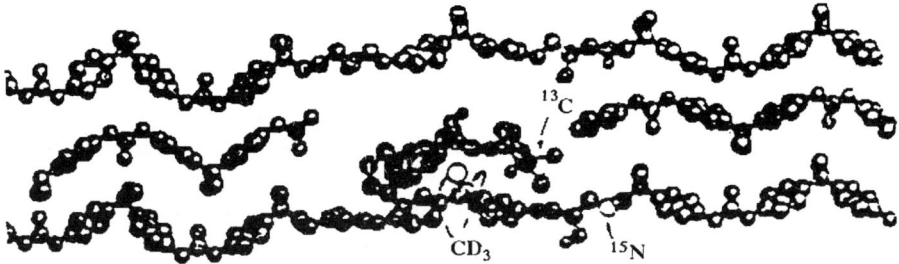

Fig. 104 Space-filling rendition of an epoxy-resin-like structure (*grey*), without crosslinks, showing the placement of the antiplasticiser (*black*) within the structure. Only the two chains directly above and below the antiplasticiser are shown in the figure. The labelled atoms are indicated by *open circles*. The antiplasticiser ^{13}C-carboxyl carbon is 4.9 Å from an amine ^{15}N (with no preference for free or crosslink sites) and 6.7 Å from a quaternary carbon of the isopropylidene moiety directly bonded to two C^2H_3 groups. The intramolecular distance from the quaternary carbon to the amine ^{15}N is approximately 10 Å (from [71])

In order to visualise the antiplasticiser placement within the epoxy network, but not its motion, molecular modelling was performed without including crosslink points. The result, presented in Fig. 104, illustrates one plausible (but not definitive) structure; in Fig. 104 only the two chains directly above and below the antiplasticiser are shown.

Regardless of the details of orientation, the positioning of the antiplasticiser shown in Fig. 104 suggests similar roles for both antiplasticiser and monoamine curing agent. In each case, antiplasticiser or monoamine interrupts the local packing of a fully crosslinked rigid lattice by insertion of a spacer. The creation of such defect sites in the epoxy network breaks down the more global interchain cooperativity involved in the high-temperature part of the β transition.

7.2.4
Conclusion on Motions in Antiplasticised Epoxy Networks

The dynamic mechanical analysis unambiguously shows that the antiplasticiser acts mainly on the cooperative motions involved in the high-temperature side of the β transition peak, implying several units. No effect is detected on the low-temperature side of the β peak.

The NMR experiments clearly show a different effect of the antiplasticiser on the mobility of either the crosslink points ($CH_2 - N$) or the hydroxypropyl ether sequence. Indeed, whereas these two groups have similar mobility in pure epoxy networks, the mobility of the crosslink points is hindered by the antiplasticiser, whereas only a slight slowing down occurs for the HPE units. Furthermore, there is no difference in mobility between the HPE sequence in the epoxy network and the one in the antiplasticiser molecule.

On the other hand, the investigations performed on pure epoxy networks unambiguously assign the β transition to motions of the HPE sequence.

These observations suggest the following scheme for the nature of the cooperativity and the effect of antiplasticiser. In the high-temperature part of the β transition, the HPE unit motions have a larger amplitude due to volume dilation and higher available thermal energy. In order to occur, correlated motions of the crosslink points are involved, either nearby along the chain sequence or spatially neighbouring (later considerations support the latter assumption). Such a cooperativity would be similar to that encountered in the low-temperature transition of bisphenol A polycarbonate, analysed in (Sect. 5).

In antiplasticised epoxy networks, the antiplasticiser interrupts the local interchain packing of a fully crosslinked rigid lattice by insertion of a spacer. Consequently, it prevents the spatial coupling with the surrounding chains and, in particular, with the spatially neighbouring crosslinks that are no longer forced to move consecutively to the motions of the chain HPE units. According to this description, a crude estimate of the number of crosslink points involved in the cooperativity that is lost by introduction of an antiplasticiser can be evaluated. Assuming a random distribution of the antiplasticisers, there is one antiplasticiser molecule per three crosslinks in the HMDA/AP19 network. Thus, the cooperativity remaining in the J'' peak of the HMDA/AP19 should involve no more than three crosslink points. Similarly, in the HMDA/AP10 system, there is one antiplasticiser molecule per six crosslinks and the cooperativity associated with the processes responsible for the J'' peak of this system cannot involve more than six crosslinks. As a consequence, the cooperativity that disappears when increasing the antiplasticiser content from 10 to 19 wt % concerns three to six units. Finally, the cooperativity suppressed by the introduction of 10 wt % of antiplasticiser deals with a number of crosslinks larger than six. From comparison of the compliances of the densely crosslinked HMDA network and of its loosely crosslinked homologue with pending monoamine residues (Sect. 7.1.1.2), the extent of cooperativity is estimated to be about six crosslink points.

Such a scheme accounts quite well for the lower cooperativity observed with the HMDA/HA networks, as well as for the lower efficiency of the antiplasticiser obtained with these systems. Indeed, the pending groups existing in such network architectures already interrupt the local interchain packing and partly avoid the propagation to the spatially neighboured crosslink points (along this line, it is interesting to note that in these systems the mobility of the crosslinks is shifted to a higher temperature than that of the HPE units). Most of the decoupling already exists in the pure epoxy systems, so the antiplasticiser brings a less significant contribution. It is worth noticing that the flexible nature of the aliphatic mono- and diamines does not play any role in the partial breakdown of the cooperativity. Indeed, quite similar behaviours are observed in networks with aromatic mono- and diamines [68].

The important feature is the occurrence of a partial breakdown of interchain cooperativity.

In the case of epoxy networks with a secondary diamine, like DMHMDA, the network architecture is such that flexible aliphatic sequences are present as chain extenders between the crosslink points. In such architectures, the motions of the HPE units can develop towards other HPE sequences (either along the chain or spatially neighbouring) without involving the crosslink points in their cooperativity. Thus, with these systems a different nature of cooperativity exists compared to the other network architectures. The introduction of an antiplasticiser in such a local packing does not affect the cooperativity as much as with the densely crosslinked architecture, for the crosslinks are not so much involved. Once more, it is important to point out that the flexible nature of the aliphatic amines does not matter since the same behaviours are observed for fully aromatic systems with identical architecture [68].

7.3
Conclusion on the Motions Involved
in the Glassy State Transitions in Aryl-aliphatic Epoxy Resins

Dynamic mechanical analysis and the ^{13}C NMR experiments performed on pure aryl-aliphatic epoxy networks, as well as on antiplasticised systems, have led to a deeper insight into the molecular motions involved in the glassy state of these epoxy resins.

The γ transition occurring around – 150 °C at 1 Hz originates from motions of aliphatic sequences of at least four methylene units within the amine moiety.

The β transition starting at 1 Hz around – 120 °C and extending to higher and higher temperatures when increasing the crosslink density, comes from motions of the hydroxypropyl ether sequence. Isolated and insensitive to crosslink density or amount of antiplasticiser in the low-temperature part of the β transition, these motions become more and more cooperative as temperature increases.

Changing the network architecture in a controlled way by inserting either monoamines (leading to aliphatic pending groups) or secondary diamines (inserted as chain extenders), using an antiplasticiser small molecule within these various networks, and looking at the consequences of these changes by dynamic mechanical and ^{13}C NMR experiments, have provided a fruitful approach for understanding the nature of the cooperativity that develops in the high-temperature part of the β transition and for accounting for the effect of the antiplasticiser.

It appears that the nature of the cooperativity that develops in aryl-aliphatic epoxy resin is similar to that encountered in bisphenol A polycarbonate: a cooperativity directly involving the local interchain packing, which can be interrupted by inserting moieties either as pending groups or as antiplasticiser small molecules.

In the absence of pending groups or antiplasticisers, the motions of the hydroxypropyl ether units in the high-temperature part of the β transition force motions of the crosslink points that are spatial neighbours of the moving HPE sequence. A crude estimate leads to an extent of cooperativity at high temperatures reaching more than six crosslink points in densely crosslinked networks.

When pending groups or antiplasticisers are present in the systems, they breakdown the local interchain packing and motions of the HPE units can occur without implying such a large number of crosslink points.

In the low-temperature part of the β transition, the motions of the HPE sequences are isolated and they remain unaffected by either changing the network architecture or adding an antiplasticiser.

Finally, it is interesting to point out that, whereas for poly(ethylene *tere*-phthalate), bisphenol A polycarbonate and the aryl-aliphatic copolyamides (Sects. 4 to 6, respectively), the phenyl ring flips plays an important role in the β transition processes. In the case of the aryl-aliphatic epoxy networks such π-flips exist but they are only indirectly involved in the β transition.

8
Poly(methyl methacrylate) and its Maleimide and Glutarimide Copolymers

This last section deals with linear polymers with side chains (as does Sect. 3). However, whereas in Sect. 3 the emphasis was on the ring motions involved in very low temperature transitions, the studies reported in this section exclusively concern the β transition and, sometimes, its coupling with the α transition. In addition to assignment of the β transition, the aim is a precise description of the involved motions as well as an analysis of the cooperativity that develops with increasing temperature.

It is worth noticing that a molecular modelling approach is used to complement the experimental techniques of dynamic mechanical analysis, dielectric relaxation, solid-state ^{13}C and ^{2}H NMR.

8.1
Poly(methyl methacrylate)

Due to the relatively simple chemical structure of poly(methyl methacrylate) (PMMA):
and the amorphous character of the material below the glass transition temperature, it was one of the earliest investigated polymers [8].

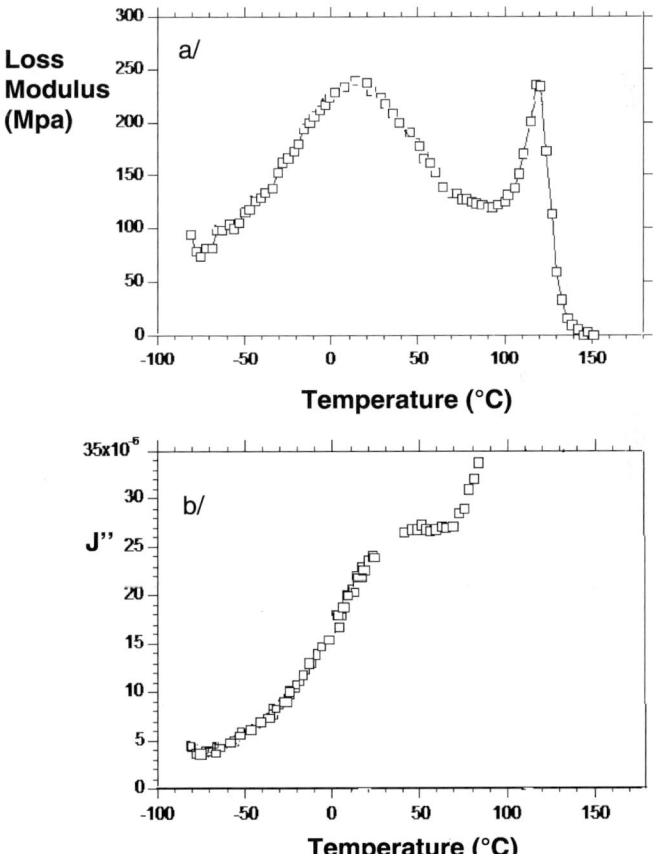

8.1.1
Dynamic Mechanical Analysis

Figure 105 shows the temperature dependence at 1 Hz of the loss modulus, E'', (Fig. 105a) and the loss compliance, J'', (Fig. 105b).

As regards the loss modulus, E'', two transitions appear:

Fig. 105 Temperature dependence for PMMA at 1 Hz of: **a** the loss modulus, E'' and **b** the loss compliance, J'' (from [75])

1. The α transition with its maximum at 116 °C, corresponding to the glass–rubber transition processes.
2. The β transition, centred around 10 °C, and corresponding to quite a broad (slightly asymmetric) peak (– 70° to 80 °C). The intensity of the peak is as large as that of the α transition.

The activation energy and entropy associated with the maximum of the β transition peak are 83 ± 5 kJ mol^{-1} and 51 ± 10 J K^{-1} mol^{-1}, respectively. This value of the activation entropy, determined according to the Stark-weather procedure (Sect. 2.3), corresponds to slightly cooperative motions at the maximum of the transition.

A detailed analysis of the temperature and frequency dependencies of the shear moduli (G' and G'') of PMMA has been reported [72]. As an example, Fig. 106 shows the frequency dependence of G'' at several temperatures.

In addition to the expected shift of the maximum of the G'' peak towards higher frequencies as temperature increases, an increase of the peak height is also observed. Furthermore, the shape of the G'' peak is asymmetric with greater broadness on the high-frequency side than on the low-frequency side (this is clearly evidenced from the dashed line drawn for an assumed symmetric shape of the spectra measured at – 34 °C). Due to these two features,

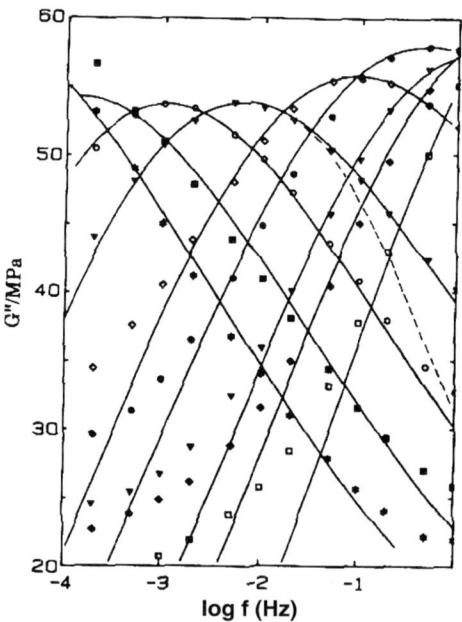

Fig. 106 Frequency dependence of the loss shear modulus, G'', for PMMA at several temperatures: ✩ – 66 °C, ■ – 55 °C, ○ – 44 °C, ▼ – 34 °C, ◇ – 17 °C, ● – 4 °C, ▽ 3 °C, ◆ + 13 °C and □ + 27 °C (from [72])

the time–temperature superposition does not properly apply and a master curve cannot be obtained. All these results indicate that the β transition of PMMA does not involve a single motional process. In order to account for the shape of the G'' peak and its evolution as a function of temperature, a distribution of correlation times, associated with a distribution of activation energies and entropies, has to be considered.

As regards the motion involved in the β transition, the dynamic mechanical response performed on a single polymer cannot lead to any assignment; the investigation of a series of polymers with a gradual change of their chemical structure is required. By performing such a systematic study, Heijboer [73, 74] attributed the β relaxation of methacrylate polymers to the motion of the COOR ester side group along their bond to the main chain.

A deeper insight can be achieved by using dielectric relaxation and solid-state NMR.

8.1.2
Dielectric Relaxation

The temperature dependence of the dielectric loss, ε'', obtained at 1 Hz for PMMA is shown in Fig. 107.

Two maxima are clearly seen:

1. The high-temperature peak, centred at 116 °C and corresponding to the glass transition
2. A peak centred around 10 °C and quite broad (– 70 to 80 °C), associated with the β transition. It is worth noticing that the intensity of the β peak is much larger than that of the α peak.

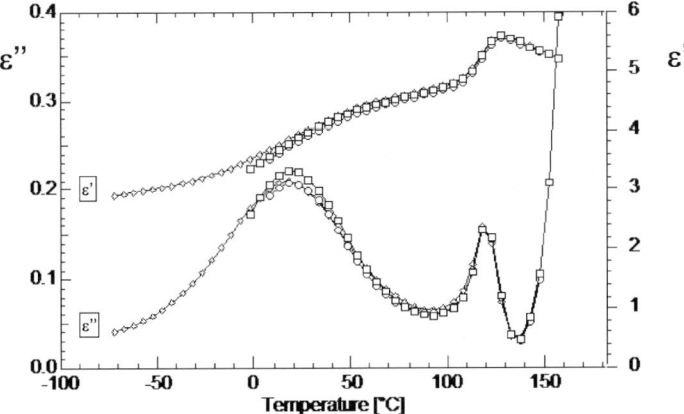

Fig. 107 Temperature dependence of the dielectric loss, ε'', for two different samples of PMMA at 1 Hz (from [75])

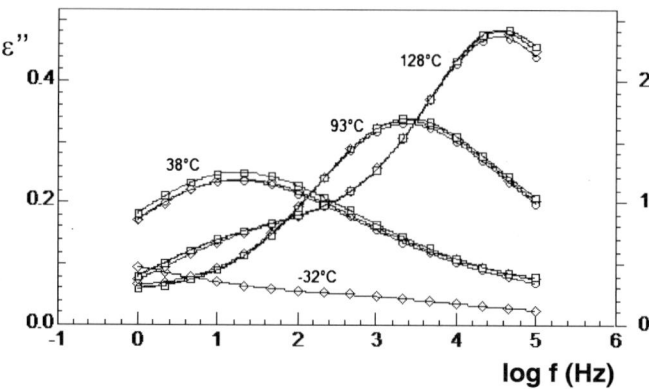

Fig. 108 Frequency dependence of the dielectric loss, ε'', for two different samples of PMMA at several temperatures (from [75])

A detailed analysis of the dielectric response of PMMA has been performed [75]. The frequency dependence of the dielectric loss, ε'', at a few temperatures is shown in Fig. 108. An increase in temperature is associated with a shift towards higher frequency, as expected, but also with quite a significant increase in the height of the peak maximum.

The activation energy, E_a, and entropy, ΔS_a, corresponding to the maximum of the β transition are $82 \pm 2 \, \mathrm{kJ \, mol^{-1}}$ and $58 \pm 3 \, \mathrm{J \, K^{-1} \, mol^{-1}}$, respectively. The activation entropy value means that the motion is slightly cooperative. In order to check the change of cooperativity with increasing temperature, activation entropies have been determined all along the β transition. A smooth increase from 50 to $60 \, \mathrm{J \, K^{-1} \, mol^{-1}}$ is observed, indicating a slight development of cooperativity with increasing temperature. It is worth noting that such a small change cannot be compared with the very large variations observed in the case of poly(ethylene *tere*-phthalate) or aryl-aliphatic epoxy resins.

The shape of the frequency dependence of ε'' has been compared in Fig. 109 in terms of reduced units $\varepsilon''/\varepsilon''_{max}$ and f/f_{max}, at various temperatures. The peak is asymmetric, being broader on the high-frequency side, especially at $10 \, ^\circ\mathrm{C}$. A gradual narrowing occurs on both the high- and low-frequency sides with increasing temperature. These results show that the motional processes involved in the dielectric β relaxation have a distribution of correlation times and that this distribution becomes narrower as temperature increases.

The chemical structure of PMMA makes assignment of the dielectric β transition easy. Indeed, the ester group $O - C = O$ is planar and its dipole moment is represented in Fig. 110.

Thus, the β transition can only originate from a rotating motion of the ester group about its bond to the chain backbone. It is interesting to point out

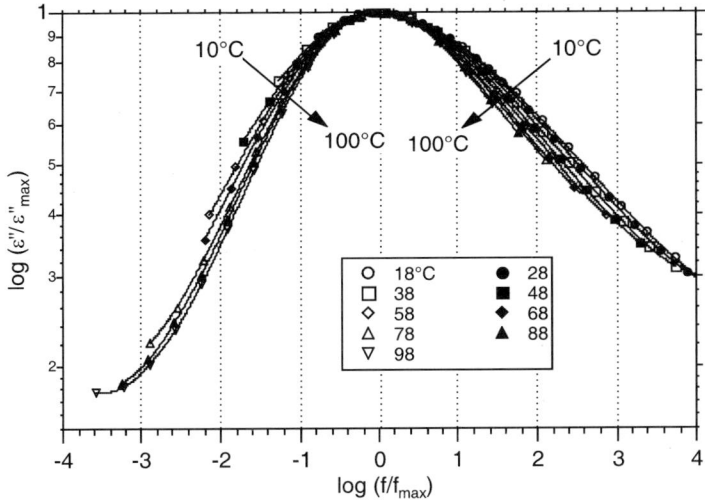

Fig. 109 Frequency dependence of ε'', in reduced units $\varepsilon''/\varepsilon''_{\max}$ and $f/f_{\max}$, at several temperatures (from [75])

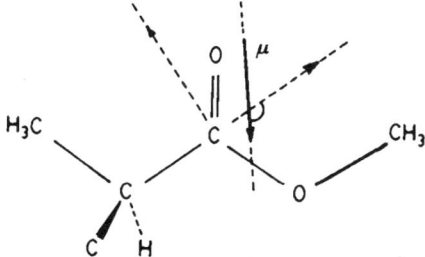

Fig. 110 Orientation of the dipole moment, μ, in a repeat unit of PMMA

that the dielectric study [76] of a series of poly(methacrylates) with various ester substituents (methyl, ethyl and butyl) shows that all these polymers have identical β transitions (the maxima occurring at the same frequency for the same temperature and the activation energies being equal). Thus, the length of the $-$R group, whenever this happens to be linear, does not increase the steric hindrance to the rotation of the ester side group.

8.1.3
Comparison of Dynamic Mechanical and Dielectric Results

The relaxation map of PMMA obtained from E'' and ε'' is shown in Fig. 111. It shows a good agreement between the two types of investigations, in agreement with the fact that the same motional groups are involved in the response of both techniques.

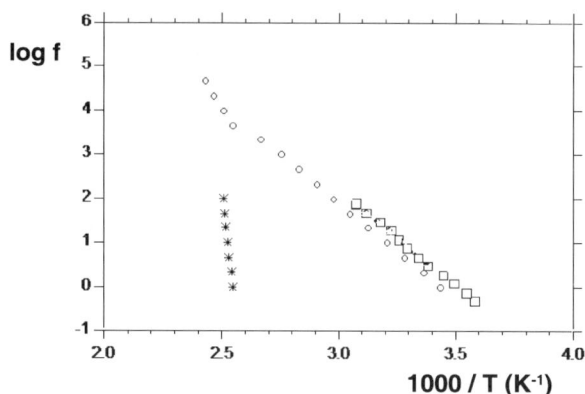

Fig. 111 Relaxation map of PMMA derived from dielectric ε'' (*, $\bigcirc$) and mechanical E'' ($\square$) data (from [75])

Furthermore, the activation energies $(83 \pm 5\,\text{kJ}\,\text{mol}^{-1}$ from mechanical analysis and $82 \pm 2\,\text{kJ}\,\text{mol}^{-1}$ from dielectric relaxation) as well as the activation entropies $(51 \pm 3\,\text{J}\,\text{K}^{-1}\,\text{mol}^{-1}$ from mechanical analysis and $58 \pm 33\,\text{J}\,\text{K}^{-1}\,\text{mol}^{-1}$ from dielectric relaxation) are identical, confirming that the processes are the same.

However, it is interesting to perform a more direct comparison of the experimental results to check whether some differences between the mechanical and dielectric behaviours could exist as a function of temperature. The appropriate quantity is E'' for the mechanics and, for the dielectric response, it is the dielectric loss modulus, m'' (defined as $\varepsilon''/(\varepsilon'^2 + \varepsilon''^2)$). Figure 112 shows the temperature dependence of E'' and m'' at 1 Hz, obtained by superposing the low-temperature part of the β transition.

The maxima occur at the same temperature, but on the high-temperature side, the mechanical loss, E'', is higher than the dielectric loss, m'', the difference increasing with temperature. Furthermore, the amplitude of the α peak observed by mechanics (Fig. 105a) is much larger than that from dielectric measurements (Fig. 107). It means that most of the dielectric relaxation, which deals specifically with the dipole moment of the repeat unit, is performed through the ester group motions involved in the β transition and only a small contribution remains to relax by the processes leading to the isotropic motions occurring at the glass–rubber transition. In contrast, the energy dissipation associated with the dynamic mechanical solicitation is more balanced between the β and α transitions. Indeed, the mechanical behaviour reflects the response of the whole material and, consequently, even if the motions of the ester groups occurring at the β transition remove local constraints, only the occurrence of the cooperative segmental motions characteristic of the α transition are able to lead to the whole set of degrees of freedom existing in the melt state above the glass transition temperature.

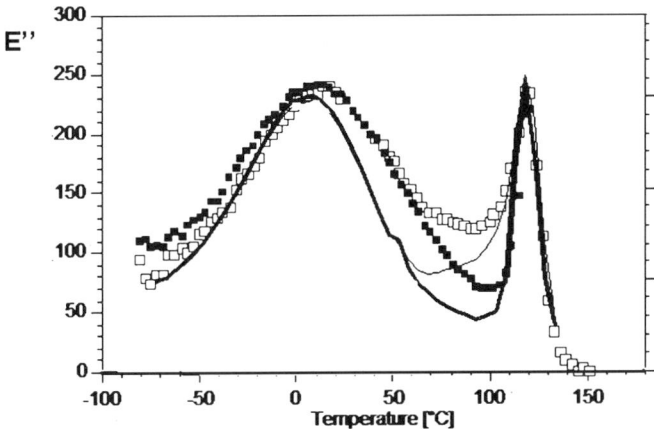

Fig. 112 Comparison of the temperature dependence at 1 Hz of mechanical E'' ($\square$, $\blacksquare$) and dielectric m'' (*thin* and *thick continuous lines*) modulus of quenched and aged PMMA, respectively, after superposition of the low-temperature part of the β transition (from [75])

Thus, the mechanical loss generated at the α transition is much larger than the dielectric one.

Another interesting feature deals with the crossover region between the β and α transitions. As shown in Fig. 112, the mechanical loss, E'', is higher than the dielectric loss, m'', in this region. Furthermore, it appears that the crossover region is sensitive to the thermal history of the PMMA sample [75], as shown in Fig. 112. Thus, the E'' and m'' curves are higher for quenched material than for a physically aged sample (sample cooled down at $5\,°\mathrm{C}\,h^{-1}$ to $87\,°C$, then maintained at this temperature for 80 h). This behaviour could be attributed to the gradual development, with increasing temperature, of a cooperativity involving the main chain more and more; cooperativity which is decreased by the higher packing developed by physical ageing. Unfortunately, the limited data available from dynamic mechanical measurements do not allow the determination of the activation entropy all along the β transition, as has been done with dielectric relaxation. However, the statement dealing with main-chain cooperativity is supported by the investigations of the maleimide and glutarimide copolymers presented in Sects. 7.2 and 7.3, respectively.

8.1.4
Solid-State NMR Investigation

The molecular motions underlying the dynamic mechanical and dielectric β transition in PMMA have been studied in detail [77] by using the 2D exchange NMR experiment. This detects slow reorientations that occur during a mixing time, t_m, by measuring the angular-dependent NMR frequencies (expressed in ppm) before and after t_m. The 2D frequency spectrum $S(\omega_1,$

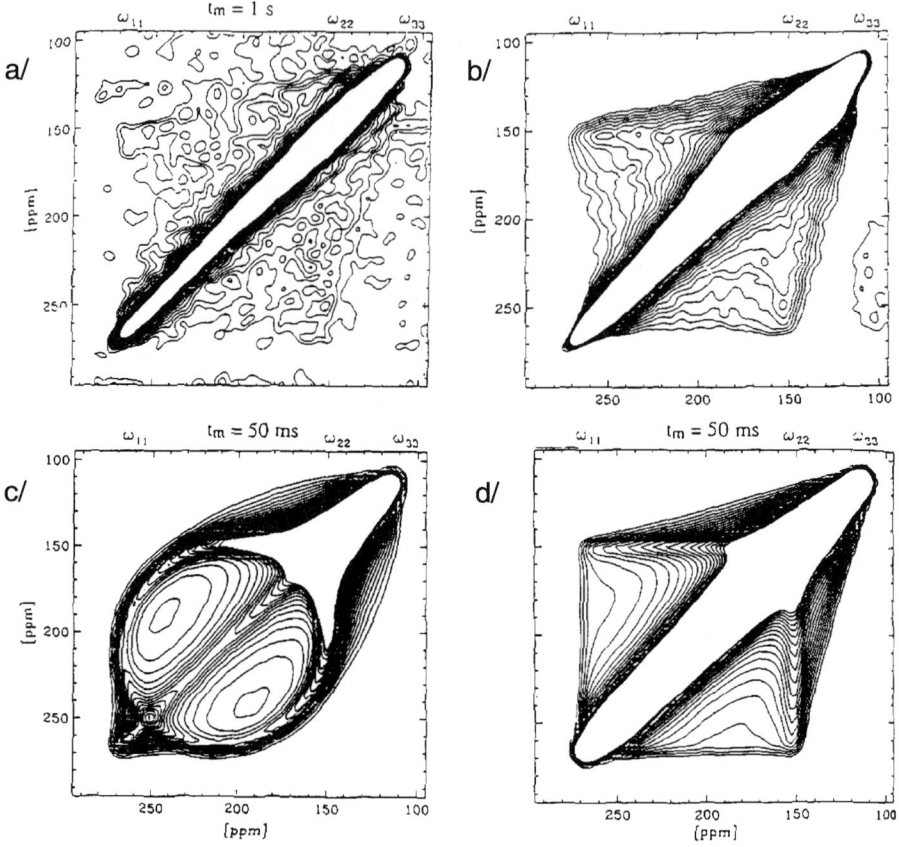

Fig. 113 2D exchange ^{13}C NMR spectrum of the carboxyl of PMMA. **a** PMMA with 20% of ^{13}COO-labelled side groups at – 40 °C and $t_m = 1$ s, the contour lines are linear between 0.5 and 5% of the maximum intensity. **b** Spectrum at 60 °C and $t_m = 50$ ms. **c** Simulation of the spectrum in **b** assuming a Gaussian distribution of side-group flip angles of $\pm 25°$ root-mean-square amplitude centred on 180°. **d** Best simulation of the spectrum in **b**, with a $180° \pm 10°$ flip angle and a concomitant rotation around the local chain axis with a root-mean-square amplitude of $\pm 20°$ (from [77])

ω_2, t_m), shown Fig. 113, represents the probability of finding a nucleus with a resonance frequency ω_1 before t_m and with a frequency ω_2 afterwards.

If no reorientation takes place during t_m, $\omega_1 = \omega_2$ and the spectrum is confined to the diagonal of the ω_1, ω_2 frequency plane. Large angle reorientations give rise to intensity in large parts of the frequency plane.

This NMR investigation has been applied to:

– ^{13}C carboxyl carbons of PMMA. In order to get stronger signals, PMMA with 20% of ^{13}C-labelled carboxyl carbons has been used.
– ^{2}H-labelled methoxy groups of PMMA.

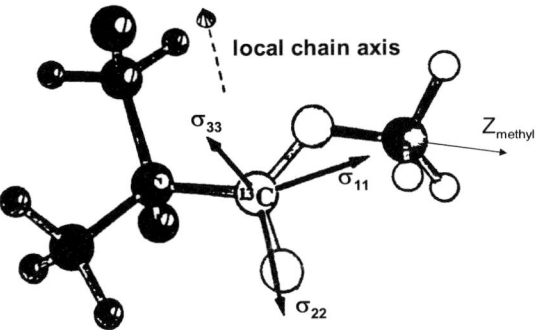

Fig. 114 Geometry of the PMMA repeat unit and the directions of the principal components of the COO chemical shift tensor (from [77])

The directions of the principal components of the COO chemical shift tensor are shown in Fig. 114. The σ_{22} axis makes an angle of 6° with the direction of the C = O bond and the σ_{33} axis is perpendicular to the COO plane. The local chain axis is approximately perpendicular to the COO plane and nearly parallel to the σ_{33} axis within about 20°. Finally, the C_3 axis of the rotating methyl group, designated as "z_{methyl}", is the unique axis of the quadrupolar interaction of the deuterons of the (O)CH$_3$ group.

It is worth noting that the OCH$_3$ groups project out of the "core" of the chain and are thus surrounded by atoms of other chains. Consequently, the environment of each side group is significantly asymmetric.

8.1.4.1
^{13}C Carboxyl Group

The ^{13}C powder NMR line shapes of the carboxyl group chemical shift tensor are shown Fig. 115 as a function of temperature. At 27 °C, a nearly regular powder spectrum is found, with $\sigma_{11} = 268$ ppm, $\sigma_{22} = 150$ ppm and $\sigma_{33} = 112$ ppm. As temperature rises, increasingly pronounced line-shape changes are observed, which are indicative of large motions with rates exceeding 10 kHz. The motional rates estimated at the various temperatures fit quite well on the relaxation map of PMMA obtained from mechanical and dielectric measurements (Fig. 111). Thus, the motions of the carboxyl groups observed by NMR are directly related to the β transition.

The 2D exchange NMR spectra obtained at – 40 °C and 60 °C are shown in Fig. 113. The purely diagonal spectrum recorded at the lower temperature of – 40 °C with $t_m = 1$ s (Fig. 113a) does not show any motion occurring in a time range of 1 s. In contrast, the spectrum obtained at 60 °C with $t_m = 50$ ms (Fig. 113b) clearly shows an exchange pattern, exhibiting intensity far off the diagonal. This indicates large frequency changes, which correspond to large reorientation angles. A large portion of the frequency plane exhibits intensity,

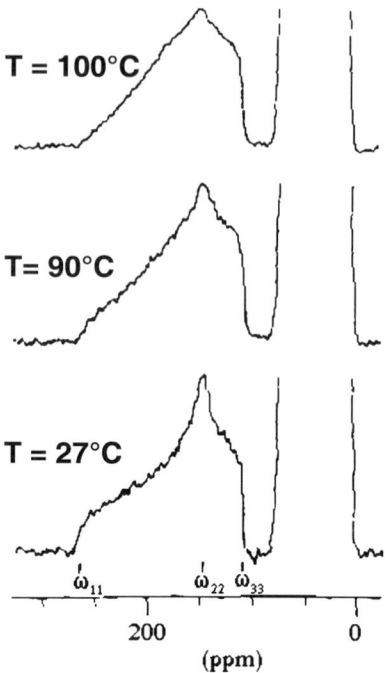

Fig. 115 Temperature dependence of carboxyl ^{13}C NMR powder spectra of PMMA (from [77])

but not all of it, which proves that the motion is not isotropic. The simplest picture of the ester group motion is a rotation of 180° around the C – C bond connecting the main chain and the side group, which corresponds to a two-site jump (π-flip). Complementary 3D exchange NMR experiments on the ^{13}C carboxyl performed at 60 °C prove that the reorientation of each moving carboxyl involves only two-sites, i.e. two relatively well-defined potential-energy minima. However, due to asymmetry of the ester side group, there is an energetic inequivalence of the two side-group orientations and, consequently, the potential-energy minima between which the ester group flip happens have unequal depths. The diffuse intensity distribution in the frequency plane arises from a broad distribution in the relative orientations of these pairs of potential-energy minima. This comes from the fact that the different side groups have different environments in the glassy state, requiring varying degrees of rearrangement of the main chain (as explained below) in the energy minimisation after the jump.

The simulated 2D exchange NMR spectrum resulting from side-group π-flips without main-chain motion, but with a Gaussian distribution of side-group flip angles of $\pm 25°$ root-mean-square amplitude, is plotted in Fig. 113c. It is clear that it cannot account for the experimental spectrum obtained at 60 °C (Fig. 113b).

Examination of this experimental spectrum reveals that the exchange maintains ω_{33} unchanged, but leads to a broadening in the region between ω_{11} and ω_{22}. Such a feature can be accounted for by considering a distribution of rotation angles around the ω_{33} axis, which is perpendicular to the OCO plane (Fig. 114) and is unchanged by a π-flip of the ester group around the C – OCO bond. Indeed, by considering that the side-group π-flips are accompanied by a restricted rotation of average amplitude $\pm 20° \pm 7°$ around the local chain axis (a rotation amplitude that can be directly derived from the 2D spectrum), quite a satisfactory simulated 2D spectrum at 60 °C is obtained, as shown in Fig. 113d.

Figure 116 shows the dynamics of the asymmetric side group in its correspondingly asymmetric environment.

This shows how a π-flip accompanied by a $\sim 20°$ rotation around the normal of the COO plane, which is approximately parallel to the local chain axis (Fig. 116c), can take place with only a limited change in the environment. A flip-back rotation (Fig. 116d) leads to a position close to its original orientation, but not exactly, due to the previous change in the environment in Fig. 116c. In contrast, a π-flip without main-chain motion would result, in general, to steric clashes of the OCH$_3$ group with surrounding units (Fig. 116b).

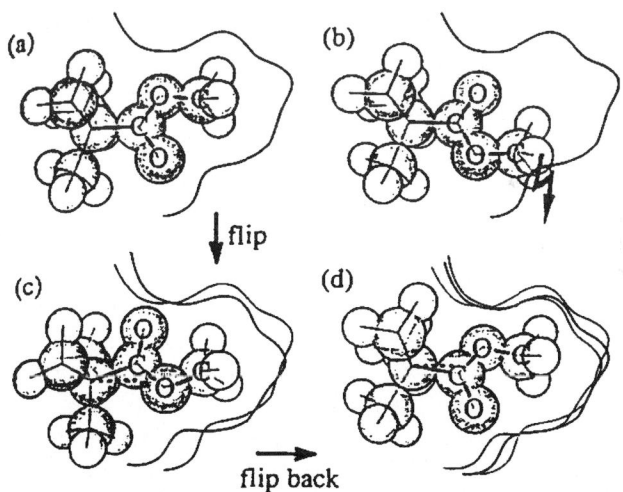

Fig. 116 Dynamics of the asymmetric side group in its asymmetric environment. **a** Initial side group orientation. **b** Steric clash with the environment if an exact 180° flip without main-chain motion is assumed. **c** To fit the asymmetric side group into the volume it had occupied before the flip, a twist around the local chain axis is required. This in turn slightly deforms the environment. **d** A second jump takes the group back close to its original orientation in **a**, but not exactly, due to the previous change in the environment in **c**, which is enhanced by rotation of other side groups that make up that environment (from [77])

Concerning the amount of ester groups undergoing a π-flip, 2D and 3D NMR at 60 °C lead to an estimate of $50 \pm 20\%$ groups participating in the exchange process on the time scale of the correlation time of the β process. The remaining $50 \pm 20\%$ ester groups, which are slow on that time scale, must be trapped in environments with higher activation barriers.

More information about the behaviour of the ester groups can be gained by looking at the line-shape changes on the diagonal in the ^{13}C NMR 2D spectra. Figure 117a,b compares the stacked plots of ^{13}C 2D spectra at – 40 °C and 60 °C. While a perfect diagonal powder pattern is observed at – 40 °C, indi-

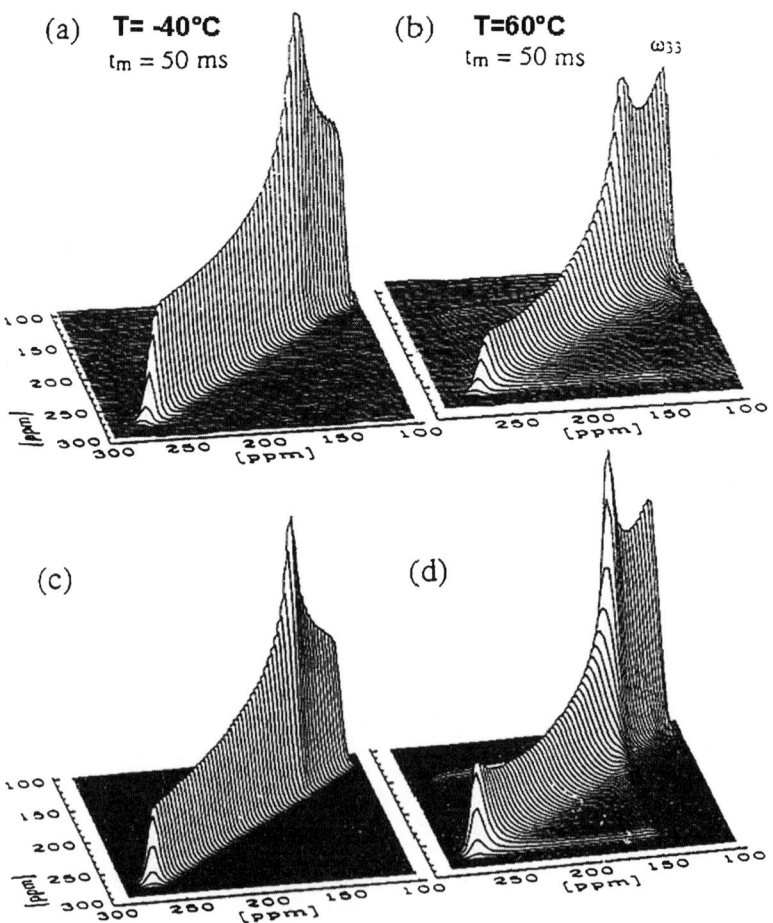

Fig. 117 Stacked plots of ^{13}C exchange NMR spectra taken with $t_m = 50$ ms. **a** $T = - 40$ °C. **b** $T = 60$ °C. **c** Simulated diagonal powder spectrum corresponding to **a**. **b** Simulated spectrum of **b** obtained by considering rotation around the σ_{33} direction with a root-mean-square amplitude of 12° (from [77])

cating the absence of carboxyl motions at a frequency higher than 10 kHz, at 60 °C the intensity of the ω_{33} end is enhanced relative to the other parts of the spectrum. Such a behaviour can be well accounted for by considering that 25% of the side groups undergo a small-angle reorientation around the local chain axis, with a mean-square-root amplitude of about 12° (compare the simulated spectrum shown in Fig. 117d with the experimental one in Fig. 117b). These small-angle reorientations result naturally when, after two (or an even number of) π-flips, a given side group does not return exactly to its original orientation due to a change in its environment (it could be called "static" reorientation), as shown in Fig. 116d. The 25% side groups concerned by these static reorientations represent half of the ester groups located in sites allowing a π-flip. It is worth noting that the static reorientations considered here have nothing to do with the 20° rotations around the local chain axis that always occur together with ester group π-flips (those could be called "dynamic").

8.1.4.2
^{2}H Methyl Group

First of all, it is worth pointing out that, due to the geometry of the PMMA repeat unit shown in Fig. 114, a rotation of the ester group around the C – OCO bond does not change the orientation of the z_{methyl} axis. As this axis is also the principal axis of the quadrupolar interaction of the rotating ^{2}H methyl group, the π-flip of the ester group does not affect the ^{2}H NMR spectra. The 2D ^{2}H NMR spectrum, obtained at 82 °C and $t_m = 500$ ms, shown in Fig. 118a, reveals that the exchange intensity is confined to the region around the diagonal.

As the 2D ^{2}H spectra are insensitive to the π-flip motion itself, the observed exchange intensity can only originate from the rotations around the local chain axis. Indeed, the 2D ^{13}C NMR spectrum of ^{13}C carboxyl (Sect. 8.1.4.1) has shown that π-flips of the ester groups are accompanied by rotations around the chain axis by about 20°.

2D ^{2}H NMR spectrum at 82 °C also provides further information on the behaviour of the 50% of ester groups which do not undergo π-flips at the involved frequency. Indeed, the absence of a spectral component (Pake powder pattern) that is very narrow perpendicular to the diagonal indicates that there are only a few, if any, completely rigid ester units. Thus, even the side groups that do not flip are undergoing restricted rotations around the local chain axis. In order to account for the experimental 2D ^{2}H NMR spectrum of Fig. 118a, an average amplitude of $\pm 7°$ has to be considered for the rotations (rocking) of the non-flipping ester groups, whereas, for flipping ester groups, the average amplitude of rotation around the local chain axis is $\pm 20°$. Under these conditions the simulated spectrum, shown in Fig. 118b, satisfactorily agrees with the experimental one.

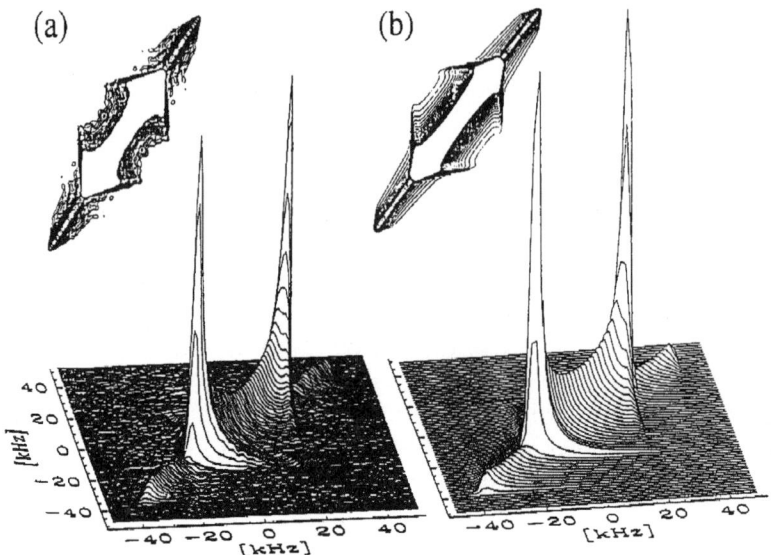

Fig. 118 2D ^{2}H NMR spectrum of PMMA deuterated in the OCH_3 group. **a** Experimental spectrum at 82° and $t_m = 500$ ms. **b** Corresponding simulation with exchange generated by motions around the local chain axis (from [77])

8.1.4.3
Summary of Ester Group Motions Derived from Solid-State NMR

The various solid-state NMR results (Sects. 8.1.4.1 and 8.1.4.2), lead to the following understanding of the motions performed by the ester groups and associated with the β transition observed in dynamic mechanical and dielectric investigations.

At 60 °C, 50% of the ester groups, trapped in constrained environments with high activation energy barriers, do not performed π-flips at a frequency higher than 10 kHz and their motions are limited to restricted rotations (rocking) around the local chain axis with an average amplitude of 7°.

At 60 °C, 25% of the ester groups undergo one (or an odd number of) π-flip at a frequency higher than 10 kHz, with no more than 25° deviation in the flip angle. The flips are accompanied by rotational readjustments with an amplitude of ca. $\pm 20°$ around the local chain axis. The flips occur between energetically inequivalent sites (thus they are active in dynamic mechanical and dielectric relaxation experiments).

At 60 °C, 25% of the ester groups which have undergone two (or an even number of) π-flips at a frequency higher than 10 kHz, return near to their original orientation with a precision of ca. 12°, corresponding to the average amplitude of the rotation around the local chain axis required by the changes in the environment of the initial position induced by the first π-flip (Fig. 116d).

8.1.5
Atomistic Modelling

The investigations of PMMA at two temperatures (– 40 and 60 °C) by multidimensional solid-state ^{13}C and ^{2}H NMR (Sect. 8.1.4) have led to quite a precise description of the ester group motions and the associated main-chain motions. However, it has not been possible to get information on the origin of the observed distribution of activation energies, nor on the extent of cooperativity along the main chain required by the π-flip of the asymmetric ester group.

The only approach presently available for getting this information deals with the atomistic modelling of the dynamics of ester groups in a glassy state cell of PMMA. Such a study has been recently performed [78].

8.1.5.1
Atomistic Modelling Method

A method for the detailed atomistic modelling of amorphous glassy polymers has been developed [50] and applied to atactic polypropylene. This method has been described in Sect. 5.4. The quasi-static modelling of chain dynamics [51] has been described in Sect. 5.4.2.

8.1.5.2
Modelling Characteristics for PMMA

In order to get more representative results, rather large cubic cells (43.9 Å edge) have been used. They are filled with 600 monomers representing either three chains of 200 monomers each or six chains of 100 monomers.

Two temperatures have been considered, – 40 and 50 °C, in order to investigate the behaviour at low temperatures (beginning of the β transition) and in the high-temperature part of the β transition where coupled motions between the ester group and the main chain have been observed by NMR (Sect. 8.1.4). The Van der Waals radius of each atom is reduced in order to reproduce the experimental densities, i.e. 1200 kg m^{-3} at – 40 °C and 1179 kg m^{-3} at 50 °C.

The tacticity of the PMMA chains was $rr = 60\%$, $rm = 35\%$, $mm = 5\%$, where r characterises the *racemic* diad and m the *meso* diad. The average structure factor, C_{400}, equal to 7.5, was in the range of the experimental values (7 to 9).

The π-flip of the ester group undergoing the quasi-static dynamics was performed by rotational steps of 2°. At each temperature 200 ester groups were been submitted to a π-flip.

8.1.5.3
Activation Energies

For each rotating ester group, the total energy of the cell is calculated. The change in energy as a function of the rotation angle is shown in Fig. 119 for three different ester groups within the cell at 50 °C.

In each case, the energy goes through a maximum whose value and rotation angle depend on the considered ester group. The activation energy, E_a, corresponds to the increase in energy associated with this maximum. As shown in Fig. 119, the activation energy strongly depends on the ester group and, consequently, over the 200 rotating ester groups there is an activation energy distribution that is almost unchanged with temperature (Fig. 120).

Values of the rotation angle corresponding to the energy maximum, $\Delta\phi_{max}$, are distributed in the same way. Figure 121 shows this distribution at – 40 and 50 °C. It is centred around 90° at the two temperatures and a slight narrowing occurs at 50 °C.

The advantage of atomistic modelling is to go deeper into the analysis of the change in energy induced by the ester rotation. In particular, it is possible to separate the inter- and intrachain changes. The various average energy changes obtained at – 40 and 50 °C are collected in Table 11. Independently of temperature, the interchain energy change is slightly negative. Furthermore, the energy change of the surroundings (defined as the interchain energy change plus the intrachain energy change associated with chains other than that of the rotating ester) is always negligible. It turns out that the activation energy originates from the change in intramolecular energy of the chain to which the rotating ester belongs. Finally, this intramolecular energy

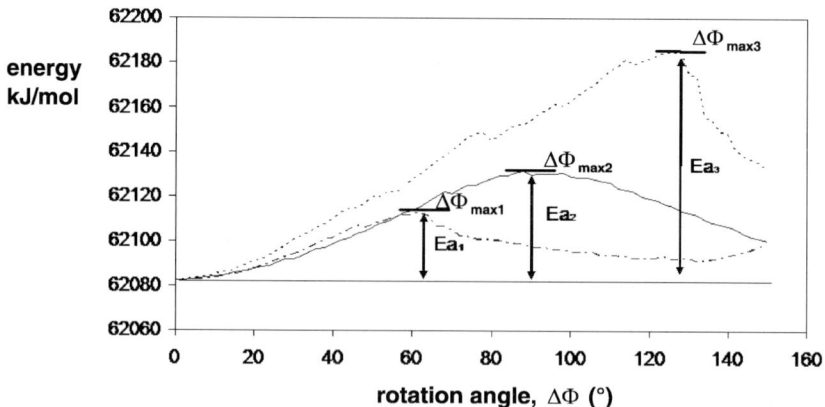

Fig. 119 Change in energy associated with the rotation angle for three different ester groups at 50 °C (from [78])

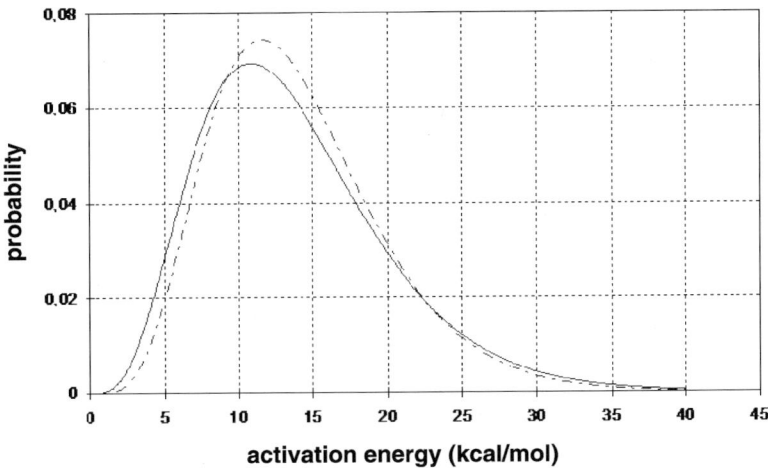

Fig. 120 Activation energy distributions obtained at − 40 °C (*broken line*) and 50 °C (*solid line*) (from [78])

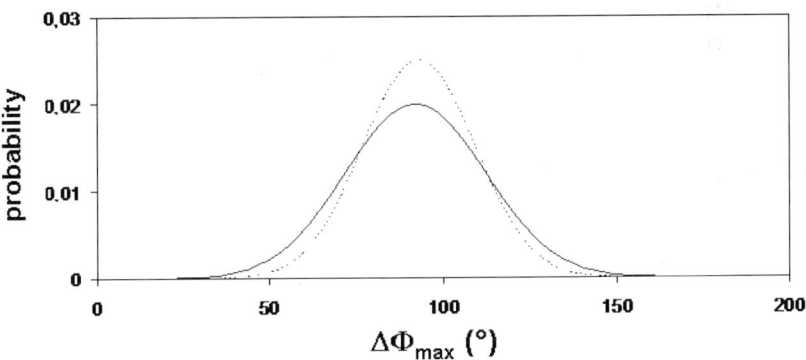

Fig. 121 Distribution of the rotation angles at the energy maximum, $\Delta\phi_{max}$, at − 40 °C (*solid line*) and 50 °C (*broken line*) (from [78])

change comes, with nearly equal contributions, from the change in energy between bonded atoms and non-bonded atoms.

It is interesting to analyse which structural features of the chain, in the vicinity of the rotating ester, are responsible for the change in intramolecular energy and, by the way, for the activation energy. The local tacticity appears not to have any effect, per se, on the values of the activation energy. The relevant parameters are related to the distances, shown in Fig. 122, between the rotating ester group and either the methyl groups or the ester groups of the nearest-neighbour monomers.

Indeed, if the activation energy distribution (Fig. 120) is split into three classes, roughly equally populated ($E_a < 42$ kJ mol^{-1}; 42 kJ mol$^{-1} < E_a < 71$ kJ mol^{-1}; and 71 kJ mol$^{-1} < E_a$), Fig. 123, corresponding to − 40 °C, shows

Table 11 Mean activation energy and average energy changes (in kJ mol^{-1}) of the various components for PMMA at $-40\,°C$ and $50\,°C$ (from [78])

Temperature [°C]	Mean activation energy	Intermolecular energy change of the chain with the rotating ester	Energy change of the surrounding
-40	56.9	53.3	3.6
50	57.6	55.9	1.7

	Intramolecular energy change of bonded atoms	Intramolecular energy change of non-bonded atoms	
-40	25.9	27.4	
50	31	24.9	

	Intermolecular energy change	Intramolecular energy change of other chains	
-40	-9.8	13.4	
50	-5.7	7.5	

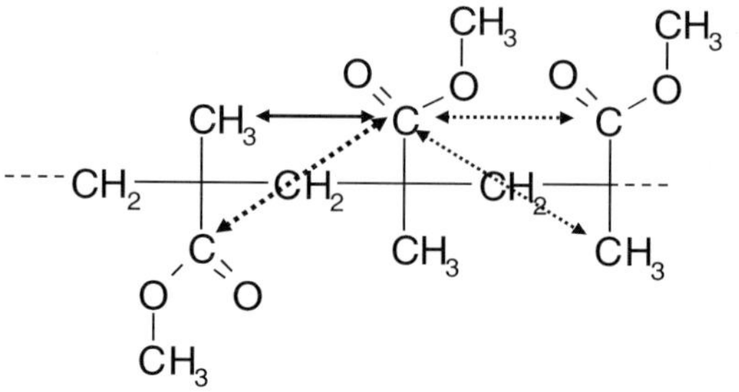

Fig. 122 Scheme of the intramolecular distances under consideration: $COOCH_3$–CH_3 (*continuous lines*) and $COOCH_3$–$COOCH_3$ (*dotted lines*) (from [78])

that the decrease of these distances leads to an increase in the probability of a high value of activation energy; the same conclusion stands at $50\,°C$.

It is worth noting that the distances considered involve both the local tacticity and the local conformation. As regards the latter, *trans* conformation corresponds to smaller distances and, thus, higher values of the activation energy.

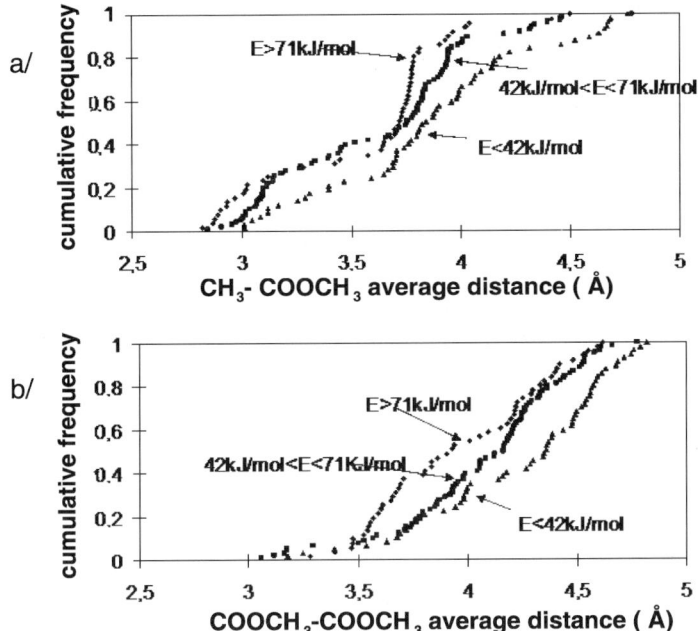

Fig. 123 Mean distances in Å as a function of the activation energy class at − 40 °C for: **a** CH$_3$–COOCH$_3$ and **b** COOCH$_3$–COOCH$_3$ (from [78])

8.1.5.4
Chain Adjustment to Ester Group Rotation

As multidimensional NMR shows, the ester group flip is accompanied by a rotation of about 20° around the local chain axis, involving a change in the
internal rotation angles of the C – C main-chain bonds. However, the extent
of the change along the chain sequence cannot be defined from NMR. In
contrast, quasi-static dynamics modelling allows one to investigate such an
extent.

Figure 124 shows the variation of the internal rotation angles of the C – C
main-chain bonds on both sides of the carbon bearing the rotating ester
group, at − 40 and 50 °C. As expected, the largest changes concern the two
nearest C – C bonds, but the deformation extends over about six to eight
bonds on each side. Thus, the perturbation of the main-chain conformation
induced by the ester flip concerns about 16 bonds, which is quite significant. However, the temperature does not seem to play any role in the mean
amplitude of extent of the changes, meaning that it is mainly driven by intramolecular release of the constraint generated by the flip of the asymmetric
ester group. Such a behaviour is quite consistent with the fact that the activation energy of the ester rotation almost entirely originates from changes in
the intramolecular energy of the chain with the moving ester (Sect. 8.1.5.3).

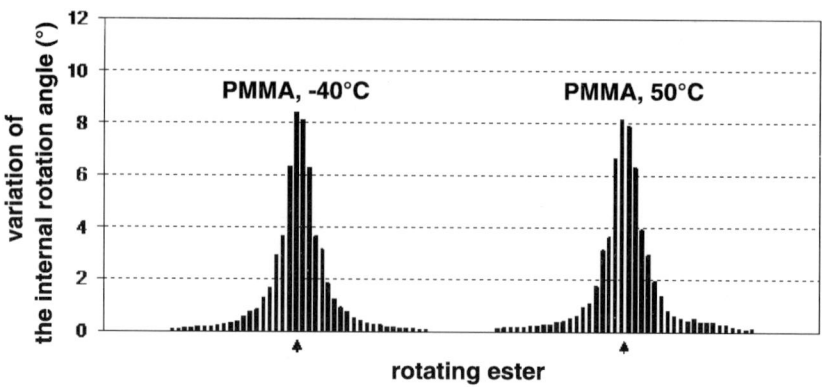

Fig. 124 Mean change of the internal rotation angles of the main-chain C – C bonds at – 40 °C and 50 °C (from [78])

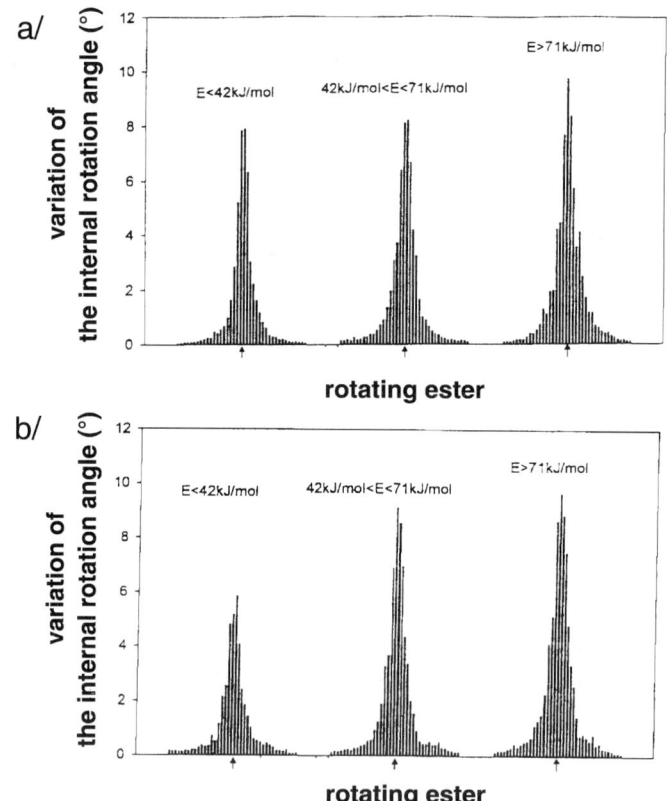

Fig. 125 Internal rotation angle changes of the main-chain C – C bonds induced by the ester group flips associated with the three different classes of activation energy: **a** at – 40 °C and **b** at 50 °C (from [78])

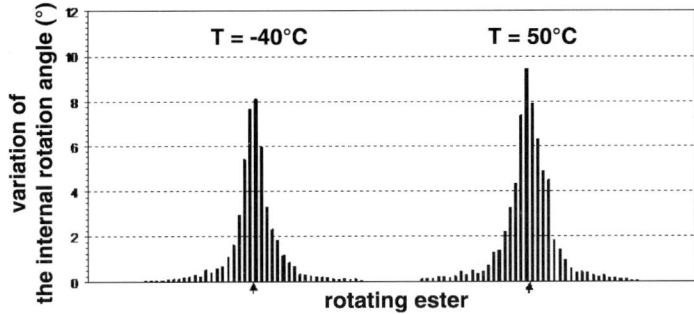

Fig. 126 Internal rotation angle changes of the main-chain C – C bonds associated with the ester group flips activated at 40, and 50 °C (from [78])

In order to go further in this approach, it is worth investigating the amplitude of the main-chain angular changes generated by ester group flip associated with different activation energies. Considering again the three classes of activation energies previously defined, the corresponding main-chain angular changes, at – 40 and 50 °C, are displayed in Fig. 125. It is clear that, at both temperatures, a higher activation energy is associated with a larger angular change.

Finally, it is interesting to combine the dielectric relaxation results and the information derived from modelling. Thus, it appears that, at – 40 °C and 1 Hz, the experimental activation energy of the involved sites is around 67 kJ mol^{-1}, whereas at 50 °C and 1 Hz, the experimental activation energy associated with the sites is 97 kJ mol^{-1}. The main-chain angular changes, corresponding to each type of site involved in the dielectric measurements at – 40 and 50 °C, are presented in Fig. 126. It appears that the amplitude of the main-chain angular changes is higher at 50 °C than at – 40 °C. More quantitatively, the total angular change is equal to 58.5° at 50 °C against 49° at – 40 °C.

8.1.5.5
Complementary Information Obtained from Modelling

The quasi-static modelling of the dynamics of the ester group flip in an amorphous cell of atactic PMMA has yielded information complementary to that derived from dynamic mechanical analysis, dielectric relaxation and, mainly, multidimensional ^{13}C and ^{2}H solid-state NMR. The main results are:

1. The activation energy of the ester group flip is distributed.
2. This activation energy originates from the change of the intramolecular energy of the chain to which the rotating ester belongs.
3. The activation energy value directly depends on the local structure through the distances between the moving ester and the methyl group, as well as the ester group of the nearest-neighbour monomers.

4. Associated with the ester group flip, there is a distortion of the internal rotation angles of the main-chain C – C bonds on both sides of the carbon bearing the rotating ester. Such a distortion concerns six to eight bonds on each side.

5. The amplitude of the main-chain angular changes is directly related to the activation energy value: a larger amplitude corresponds to a higher activation energy.

6. The ester group flips occurring at – 40 °C induce a smaller main-chain angular change (a total change of 49°) than those occurring at 50 °C (ca. 65°).

8.1.6
Conclusion on the Motions Involved in the β Transition of PMMA

The various investigations (dynamic mechanical analysis, dielectric relaxation, multidimensional ^{13}C and ^2H solid-state NMR, and atomistic modelling) performed on the β transition of PMMA have led to a very good understanding of the motions involved in this transition.

The β transition clearly originates from π-flips of the ester group around the bond with the main chain. These flips are accompanied by a change in the internal rotation angles of the main chain. The amplitude of this change gradually decreases over a scale of six to eight bonds on both sides of the carbon bearing the flipping ester group. The amplitude of the main-chain adjustment is larger in the high-temperature part of the β transition (typically 50 °C) than in the low-temperature part (as – 40 °C). These effects are reflected in the moderate cooperativity of the motions indicated by the activation entropy values along the β transition.

The observed activation energy comes from a change of the intramolecular energy of the chain with the rotating ester group and it increases with the amplitude of the main-chain internal rotation angle changes.

It is worth noting that, as for bisphenol A polycarbonate (Sect. 5), the investigation of the β transition of PMMA constitutes quite an interesting illustration of the level of understanding that can be reached when the various experimental and modelling approaches are applied together.

8.2
Methyl Methacrylate-co-N-cyclohexyl Maleimide Random Copolymers

The methyl methacrylate-co-N-cyclohexylmaleimide copolymers are based on methyl methacrylate and N-cyclohexylmaleimide repeat units with chemical formula:

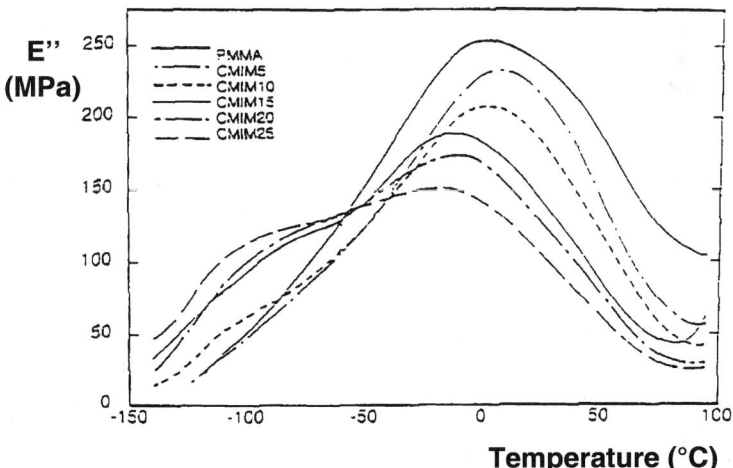

The maleimide cycle is a rigid structure whose C – C bond participates in the chain backbone. On each side of the rigid cycle are flexible sequences of MMA linked to the cycle either directly by the quaternary carbon or through the methylene group.

These copolymers will be coded CMIMx, where x indicates the mol% of cyclohexylmaleimide (CMI) units in the copolymer. Prepared by radical polymerisation of MMA and maleimide, the copolymers have a random distribution of the maleimide units.

8.2.1
Dynamic Mechanical Analysis

The viscoelastic response of a series of CMIMx copolymers has been determined as a function of temperature [79]. The loss modulus, E'', at 1 Hz, of a series of copolymers with molar content of maleimide units varying from 5 to 25% is plotted in Fig. 127; for comparison the curve for pure PMMA is also drawn.

Fig. 127 Temperature dependence of the loss modulus, E'', at 1 Hz, for PMMA and various CMIMx copolymers (from [79])

The α transition is not included in the considered temperature range. For CMI contents equal to or higher than 10%, two transitions are observed. At low temperatures a shoulder is present, whose extent increases with increasing CMI (γ transition). Studies performed on copolymers with maleimide unit N-substituted by isopropyl or phenyl groups [79] do not show this low-temperature transition, which appears to be specific for cyclohexylmaleimide. Such a situation is analogous to the one encountered with poly(cyclohexyl methacrylate) described in Sect. 3. Consequently, this low-temperature transition is assigned to the internal motion of the cyclohexyl ring, i.e. the chair–chair inversion represented in Fig. 7.

The second transition occurs at a higher temperature, in the range of the β transition of PMMA, and will be referred to as the β transition of the CMIMx copolymers. Two main features appear. The first concerns the decrease of intensity of the transition when increasing the CMI content. Actually, there is quite a linear decrease, strongly suggesting that the MMA units are the only ones contributing to the β peak.

The second feature deals with the asymmetry of the effect of CMI on the β peak. Indeed, when increasing the amount of CMI in the copolymer, one observes a rather large decrease of the high-temperature side of the β peak, whereas the low-temperature side is only slightly affected. As a consequence, the temperature at the maximum of the peak is shifted from 0 °C for PMMA to – 25 °C for CMIM25.

The investigations of the β transition of PMMA (Sect. 8.1) show that the high-temperature part of the peak involves π-flip motions of the ester group, which are coupled with a 20° rotation around the local chain axis and an adjustment of the internal rotation angles of the main-chain C – C bonds over six to eight bonds on each side of the carbon bearing the rotating ester. From this picture, it is clear that the CMI unit hinders any rotation required by the π-flips of ester groups located in its neighbourhood along the chain and, thus, breaks down the cooperativity that occurs in the high-temperature part of the β transition. Only ester groups belonging to MMA units far enough from the rigid CMI cycle (two or three repeat units farther) can undergo cooperative ester group π-flips at the same temperature as in pure PMMA. The probability of finding such sequences decreases by increasing the CMI content, leading to the observed decrease of the intensity of the high temperature side of the β peak.

Another interesting temperature range concerns the crossover between the β and α transitions. The temperature dependence of the loss modulus, E'', at 1 Hz in this region is shown in Fig. 128 for CMIMx copolymers with a low CMI content. A decrease of E'' value clearly appears in the crossover region. With 3% CMI the effect has an amplitude comparable to that obtained by physical ageing of PMMA (Fig. 112). The effect increases with 5% but it seems to level off for higher contents. As in this concentration range (3 to 5%) there is a small increase only of the α transition temperature, the observed decrease of E'' cannot be attributed to a change in the β – α temperature difference.

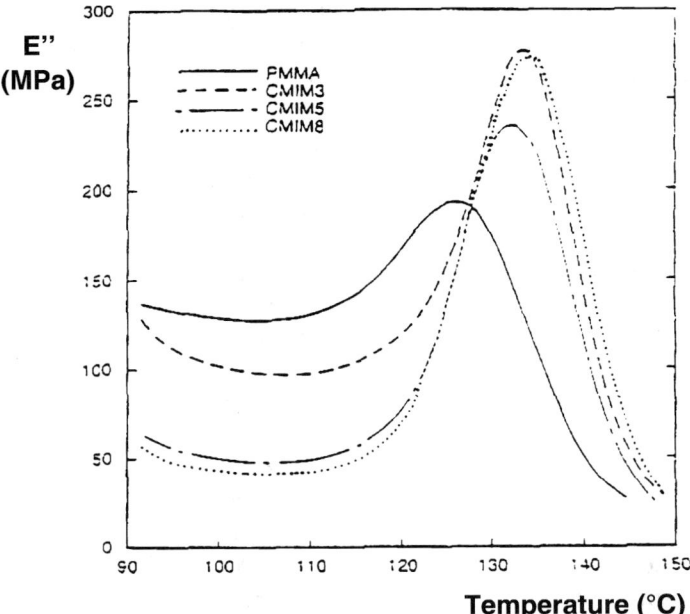

Fig. 128 Temperature dependence of the loss modulus, E'', in the β–α crossover region for PMMA and various CMIMx copolymers (from [79])

Thus, what is observed is a decoupling between the motions of the β transition and those involved in the α transition. Indeed, in PMMA, along the high-temperature side of the β peak, more and more cooperativity between ester flips and chain motions is involved, in such a way that the chain motions in this temperature region could be considered as the beginning of cooperative chain motions characteristic of the α processes. The presence of rigid CMI units within the chain backbone interrupts this cooperativity and, consequently, the motions involved in the β transition are confined to small MMA sequences and cannot reach the extent required by the chain motions occurring at the onset of the α transition.

8.2.2
Dielectric Relaxation

The dielectric relaxation of CMIMx copolymers has been studied and compared to the PMMA dielectric behaviour [75]. Figure 129 shows the temperature dependence of the dielectric loss, ε'', at 1 Hz.

In the temperature range of the α transition, a shift towards higher temperature is associated with the introduction of increasing amounts of CMI; it reflects the effect of the rigid maleimide cycles which hinder the main-chain motions, comparative to PMMA.

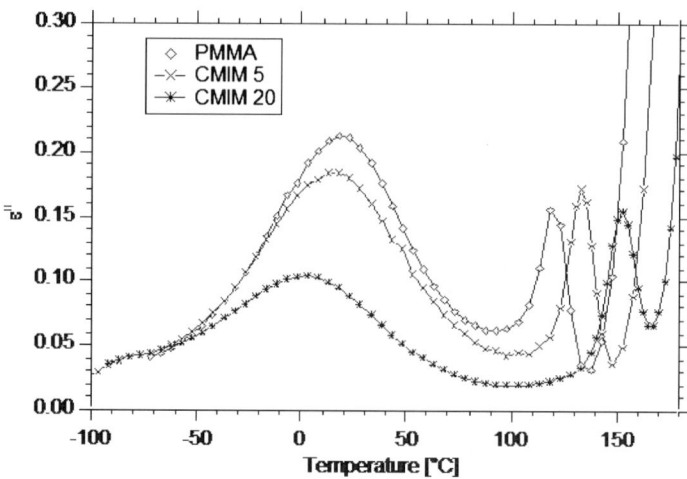

Fig. 129 Temperature dependence of the dielectric loss, ε'', at 1 Hz, for PMMA, CMIM5 and CMIM20 copolymers (from [75])

Concerning the β transition, the effect obtained with 5% CMI is rather small and so only the copolymer containing 20% CMI will be considered. With this latter copolymer, one observes quite a large decrease of the peak intensity, as well as a shift of ca. 20 °C of the temperature of the peak maximum towards a lower temperature. Such a shift comes from the fact that the high-temperature side of the β peak is more affected by the presence of the CMI units than is the low-temperature side. In the chemical structure of the copolymer, two types of dipoles are present:

– The dipole associated with the ester group of the MMA unit, whose π-flips are responsible for the β transition in PMMA.
– The dipole associated with the maleimide unit, which is located in the plane of the maleimide cycle and perpendicular to the C – C bond taking part to the chain backbone. The motion of this resultant dipole implies a rotation of the rigid maleimide unit around the local chain axis. This structure is, in some way, analogous to that encountered with *iso*-phthalic units in aryl-aliphatic polyamides (Sect. 6) where the *iso*-phthalic rings only undergo small amplitude oscillations but no flips at temperatures below the glass transition temperature. Thus, it is unlikely that the maleimide dipole could be involved in the β transition motions of CMIMx copolymers.

The larger effect of CMI units on the high-temperature side of the β peak indicates that the rigid CMI units hinder the main-chain internal rotation angle adjustment required by the π-flip of the ester groups of MMA occurring in this temperature range. Consequently, these cooperative motions cannot happen in the sequences of MMA repeat units, which are not long enough (typically smaller than six repeat units). Even for such rather long sequences, the cooperativity for π-flips of ester groups near the CMI units cannot develop. In contrast, the more localised ester group π-flips performed in the low-temperature side of the β transition are not affected by the CMI units for they occur in sites with lower surrounding constrains and do not require main-chain cooperativity.

This conclusion is also supported by a quantitative analysis of the strength of the dielectric β relaxation, expressed through the change of permitivity, ε', determined from the frequency response. Indeed, at 23 °C, ε' is consistent with the amount of ester groups present in the CMIMx copolymer, whereas at 98 °C, ε' is much weaker than expected from the ester content, showing that CMI units hinder cooperative π-flip ester motions.

The more localised character of the ester motions occurring in the CMIM20 copolymer is reflected in the lower value of the activation entropy determined for the maximum of the β peak: 40 J K^{-1} mol^{-1} for CMIM20 instead of 58 J K^{-1} mol^{-1} for PMMA.

It is interesting to point out the change in the relative heights of the β and α peaks between PMMA and CMIM20 (Fig. 129). In the latter polymer, only a small part of the dielectric relaxation happens through the β motional processes. The cooperative motions involved in the α transition are required for achieving an important relaxation, whereas it is the opposite for PMMA. Such a behaviour is consistent with the hindrance of main-chain cooperativity by the rigid CMI units.

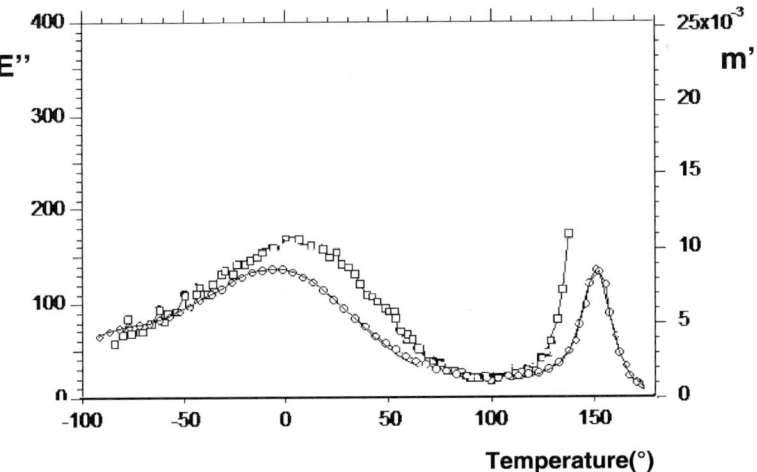

Fig. 130 Temperature dependence of the mechanical loss modulus, E'' ($\square$) and the dielectric loss modulus, m'' ($\bigcirc$) for CMIM20 at 1 Hz, obtained by superposing the low-temperature part of the β peak (from [75])

Finally, it is worth comparing the dynamic mechanical and dielectric results. For this purpose, Fig. 130 shows the temperature dependence of the mechanical loss modulus, E'', and the dielectric loss modulus, m'', at 1 Hz, obtained by superposing the low-temperature part of the β transition.

Whereas for PMMA (Fig. 113) the two β peaks are quite well superposed, except in the $\beta - \alpha$ crossover region, in the case of the CMIM20 copolymer, the dielectric loss is weaker than the mechanical loss in the high-temperature part of the β transition. In contrast, in the $\beta - \alpha$ crossover region, the same behaviour is observed for the mechanical and dielectric responses, showing that, in CMIM20, the CMI units lead to a complete decoupling between the β and α transitions, which is not the case for PMMA.

8.2.3
^{13}C NMR Investigations

In the case of the CMIM20 copolymer, the NMR investigations have been performed both in the solid-state and in solution [75].

8.2.3.1
Solid-State ^{13}C NMR

Unlike the NMR study on PMMA, ^{13}C- and ^{2}H-labelled CMIM20 copolymers were not available, so investigation has been limited to analysis of the chemical shift anisotropy of the carboxyl group [75].

From the relaxation map of the copolymer and the frequency range involved in the NMR experiment, the motional processes involved in the β transition should be detected by NMR in the temperature range 20–140 °C.

For comparison, the various NMR spectra are given for both PMMA and CMIM20.

Figure 131 shows the powder NMR spectra of the carboxyl group at 25, 90 and 115 °C.

At 25 °C, the chemical shift anisotropy parameters (Fig. 114) are $\sigma_{33} = 112$ ppm, $\sigma_{22} = 150$ ppm and $\sigma_{11} = 268$ ppm. When increasing temperature, several differences are revealed:

- In the σ_{33} region, the intensity becomes higher and higher for PMMA, compared to CMIM20
- In the σ_{22} region, the signal of CMIM20, broader at 25 °C than that from PMMA, becomes narrower at 115 °C
- In the σ_{11} region, the signal of CMIM20, smaller at 25 °C than for PMMA, turns out to be larger at 115 °C

All these results indicate that the C = O motions develop more rapidly in PMMA with increasing temperature than they do in CMIM20. At 115 °C, there are less motions in CMIM20 than in PMMA.

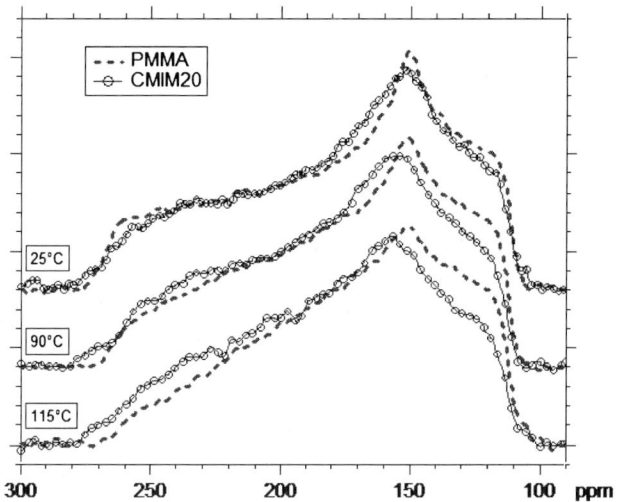

Fig. 131 Powder spectra of $C = O$ for PMMA and CMIM20 at 25, 90 and 115 °C (from [75])

Such a behaviour is consistent with the weaker intensity, observed in the high-temperature part of the dielectric β peak, for CMIM20 compared to PMMA.

8.2.3.2
Solution ^{13}C NMR

^{13}C Spin-lattice relaxation times T_1 at 25 and 50 MHz have been determined for PMMA and CMIM20 in deuterated chloroform solution over a temperature range of – 50 to 50 °C.

In CMIM20, MMA sequences contain an average of four consecutive repeat units. The CH$_2$(MMA) groups whose relaxation has been measured are indicated by circles in Fig. 132, whereas the CH(CMI) groups also studied are indicated by squares.

The plot of T_1 as a function of $1000/T$ (K^{-1}) exhibits a minimum at the temperature where the observed segmental motion happens at the frequency

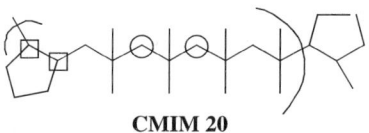

CMIM 20

Fig. 132 Simplified scheme of a CMIM20 chain. The studied CH$_2$(MMA) and CH(CMI) carbons are indicated by *circles* and *squares*, respectively

of the NMR experiment (25 or 50 MHz). The absolute value of $T_{1\,min}$ reflects the amplitude of the libration motion of the ^{13}C group, but not its reorientation due to segmental motions of the chain. For this reason, it is more appropriate for investigating chain dynamics to consider the $T_1/T_{1\,min}$ ratio, which only depends on the segmental motions.

The temperature dependence of $T_1/T_{1\,min}$ for the CH$_2$(MMA), both in PMMA and CMIM20, is plotted in Fig. 133 at 25 and 50 MHz.

It clearly appears that, at both frequencies, the minimum is reached at a higher temperature for CMIM20 than for PMMA. This indicates a slowing down of the motions of the MMA units by the rigid CMI units within the chain.

Looking at the dynamics of the CH$_2$(MMA) and CH(CMI) groups within the CMIM20 chain, the results presented in Fig. 134 show, particularly at 50 MHz, a minimum occurring at a higher temperature for the CH(CMI) than for CH$_2$(MMA). Thus, within the same CMIM20 chain, the MMA units are more mobile than the CMI units.

These ^{13}C NMR investigations on PMMA and CMIM20 in solution unambiguously demonstrate the hindrance of the mobility of the MMA repeat units by the rigid maleimide cycle of the CMI group. This increase of rigidity is, of course, reflected in the higher glass transition temperatures observed in the CMIMx copolymers, but also in the hindrance of cooperativity of the ester group π-flips observed in the high-temperature part of the β transition.

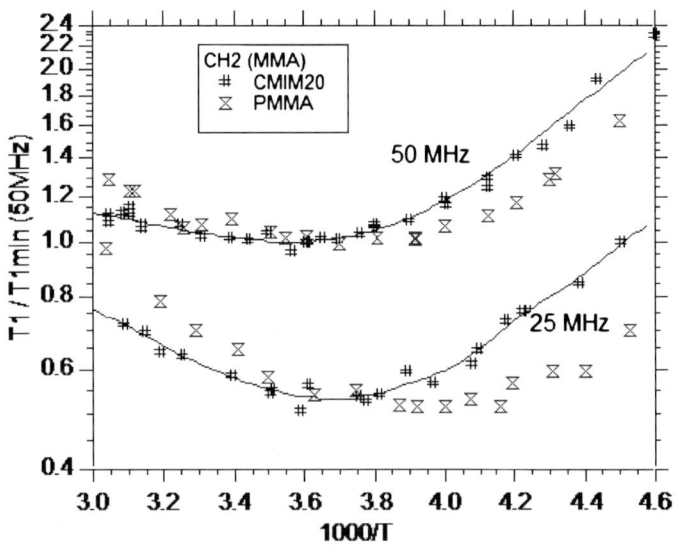

Fig. 133 Temperature dependence of the $T_1/T_{1\,min}$ ratio of CH$_2$(MMA) for PMMA and CMIM20 at 25 and 50 MHz. The curves are guides for the eye (from [75])

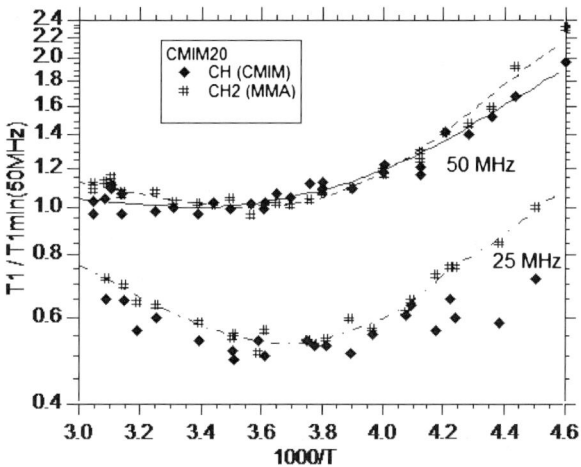

Fig. 134 Temperature dependence of the ratio $T_1/T_{1\,min}$ of CH_2(MMA) and CH(CMI) for CMIM20 at 25 and 50 MHz. The curves are guides for the eye (from [75])

8.2.4
Atomistic Modelling

A molecular modelling using the quasi-static dynamics, analogous to that described for PMMA (Sect. 8.1.5), has been applied to the CMIM20 copolymer [78].

8.2.4.1
Modelling Characteristics for CMIM20

The amorphous glassy cells of CMIM20 have been constructed from the PMMA cells according to the following procedure:

- 20% of MMA repeat units are randomly selected by pairs, then each pair is replaced by a CMI unit in the original cell
- In order to remove the distortions created by such substitutions, a first energy minimisation is performed
- An annealing at 1727 °C, followed by a cooling down to 50 °C in 100 steps, with energy minimisation at each step, allows the chain to reach the required equilibrium

The density of the CMIM20 cell is adjusted to the experimental one at 50 °C, i.e. 1179 $kg\,m^{-3}$, identical to the PMMA cell.

The quasi-static dynamics is applied to the π-flip of the ester groups. The CMI units are not submitted to any forced rotation, but small rotations around the local chain axis can be induced by the π-flip of the ester group born by a neighbouring MMA monomer.

8.2.4.2
Activation Energies

The mean activation energy for CMIM20 is 53 kJ mol^{-1}, significantly smaller than that obtained for PMMA (58 kJ mol^{-1}). Such a trend agrees with the dielectric data (Sect. 8.2.2), in spite of the larger change observed experimentally.

As regards the intra- or intermolecular origin of the activation energy, the relative contributions are similar in CMIM20 and PMMA. The activation energy comes essentially from the intramolecular potential energy change of the chain bearing the moving ester.

An interesting feature deals with the effect of the position of the CMI unit relatively to the MMA with the rotating ester group. Indeed, Fig. 135 shows that the activation energy, associated with the π-flip of an ester group, gradually increases when the corresponding MMA repeat unit is located farther and farther from the CMI unit. Thus, five C – C bonds from the CMI cycle are enough to recover the behaviour observed in PMMA. This result is consistent with the lower local density (at a scale of 5.5–6.5 Å) observed in CMIM20 than in PMMA. The bulkiness of the CMI comonomer locally creates a lower density site in which an ester group π-flip can occur with a weaker intramolecular adjustment and, consequently, a lower activation energy.

8.2.4.3
Chain Adjustment to Ester Group Rotation

Comparison of the chain adjustments induced by the ester π-flips in PMMA and CMIM20 reveals quite significant differences.

The total change of internal rotation angles of the main-chain C – C bonds in the case of CMIM20 is weaker (55°) than for PMMA (58.5°). Furthermore, one observes a gradual increase of this total angular change when the MMA

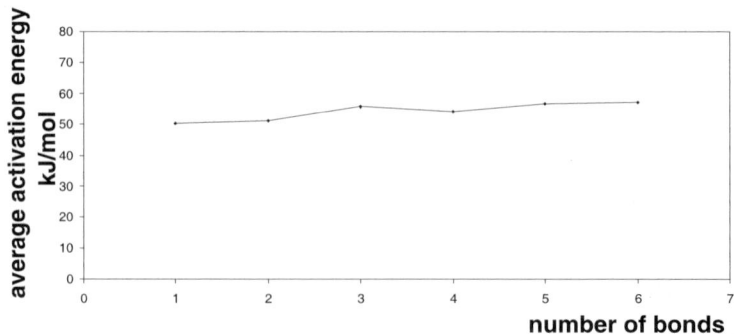

Fig. 135 Average activation energy as a function of the number of bonds between the MMA unit bearing the moving ester and the CMI cycle (from [78])

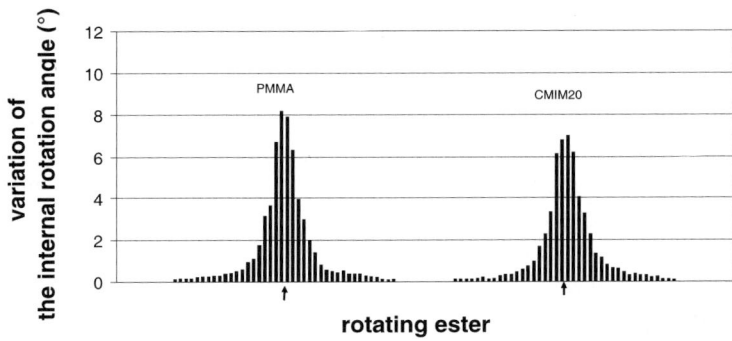

Fig. 136 Internal rotation angle changes of the main-chain C – C bonds associated with the ester group π-flips occurring in PMMA and CMIM20 at 50 °C (from [78])

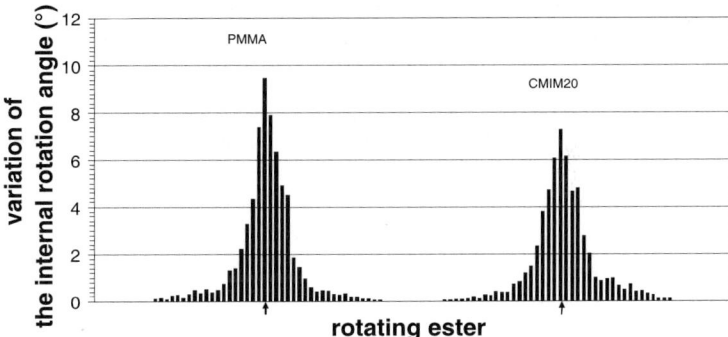

Fig. 137 Internal rotation angle changes of the main-chain C – C bonds associated with the ester group π-flips activated in the dielectric relaxation of PMMA and CMIM20 at 50 °C. (from [78])

unit with the rotating ester is located farther and farther away from the CMI unit (the PMMA value is reached when five bonds separate the two units).

Figure 136 shows the amplitude and extent of the internal rotation angle changes for PMMA and CMIM20.

It is clear that a lower amplitude change occurs for the four to five nearest bonds on each side of the rotating ester in the case of CMIM20. This effect is even more pronounced when the comparison is performed on the sites that are active in dielectric relaxation at 50°, as shown in Fig. 137.

8.2.5
Conclusion on the Motions Involved in the Solid-State Transitions of Methyl Methacrylate-*co*-*N*-cyclohexyl Maleimide Random Copolymers

In this particular type of maleimide copolymer, due to the cyclohexyl ring, a very low-temperature transition exists that originates from the chair–chair inversion of the cyclohexyl ring.

The β transition comes from the ester group π-flip of the MMA units, whereas the maleimide cycles are not able to undergo large amplitude motions at temperatures below the glass transition temperature. However, due to the rigid structure of the maleimide unit, the ester group motions of the neighbouring MMA monomers are perturbed in different ways.

At first, the bulkiness of the CMI group creates sites with a lower local density, in which ester group π-flips with a smaller activation energy occur. This effect gradually disappears for MMA units located farther away from the CMI group along the chain, in such a way that the pure PMMA behaviour is recovered for MMA units separated from CMI cycle by five or more C – C bonds.

The rigidity of the CMI unit also hinders the internal rotation angle change of the main-chain MMA sequences. Consequently, the coupling between the ester group π-flip and the main-chain adjustment occurring in the high-temperature part of the β transition peak of pure PMMA is limited to long-enough MMA sequences whose probability sharply decreases when the CMI content increases. As a result, in the dynamic mechanical and dielectric β peak, the high-temperature part gradually disappears with increasing the amount of CMI units in the random copolymers. This leads to a progressive shift of the temperature at the peak maximum towards lower temperatures. In other words, one can say that the CMI units hinder the cooperativity which exists in the high-temperature part of the β transition of pure PMMA.

8.3
Methyl Methacrylate-*co*-*N*-methyl Glutarimide Random Copolymers

Another series of methyl methacrylate copolymers is constituted by the methyl methacrylate-*co*-*N*-methyl glutarimide random copolymers. The chemical structure of the *N*-methyl glutarimide repeat unit is:

These copolymers will be coded MGIM*x*, where *x* indicates the mol% of glutarimide units in the copolymer.

The copolymers are obtained by the reaction, in solution, of methylamine with PMMA, leading to methyl glutarimide (MGI) cycles randomly distributed along the original PMMA chain backbone. As the content of MGI units varies from 4 to 76%, it is worth considering the percentages of the different triads that can be encountered (the calculations are performed as-

Table 12 Percentages of the different types of triads as a function of copolymer composition

Copolymer	GIM-GIM-GIM	MMA-GIM-GIM GIM-GIM-MMA MMA-GIM-MMA	MMA-MMA-MMA	GIM-MMA-MMA MMA-MMA-GIM GIM-MMA-GIM
Calculations performed on the basis of a random distribution of the MGIM and MMA units				
GIM76	43.9	32.1	1.4	22.6
GIM63	25.0	38.0	5.1	31.9
GIM58	19.5	38.5	7.4	34.6
GIM36	4.7	31.3	26.2	37.8
GIM21	0.9	20.1	49.3	29.7
GIM8	0.1	7.9	77.9	14.1
GIM4	0.0	4.0	88.5	7.5

suming a random distribution of the MGI and MMA units). The results are reported in Table 12. Whereas, with increasing the MGI content, there is a gradual change in the probability of finding a MGI–MGI–MGI triad or a triad with MMA–MMA linkages, the probability of finding a triad involving MGI–MMA linkages does not change significantly when the composition is modified within the range 36–76% of MGI units.

Compared to the CMIMx chemical structure, the main differences are:

- The glutarimide cycle, due to the additional CH_2 group, is less rigid than the maleimide one and can undergo distortions of the internal rotation angles of its $C - CH_2 - C$ part.
- The glutarimide unit is linked, on each side, to the neighbouring MMA units through a CH_2 group. Such a linkage is less constraining for the MMA chain backbone than for the maleimide cycle which, on one side, is directly linked to the quaternary carbon of the MMA monomer.

Due to the introduction of a cyclic structure within the PMMA chain backbone, the glass transition temperature of the MGIMx copolymers increases with the amount of MGI units; for MGIM76 a glass transition temperature as high as 151 °C is reached.

Though MGI units have similar effects on the glass transition temperature as CMI comonomer, the specific features of the MGIMx copolymer chemical structure yield mechanical behaviours, in particular in fracture, quite different to the CMIM series (as described in [1]). For this reason, it is interesting to investigate the effect of MGI unit content on the β transition of the copolymers and compare it to that observed in the CMIM series.

8.3.1
Dynamic Mechanical Analysis

The β transition of a series of MGIMx copolymers has been investigated by dynamic mechanical analysis [80]. The temperature dependence of the loss modulus, E'', at 1 Hz is shown in Fig. 138. In the region of the α transition, when increasing the MGI content, a shift towards a higher temperature is observed.

8.3.1.1
β Transition of MGIMx Copolymers

In the β transition region, Fig. 138 reveals a decrease of the peak amplitude with increasing MGI content. For MGIM21, MGIM36 and MGIM58, the decrease of E'' in the temperature range from -20 to $100\,°C$ is accompanied by an increase of E'' in the low-temperature part. In contrast, for MGIM76, one observes quite a weak and broad peak over the whole temperature range from $-80\,°C$ to the beginning of the α transition at $120\,°C$.

In order to go further in the analysis, it is worth considering the loss compliance, J'', shown at 1 Hz in Fig. 139. Indeed, the loss compliance allows one to perform quantitative comparisons based on additive contributions of the various units, as already mentioned (Sects. 6.3.1 and 7.1.1). Such an approach is illustrated as a function of temperature in Fig. 140 for MGIM36.

Thus, assuming that the contribution of the MMA unit to J'' is the same in MGIM36 as in pure PMMA, the contribution to J'' of the MMA units contained in MGIM36 is given by $0.64\,J''(\text{PMMA})$ and, consequently, the effect of

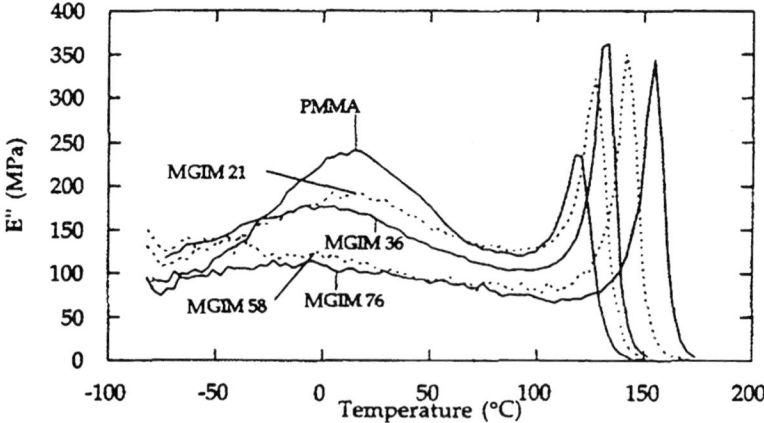

Fig. 138 Temperature dependence of the loss modulus, E'', at 1.2 Hz for some typical MGIMx copolymers (from [80])

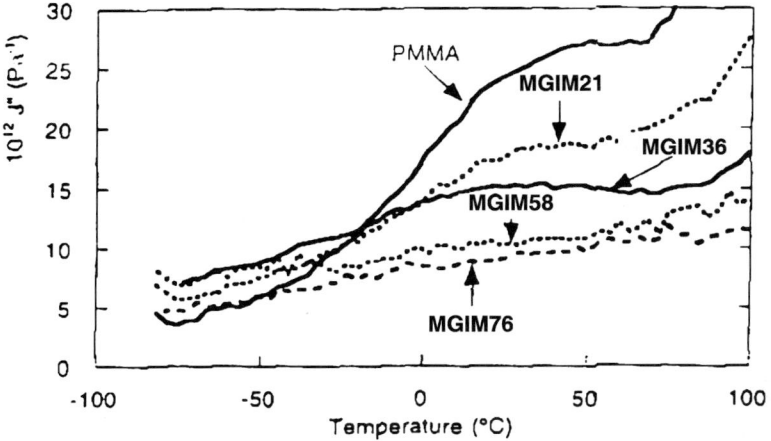

Fig. 139 Temperature dependence of the loss compliance, J'', at 1.2 Hz for some typical MGIMx copolymers (from [80])

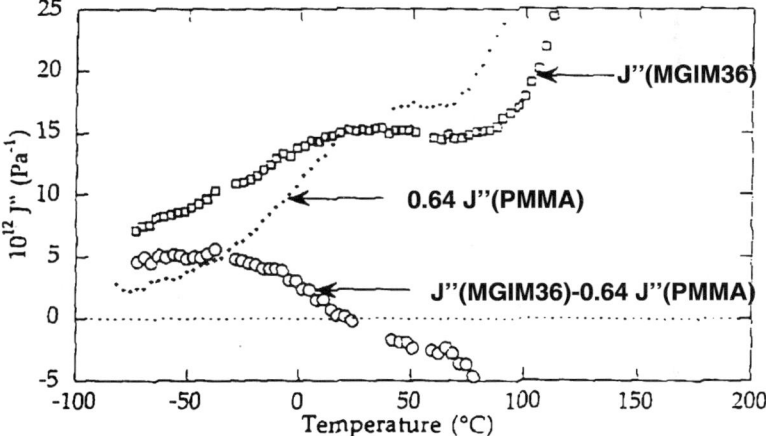

Fig. 140 Temperature dependence, for MGIM36, of the loss compliance, J'', and of the contribution to J'' of the MMA units, as well as the difference $\Delta J''$(MGIM36) (from [80])

the MGI units is given by the difference:

$$\Delta J'' \,(\text{MGIM36}) = J'' \,(\text{MGIM36}) - 0.64\, J''(\text{PMMA}) \tag{6}$$

The results, plotted in Fig. 140, clearly show two different behaviours, depending on the considered temperature range:

– At temperatures below 20 °C, the J'' values for MGIM36 are larger than what should be expected from the corresponding response of MMA units in PMMA; the effect is larger and larger as temperature decreases

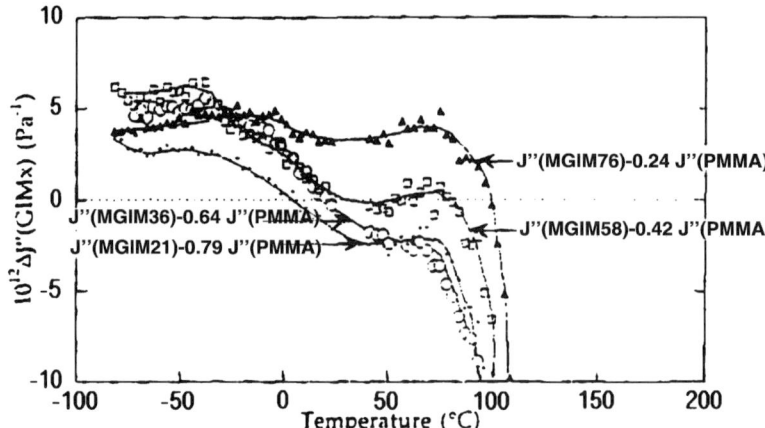

Fig. 141 Temperature dependence of the difference, $\Delta J''$ (MGIMx), for the various MGIMx copolymers (from [80])

- At temperatures above 20 °C, the opposite effect is observed, J''(MGIM36) being smaller than 0.64 J''(PMMA)

This change in behaviour is reflected on the difference curve, $\Delta J''$(MGIM36), which goes from positive values at low temperatures to negative ones above 20 °C.

The difference curves, $\Delta J''$(MGIMx), corresponding to the various MGIMx copolymers are shown in Fig. 141. At low temperatures, whatever the MGIM content, $\Delta J''$(MGIMx) is positive, indicating a higher dissipation than expected from additive MMA contributions. More precisely, this effect passes through a maximum for MGIM58.

As regards the high-temperature part of the β transition (20–100 °C), it clearly appears in Fig. 141 that MGIM21 and MGIM36 lead to negative values for the difference, $\Delta J''$(MGIMx), indicating that the dissipation of these copolymers is smaller than expected from additive MMA contributions. In contrast, values of $\Delta J''$(MGIMx) close to zero are found for MGIM58 up to 80 °C and MGIM76 leads to positive values. It means that another process develops in these MGI-rich copolymers.

8.3.1.2
Comparison of CMIMx and MGIMx Copolymers

It is interesting to compare the dynamic mechanical loss compliance, J'', of a MGIM copolymer of intermediate glutarimide content with that of a CMIM copolymer of similar composition, taking as a reference the PMMA response weighted by the %(mol/mol) of MMA in each considered copolymer. Thus, the temperature dependencies of J''(MGIM21) and 0.79 J''(PMMA)

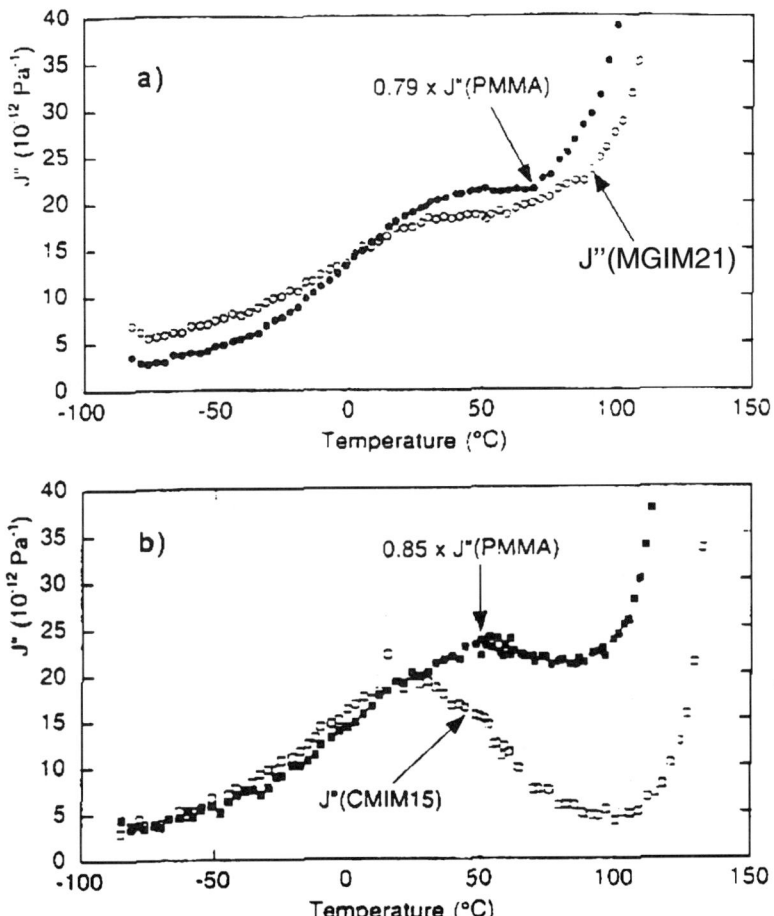

Fig. 142 Comparison of the loss compliance of MMA-*co*-MGIM and MMA-*co*-CMI copolymers in the β transition region. **a** J''(MGIM21) (*open circles*) and 0.79 J''(PMMA) (*filled circles*) vs temperature. **b** J''(CMIM15) (*open squares*) and 0.85 J''(PMMA) (*filled squares*) vs temperature (from [80])

are shown in Fig. 142a and that of J''(CMIM15) and 0.85 J''(PMMA) in Fig. 142b.

The most interesting feature deals with the high-temperature part of the β transition where cooperativity of ester group π-flips and chain motions occurs in PMMA. Indeed, whereas CMI units clearly hinder such a cooperativity, MGI units, at this low content (21%), only slightly affect the PMMA cooperativity. Such a difference can be explained by the difference of rigidity of the CMI and MGI units already pointed out.

8.3.2
Dielectric Relaxation

Dielectric relaxation measurements have been performed on a series of MGIMx copolymers with x varying from 21 to 76% [75].

8.3.2.1
Effect of MGI Content

The temperature dependence of the dielectric loss, ε'', at 1 Hz, for the various MGIMx copolymers is shown in Fig. 143.

It is clear that introducing MGI cycles in the PMMA chain backbone strongly decreases the intensity of the β peak. Interestingly, at first sight the decrease of intensity appears almost identical for the low- and high-temperature parts of the β peak for MGIM21 and MGIM36. However, for MGIM58 and, even more for MGIM76, a levelling of the ε'' values takes place in the high temperature part, which extends till the α transition. Nevertheless, a small shift of the peak maximum temperature towards a lower temperature is observed when increasing the MGI content up to 58%. This effect is reversed in the case of MGIM76 whose peak maximum occurs at nearly the same temperature as for PMMA.

The activation energy, E_a, shown in Fig. 144, smoothly decreases with increasing MGI content and seems to level off above 58%.

A similar behaviour is observed for the activation entropy, ΔS_a, up to 58%. MGIM76 leads to a slightly higher value, indicating that a more cooperative motion happens within this copolymer chain, which is very rich in MGI units.

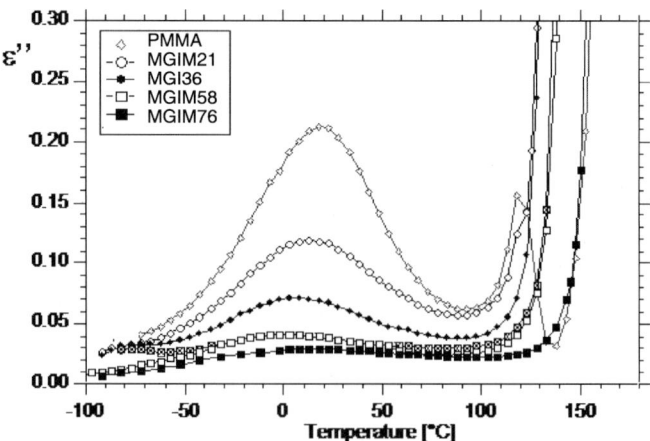

Fig. 143 Temperature dependence of the dielectric loss, ε'', at 1 Hz, for some MGIMx copolymers (from [75])

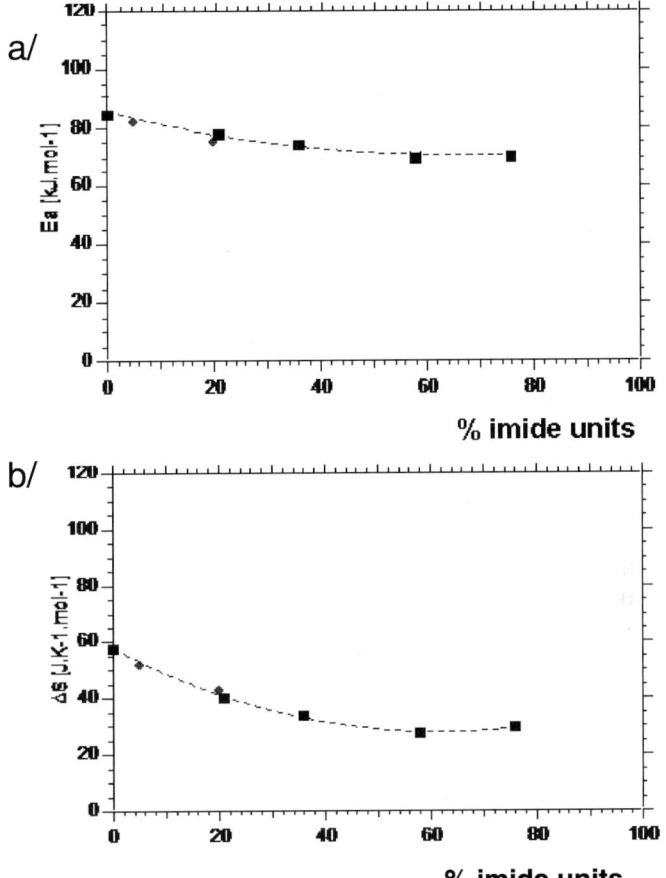

Fig. 144 Activation energy, E_a (**a**) and entropy, ΔS_a (**b**) versus the %(mol/mol) of comonomer for a series of MGIMx copolymers (■) and CMIM20 (●) (from [75])

It is worth noticing that similar behaviours are observed whatever part of the β transition is considered.

In order to go further in the analysis of the change in the shape of the β peaks within the MGIMx series, it is convenient to compare the shapes of the corresponding dielectric loss peaks after rescaling them to get the same integrated value. The results are shown in Fig. 145. In the low temperature part, the same shape is observed, whereas in the high temperature range quite a different shape is obtained for MGIM58 and, even more, for MGIM76. This clearly indicates that, in the case of MGIM58 and MGIM76, new dissipative processes occur at high temperature that do not exist in PMMA.

It is interesting to take advantage of the additivity of contributions to the dielectric loss modulus, m'', to compare the m'' values of any MGIMx copolymer with the contribution to m'' of the MMA units contained in the

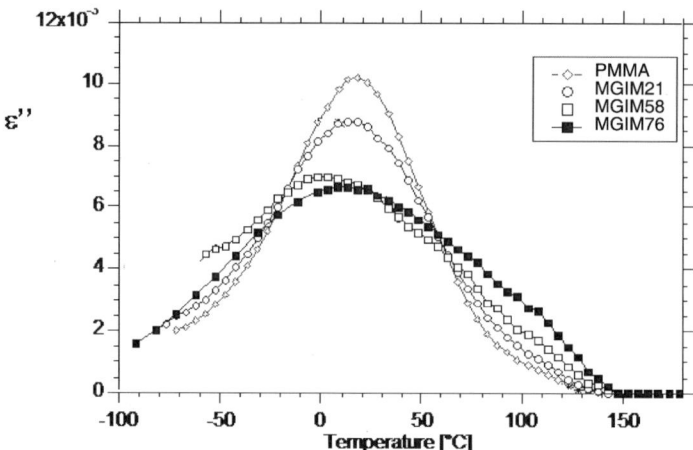

Fig. 145 Temperature dependence of the dielectric loss, ε'', rescaled to get the same integral value, for the various MGIMx copolymers (from [75])

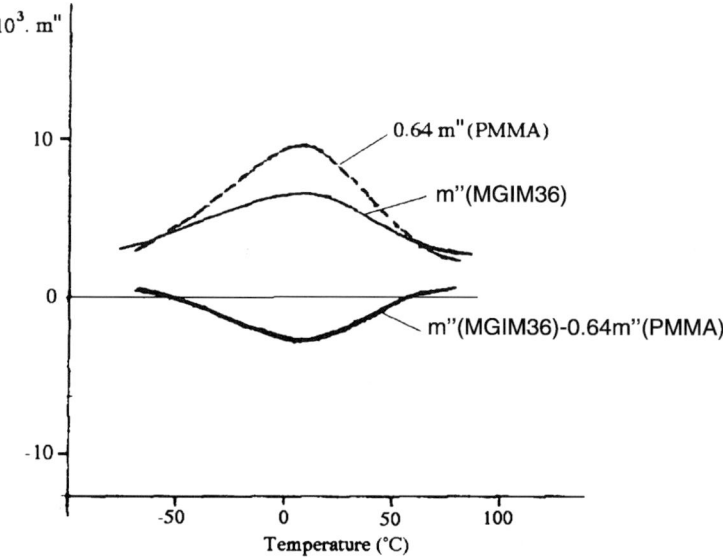

Fig. 146 Temperature dependence, for MGIM36, of the dielectric loss modulus, m'', and of the contribution to m'' of the MMA units, as well as the difference, $\Delta m''$(MGIM36)

considered MGIMx copolymer and, consequently, to examine the effect of the MGI units given by the difference:

$$\Delta m''(\text{MGIM}x) = [m''(\text{MGIM}x) - xm''(\text{PMMA})]/100 \qquad (7)$$

To illustrate the above approach, the results for MGIM36 are shown in Fig. 146.

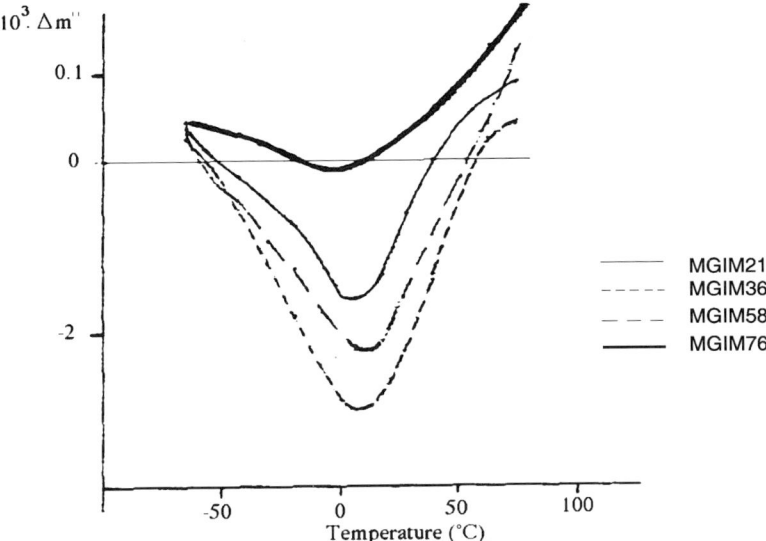

Fig. 147 Temperature dependence of the difference, $\Delta m''$ (MGIMx), for the various MGIMx copolymers

It is clear that in the low-temperature part, the introduction of MGI units decreases the dissipation relative to what should be expected from the MMA content. However, this effect gradually disappears along the high-temperature part to reach, at the end of the β transition, values equal to or slightly higher than those expected.

The temperature dependence of the difference, $\Delta m''$, for the various MGIMx copolymers is shown in Fig. 147. The curves have similar shapes for MGIM21, MGIM36 and MGIM58. However, the amplitude of the effect changes with the MGI content, being maximum (the largest negative values of $\Delta m''$) for MGIM36, then decreasing for MGIM58. Such a trend is even more pronounced in the case of MGIM76 for which positive values of $\Delta m''$ are obtained over almost the whole temperature range, indicating a larger dissipation than estimated from the MMA unit content.

8.3.2.2
Comparison of Dielectric Relaxation of CMIM20 and MGIM21 Copolymers

The temperature dependence of the dielectric loss, ε'', at 1 Hz for CMIM20 and MGIM21 and PMMA, is shown in Fig. 148. It appears that the two imide groups have different effects on the dielectric response of PMMA. In the low-temperature region, the MGI unit leads to ε'' values lower than the CMI unit. However, the largest effect arises in the high-temperature part of the β transition. Indeed, whereas the CMI units strongly hinder the cooperativity of the

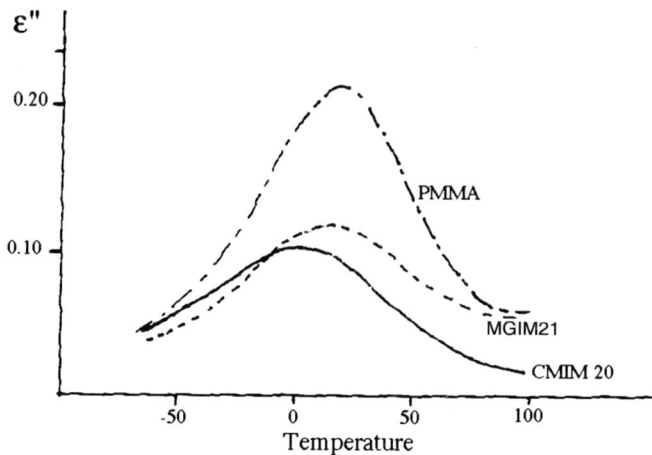

Fig. 148 Temperature dependence of the dielectric loss, ε'', at 1 Hz, for CMIM20, MGIM21 and PMMA

ester group π-flips with the chain backbone leading to quite low values of ε'', the effect of MGI units gradually vanishes when temperature increases, in such a way that almost the same ε'' values are obtained at the end of the β transition.

8.3.2.3
Comparison of Dielectric and Mechanical Results

The mechanical loss modulus, E'', and the dielectric loss modulus, m'', at 1 Hz, are compared for the various MGIMx copolymers in Fig. 149, using the same scales which lead in PMMA to superposition of the low-temperature parts of the β transition (Fig. 112). In all cases, the mechanical loss, E'', is higher than the dielectric loss, m'', over the whole temperature range.

A useful approach for comparing the effects of introducing MGI units within a PMMA chain is by taking as a reference the response expected from the MMA content in the MGIMx copolymers. This refers to Fig. 142 for the mechanics and Fig. 147 for the dielectric experiments.

In the low-temperature part of the β transition, the MGI units lead to an increase of mechanical dissipation, whatever the MGI content. In contrast, except for MGIM76 copolymers, one observes for all the other MGIMx a decrease of the dielectric dissipation. In both cases, the effect develops when increasing the MGI content, but passes through a maximum for either MGIM58 in mechanics or MGIM36 in dielectrics. The lowest effect in mechanics, even an opposite effect in dielectrics, are obtained for the MGI-richest copolymer, MGIM76.

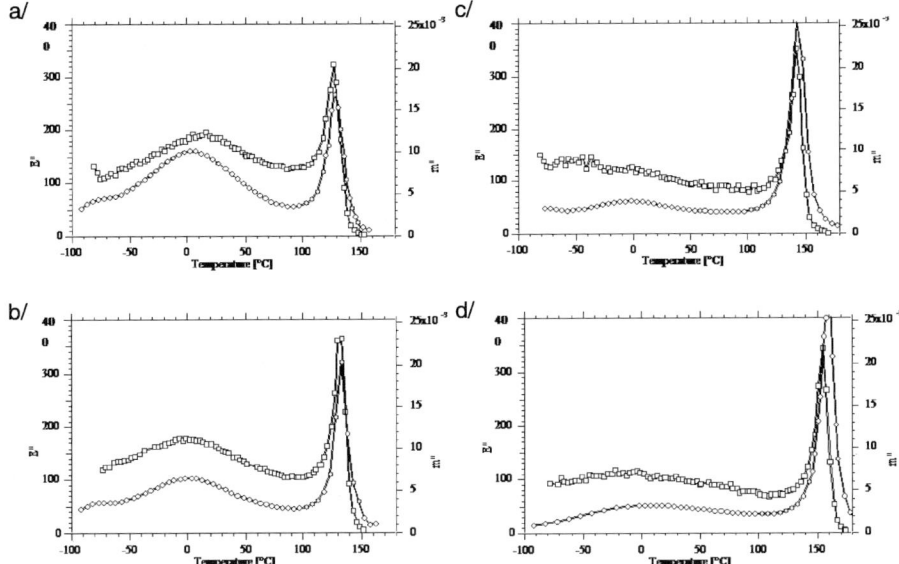

Fig. 149 Temperature dependence of the mechanical loss modulus, E'', ($\square$) and the dielectric loss modulus, m'', ($\bigcirc$) at 1 Hz, for the various MGIMx copolymers: **a** MGIM21, **b** MGIM36, **c** MGIM58 and **d** MGIM76 (from [75])

In the high-temperature part, opposite effects are observed in mechanics and dielectrics, except for MGIM76. Indeed, the decrease of mechanical dissipation (relatively to the MMA content response) increases with increasing temperature, but it is the opposite with dielectric dissipation. For MGIM76, in both experiments, the dissipation is larger than expected from the MMA content.

8.3.3
Discussion of the Origins of the Effect of MGI Units

It clearly appears from the results that one has to consider differently the low- and high-temperature parts of the β transition.

In the low-temperature part, the motions involved in PMMA are isolated ester group π-flips occurring in sites with a lower local density, in such a way that the π-flip can be performed without any cooperative adjustment of the main chain. Dealing with the MGIMx copolymers, the amplitude of the effect of MGI units passes through a maximum for either 36 (dielectric) or 58 (mechanics) %, which suggests that the effect is related to the occurrence of MMA–MGI linkages. Indeed, as shown in Table 12, the probability of such linkages is maximum around 36–58%. The associated increase of mechanical dissipation cannot be attributed to a larger number of isolated ester flips, for it should also lead to a higher dielectric dissipation, whereas the opposite effect is observed. Thus, it suggests that what could happen is that the MGI unit next to a flipping ester group takes part in a local adjustment, which generates a larger mechanical dissipation. However, this coupling would slow down the ester flip frequency, resulting in a lower number of ester flips contributing at this temperature to the dielectric dissipation at 1 Hz. Thus, a higher temperature is required for their contribution at the investigation frequency of 1 Hz.

In the high-temperature part, the concerned motions in PMMA are ester group π-flips coupled to chain backbone adjustments that extend over six to eight bonds on both sides of the carbon bearing the rotating ester group. Thus, the introduction of glutarimide cycles within the chain backbone can hinder this cooperativity and decrease the relative dissipation due to the MMA motions, as observed with CMI units (Fig. 148). However, according to this assumption, one should expect an effect increasing with the MGI content, whereas the opposite is observed both in mechanics (Fig. 141) and dielectrics (Fig. 147). Actually, if this hindrance effect occurs at low MGI contents (21 and 36%), it is counterbalanced by another process, which contributes more and more when the MGI content increases (58 and 76%) and leads to an additional dissipation both in mechanics and dielectrics. Considering the MGI content dependence of the effect, the more likely process would be MGI unit cooperative motions occurring in MGI sequences. Indeed, this hypothesis is supported by the fact that, as shown in Table 12, the probability of finding the MGI–MGI–MGI triad is much larger in MGIM58 and, particularly, in MGIM76 than in the other copolymers, which agrees with the observations. Thus, in MGIM76 a new type of cooperative motion takes place in the high-temperature part of the β transition.

8.3.4
^{13}C NMR Investigation

In the case of MGIMx copolymers, the NMR investigation has been performed both in solid-state and in solution [75].

8.3.4.1
^{13}C Solid-State NMR

^{13}C chemical shift anisotropy NMR measurements deal with frequencies of the order of or higher than 10^4 Hz and, consequently, the temperature range at which the motions involved in the β transition contribute to the NMR response are considerably shifted towards higher temperatures. The dielectric relaxation map of the β transition allows one to determine the following temperature ranges for the considered MGIMx copolymers:

MGIM21 30–130 °C
MGIM76 20–160 °C

At higher temperatures the processes observed correspond to the α transition.

As previously mentioned (Sect. 8.1.4.1) the rigid-lattice components of the chemical shift anisotropy tensor of the carboxyl group in PMMA are: $\sigma_{11} = 268$ ppm, $\sigma_{22} = 150$ ppm and $\sigma_{33} = 112$ ppm. An analysis of the MGIM76 $C = O$ spectrum, after subtraction of the $C = O(MMA)$ contribution, leads for the $C = O(MGIM)$ to the following values: $\sigma_{11} = 250$ ppm, $\sigma_{22} = 170$ ppm and $\sigma_{33} = 105$ ppm.

MGIM21 Copolymer

The chemical shift anisotropies of carboxyl groups of MGIM21 at 25, 90 and 115 °C are shown in Fig. 150. For comparison, the spectra of PMMA are also plotted.

The spectra of MGIM21 and PMMA are quite similar, whatever the temperature. In particular, in the chemical shift range from 200 to 270 ppm, there is no indication of motionless $C = O(MGI)$, although the copolymer contains 36 wt % of these groups. This indicates that, in the β transition region, the $C = O(MGI)$ groups undergo some motions.

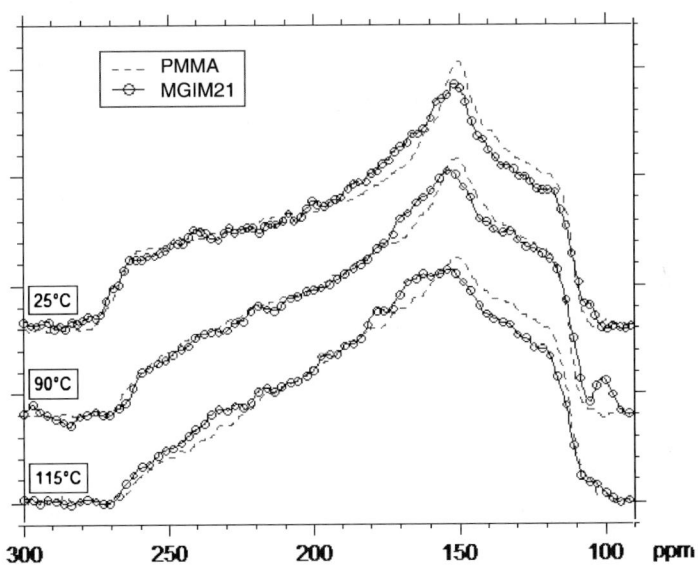

Fig. 150 Chemical shift anisotropy of $C = O$ groups for PMMA (*broken lines*) and MGIM21 (○) at different temperatures (from [75])

MGIM76 Copolymer

Figure 151 shows the $C = O$ chemical shift anisotropies of MGIM76 at several temperatures corresponding to the β transition. Furthermore the contribution of the $C = O(MGI)$ obtained by subtracting the 14 wt % of $C = O(MMA)$ is plotted. The spectrum relative to $C = O(MGI)$ does not change from 25 to 115 °C, which indicates that these groups do not perform any motion at 10^4 Hz in this temperature range. In contrast, at 160 °C, quite a different spectrum shape is observed. There is a decrease of intensity in the σ_{33} region, compensated by an intensity increase in the σ_{22} region. Such an effect could be attributed to a reorientation motion of the σ_{33} axis (Fig. 152), mainly in

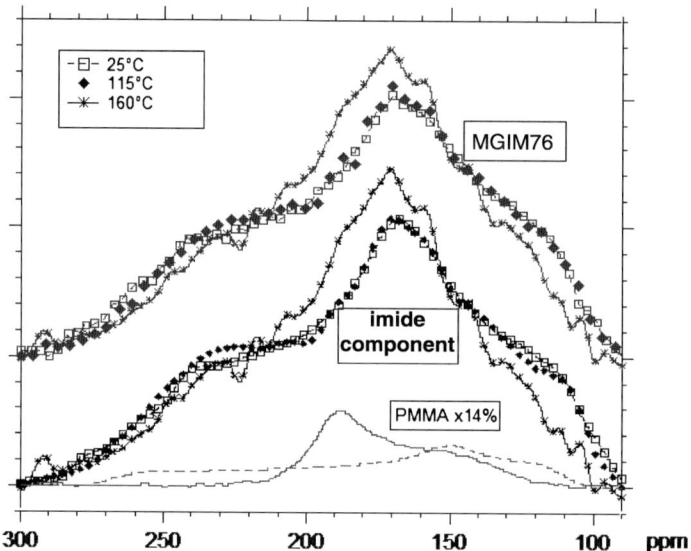

Fig. 151 Chemical shift spectrum of total $C = O$ groups and of the $C = O(MGIM)$ contribution for MGIM76 at the indicated temperatures (from [75])

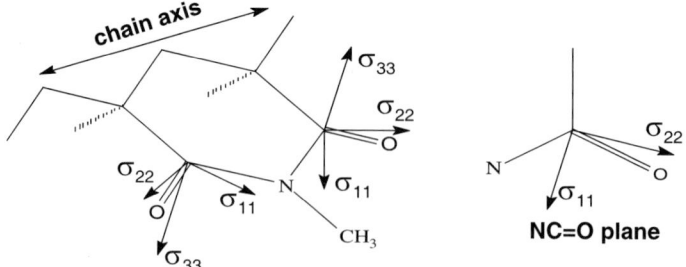

Fig. 152 Geometry of the chemical shift anisotropy tensors of $C = O$ groups in MGIM units

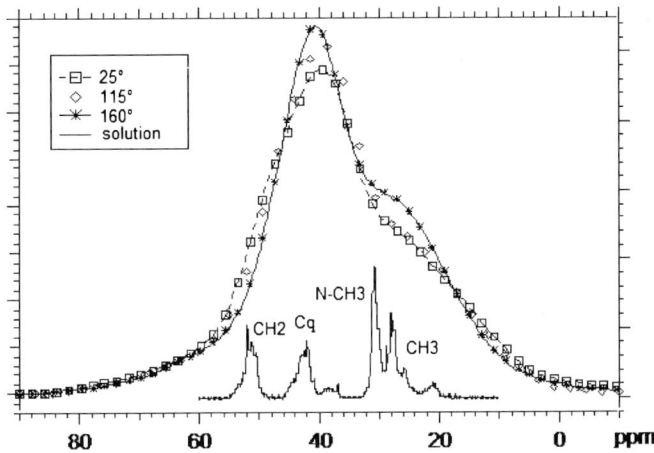

Fig. 153 Chemical shift spectrum of the aliphatic carbons of MGIM76 at the indicated temperatures and NMR spectrum of this copolymer in solution (from [75])

the plane perpendicular to the local chain axis. It is worth noticing that this reorientation motion happens in the highest temperature part of the β transition.

As regards the aliphatic carbons, the chemical shift spectra of MGIM76 at various temperatures are shown in Fig. 153, as well as the NMR spectrum obtained for MGIM76 in solution. As for C = O(MGI) groups, spectrum shape changes only occur between 115 and 160 °C. Two spectrum regions are concerned:

- α-CH$_3$ and N-CH$_3$ region (20–35 ppm)
- Quaternary carbons, C$_q$, and CH$_2$ region (40–55 ppm)

Concerning the α-CH$_3$ or N-CH$_3$ groups: the α-CH$_3$ undergoes a very fast rotation around its own C$_3$ axis and should not participate in the spectrum change above 115 °C. It is the same for the rotation of CH$_3$ groups around the C – N bond. However, in the latter case, a reorientation of the C$_3$ axis in a plane perpendicular to the local chain axis (Fig. 152) can lead to the narrowing observed at 160 °C in this spectrum region. Finally, the narrowing occurring at 160 °C in the CH$_2$ and C$_q$ region comes from a reorientation motion of the C$_q$ – CH$_2$ – C$_q$ part of the glutarimide cycle (Fig. 152).

Thus, in the case of MGIM76 copolymers, the solid-state NMR investigation indicates the existence, in the highest temperature part of the β transition, of a motion of the MGI units involving both the aliphatic and carboxyl groups, which could be assigned to an intracycle conformational change. It is likely that such a conformational change, performed within a MGI sequence, involves an adjustment of the nearest-neighbour MGI units, leading to some cooperativity.

8.3.4.2
¹³C NMR of MGIMx in Solution

The chain dynamics of the MGIMx copolymers has been studied in deuterated chloroform solutions through spin-lattice relaxation time, T_1, or more precisely through the $T_1/T_{1\,min}$ ratio, in order to avoid the contribution of libration motions (Sect. 8.2.3.2). The temperature range considered extends from – 50 to 50 °C.

In the case of MGIM21, the results are identical to those of PMMA over the whole temperature range.

As regards the $CH_2(MMA)$, in MMA sequences, the results, plotted in Fig. 154, show that identical dynamics are observed for PMMA, MGIM36 and MGIM58.

For the latter copolymer, in which MMA units are frequently next to MGI units, this result means that the internal rotation changes associated with a conformational jump involving a MMA unit occur similarly for either MMA or MGI backbones.

Concerning the $CH_2(MGI)$ dynamics, it appears to be the same for MGIM36, MGIM58 and MGIM76 copolymers.

Another interesting feature deals with the comparison of the dynamics of $CH_2(MMA)$ and $CH_2(MGI)$ in MGIM58 copolymer. The temperature dependence of the $T_1/T_{1\,min}$ ratio for each CH_2 is shown in Fig. 155. It is clear that for $CH_2(MGI)$ the minimum occurs at a higher temperature, meaning that

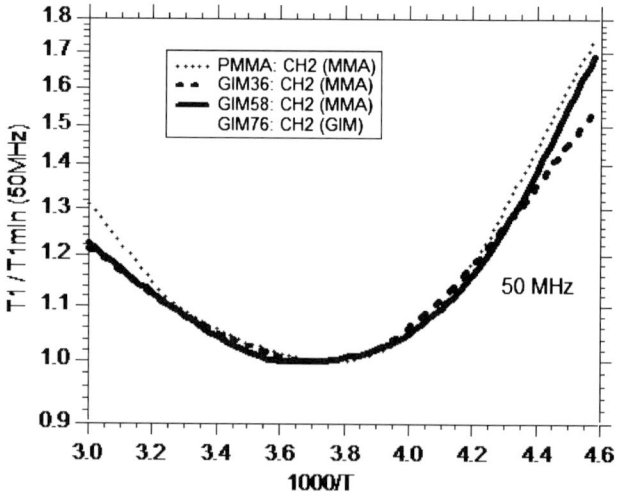

Fig. 154 Temperature dependence of the $T_1/T_{1\,min}$ ratio relative to $CH_2(MMA)$, for various MGIMx copolymers and PMMA (for clarity, only the interpolated curves are presented) (from [75])

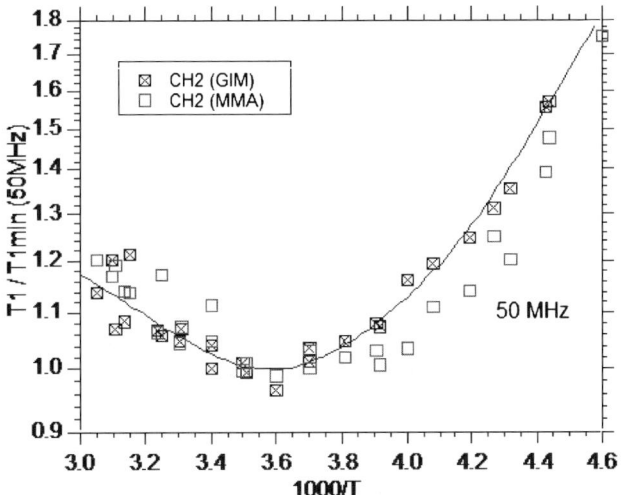

Fig. 155 Temperature dependence of the $T_1/T_{1\,min}$ ratio for $CH_2(MMA)$ and $CH_2(MGIM)$ in MGIM58 copolymer (from [75])

the $CH_2(MGI)$ motions are slower than the $CH_2(MMA)$ ones within the same MGIM58 chain backbone.

8.3.5
Atomistic Modelling

A molecular modelling approach to the ester group of MMA units, using the quasi-static dynamics, as already described for PMMA (Sect. 8.1.5) and CMIM20 (Sect. 8.2.4), has been applied to MGIMx copolymers [78].

Two MGIMx copolymers have been considered, i.e. MGIM21 and MGIM58. However, as the results obtained for MGIM21 are very close to the ones for PMMA, only those concerning MGIM58 will be presented.

8.3.5.1
Modelling Characteristics for MGIM58

The amorphous glassy cell of MGIM58 is constructed according to the procedure described in Sect. 8.2.4.1 for CMIM20; it contains 168 MMA units.

The quasi-static dynamics is applied to the π-flip of MMA ester groups. The MGI units are not submitted to forced rotations, but rotations around the local chain axis can be induced by the π-flip of an ester group born by a neighbouring MMA unit.

The density of the cell at $50\,°C$ is $1160\,kg\,m^{-3}$, close to the experimental value of $1190\,kg\,m^{-3}$, which is slightly higher than that of PMMA ($1180\,kg\,m^{-3}$). As regards the local density, whatever the size of the cell from

7.5 Å to 15.5 Å, it is higher than for PMMA and the density distribution is narrower.

8.3.5.2
Activation Energies

The mean activation energy for MGIM58 is 60 kJ mol^{-1}, significantly higher than for PMMA (58 kJ mol^{-1}). Concerning the intra- or intermolecular origin of the activation energy, Table 13 shows, by comparison with Table 11 data dealing with PMMA at 50 °C, that there is a higher contribution of the surrounding and that the inter- and intramolecular energy changes related to the surrounding are twice those for PMMA.

The quite large negative value of the intermolecular energy change of the surrounding between the equilibrium state and the transition state indicates a better local packing in the latter polymer. In contrast, a large positive value is found for the intramolecular energy change of the surrounding chains, which corresponds to a larger chain distortion in the transition state, compared to PMMA.

It is interesting to examine the effect of the MMA or MGI nature of the nearest neighbours on the activation energy, E_a, of a rotating ester. In a MMA sequence, E_a is equal to 58.8 kJ mol^{-1}. When the involved MMA is next to a MGI unit, the value decreases to 57.2 kJ mol^{-1}, whereas when it has a MGI unit on each side, E_a increases to 62 kJ mol^{-1}.

Table 13 Mean activation energy and average energy changes (in kJ mol^{-1}) of the various components for MGIM58 at 50 °C (from [78])

Mean activation energy	Intermolecular energy change of the chain with the rotating ester	Energy change of the surrounding
59.9	57.6	2.3
Intramolecular energy change of bonded atoms	Intramolecular energy change of non-bonded atoms	
29.2	28.4	
Intermolecular energy change	Intramolecular energy change of other chains	
– 11.4	13.7	

8.3.5.3
Chain Adjustment to Ester Group Rotation

The ester group π-flips in MGIM58 lead to average changes in the internal rotation angles of the chain backbone C – C bonds that are very close to those obtained in PMMA, as shown in Fig. 156. The average total angular changes are 60.8° for MGIM58 and 58.5° for PMMA. Similar results are found when considering the sites that are active in dielectric relaxation at 50°. Thus, the presence of glutarimide cycles does not hinder the chain backbone adjustment and, consequently, the cooperativity which exists in PMMA. Therefore, the glutarimide cycle is able to undergo distortions of a few degrees (2– 6°) of the $C_q – CH_2 – C_q$ bonds, unlike the rigid imide group.

Concerning the effect of the distance to a MGI unit on the chain adjustment, it appears that, with a MGI as nearest neighbour of the rotating ester, the total angular change is 59.9° and with a MGI unit on each side it rises to 61.9°. This effect is directly reflected in the activation energies.

Finally, it is worth noting that the behaviour observed for MGIM58 is quite different from that for CMIM20 (Sect. 8.2.4.3). The MGI unit takes part in the chain adjustment, whereas the CMI unit hinders it.

8.3.6
Conclusion on the Motions Involved
in the β Transition of MGIMx Copolymers

In the case of MGIMx copolymers, the motions involved in the β transition are more complex than for CMIMx copolymers.

Indeed, in the low-temperature part, MMA ester group π-flips coupled with MGI distortions arise in sites of lower local density. This coupling in-

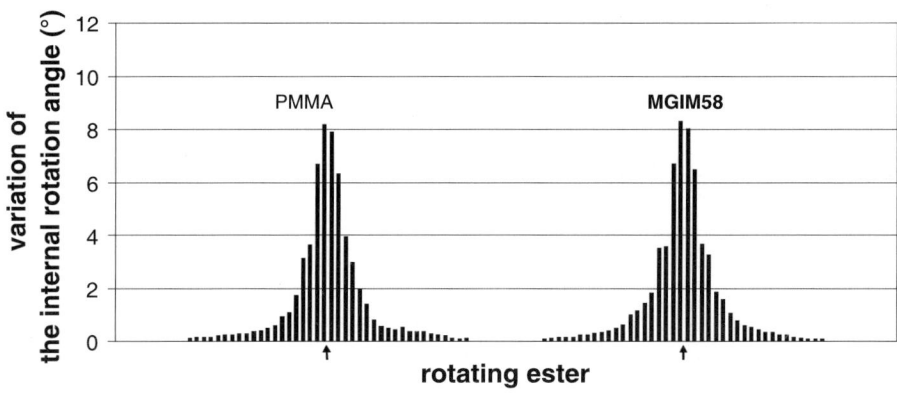

Fig. 156 Internal rotation angle changes of the main-chain C – C bonds associated with the ester group flips occurring in MGIM58 and PMMA at 50 °C (from [78])

duces a higher mechanical loss and also induces a slowing down of the π-flips.

In the high-temperature part, two processes occur:

1. MMA ester group π-flips, associated with internal rotation angle changes of the backbone C – C bonds, including both the MMA backbone bonds and the C_q – CH_2 – C_q sequence of the glutarimide cycles
2. In the highest temperature range, reorientation of the glutarimide cycle around the local chain axis occurring within a MGI sequence, a reorientation that is associated with some MGI–MGI cooperativity, as in MGI-rich copolymers

8.4
Conclusion on the β Transition of PMMA
and its Maleimide and Glutarimide Copolymers

By combining both experimental techniques and atomistic modelling, the investigation into the motions involved in the β transition of PMMA, CMIMx and MGIMx copolymers illustrates the detailed description that can be achieved (as for bisphenol A polycarbonate in Sect. 5).

In particular, the intramolecular cooperativity considered from experimental data on PMMA and its copolymers, is well supported by atomistic modelling results. The ester group π-flips, coupled with main-chain adjustment of torsion angles of backbone C – C bonds in the case of PMMA, remain isolated in the CMIMx copolymers. Indeed, the rigidity of the maleimide cycle hinders the propagation of main-chain adjustment. For the MGIMx copolymers, the more easily deformable glutarimide cycle allows the propagation of main-chain adjustment, although with a lower amplitude. Furthermore, for MGI-rich copolymers, the glutarimide cycle motions develop an intramolecular cooperativity between MGI units.

It is worth noting that the cooperativity of β transition motions, observed in PMMA and MGIMx copolymers, is intramolecular, unlike the case of bisphenol A polycarbonate where intra- and intermolecular cooperativities exist. The consequences of such a difference on the mechanical properties (in particular the strain softening) of these two types of polymers are investigated in a second paper [1].

9
Conclusion

In this paper, the investigation of solid-state transitions has been performed on quite a large series of amorphous polymers. Various types of chemical structures have been considered:

– Polymers without side-chains, containing phenyl rings, either linear like poly(ethylene *tere*-phthalate), bisphenol A polycarbonate and aryl-aliphatic copolyamides, or crosslinked as epoxy resins. For the linear polymers, the phenyl ring π-flips, occurring in the high-temperature part of the mechanical β transition, are associated with intramolecular cooperativity and, for polycarbonate, also with intermolecular cooperativity. In this type of polymer, small molecule antiplasticisers exist and comparison of pure and antiplasticised polymers provides a powerful way of analysing the nature of the cooperativity involved.

– Polymers with side-chains, like poly(methyl methacrylate) and its male-imide and glutarimide copolymers. The ester group π-flips, resulting in the β transition of these polymers, are isolated in the low-temperature part of the mechanical β peak, whereas, in the high-temperature part, they have an intramolecular cooperativity with main-chain adjustments. This cooperativity is hindered by the rigid maleimide cycle, but remains active with the more flexible glutarimide cycle. It is worth noting that small molecule antiplasticisers have never been reported for this type of polymer.

The approach developed in this paper, combining on the one side experimental techniques (dynamic mechanical analysis, dielectric relaxation, solid-state ^{1}H, ^{2}H and ^{13}C NMR on nuclei at natural abundance or through specific labelling), and on the other side atomistic modelling, allows one to reach quite a detailed description of the motions involved in the solid-state transitions of amorphous polymers. Bisphenol A polycarbonate, poly(methyl methacrylate) and its maleimide and glutarimide copolymers give perfect illustrations of the level of detail that can be achieved.

The investigation reported in this paper shows that, in spite of the detailed information provided by solid-state NMR or atomistic modelling, combination with mechanical and dielectric measurements is necessary.

The consequences of secondary transition motions on the mechanical properties (deformation, yield and fracture) of amorphous polymers are addressed in a second paper in this volume [1].

Acknowledgements The authors are greatly indebted to the Ph.D. students D. Bauchiere, F. Beaume, B. Bordes, B. Brule, S. Cukierman, L. Heux, B. Lousteau, L. Teze and P. Tord-jeman; to the post-docs P.L. Lee and A.S. Maxwell who have been involved in many of these studies; and to their colleagues S. Choe, G. Davies, J. Schaefer, U. Suter, J. Virlet, I.M. Ward for their fruitful collaboration on various topics. They thank C. Gaillet-Sulpice for her help in producing the figures given in this paper.

References

1. Monnerie L, Halary J-L, Kausch H-H (2005) Adv Polym Sci (this volume) Springer, Berlin Heidelberg New York
2. Starkweather HW (1981) Macromolecules 14:1277
3. Starkweather HW (1988) Macromolecules 21:1798
4. Starkweather HW (1991) Polymer 32:2443
5. Heijboer J (1972) PhD Thesis, University of Leyden
6. Lauprêtre F, Monnerie L, Virlet J (1984) Macromolecules 17:1397
7. Lauprêtre F, Virlet J, Bayle J-P (1985) Macromolecules 18:1846
8. McCrum NG, Read BE, Williams G (1967) Anelastic and dielectric effects in polymer solids. Wiley, London
9. English AD (1985) Macromolecules 17:2182
10. Allen RA, Ward IM (1992) Polymer 33:5191
11. Wilhem M, Spiess HW (1996) Macromolecules 29:1088
12. Maxwell AS, Ward IM, Lauprêtre F, Monnerie L (1998) Polymer 39:6835
13. Maxwell AS, Ward IM, Monnerie L (1998) Polymer 39:6851
14. Jackson WJ, Caldwell JR (1967) J Appl Polym Sci 11:211
15. Jackson WJ, Caldwell JR (1967) J Appl Polym Sci 11:227
16. Robeson LM, Faucher JA (1969) J Polym Sci Lett Ed 7:35
17. Fischer EW, Hellmann GP, Spiess HW, Wehrle M (1985) Macromol Chem Phys Suppl 12:189
18. Matsuoka S, Ishida Y (1966) J Polym Sci Part C 14:247
19. Katana G, Kremer F, Fischer EW, Plaetschke R (1993) Macromolecules 26:3075
20. Aoki Y, Brittain JO (1977) J Polym Sci Polym Phys 14:1297
21. Brûlé B (1999) PhD Thesis, Université Pierre et Marie Curie, Paris
22. Illers KH, Breuer H (1963) J Colloid Sci 18:31
23. Legrand DG, Erhardt PF (1969) J Appl Polym Sci 13:1707
24. Goldstein M (1972) In: Douglas RW, Ellis B (eds) Amorphous materials. Wiley, London, p 23
25. Yee AF, Smith SA (1981) Macromolecules 14:54
26. Jho JY, Yee AF (1991) Macromolecules 24:1905
27. Xiao C, Yee AF (1992) Macromolecules 25:6800
28. Kim CK, Aguilar-Vega M, Paul DR (1992) J Polym Sci Polym Phys 30:1131
29. Davenport RA, Manuel AJ (1977) Polymer 18:557
30. Ingelfield PT, Jones AA, Lubianez RP, O'Gara JF (1981) Macromolecules 14:288
31. O'Gara JF, Desjardins SG, Jones AA (1981) Macromolecules 14:64
32. O'Gara JF, Jones AA, Hung C-C, Ingelfield PT (1985) Macromolecules 18:1117
33. Williams G, Watts DC (1970) Trans Farad Soc 66:80
34. Roy AK, Jones AA, Ingelfield PT (1986) Macromolecules 19:1356
35. Schmidt C, Kuhn KJ, Spiess HW (1985) Progr Coll Polym Sci 71:71
36. Wehrle M, Hellmann GP, Spiess HW (1987) Coll Polym Sci 265:815
37. Spiess HW (1983) Coll Polym Sci 261:193
38. Schaefer D, Hansen M, Blümich B, Spiess HW (1991) J Non-crystalline Solids 131–133:777
39. Steger TR, Schaefer J, Stejskal EO, McKay RA (1980) Macromolecules 13:1127
40. Schaefer J, Stejskal EO, McKay RA, Dixon WT (1984) Macromolecules 17:1479
41. Ingelfield PT, Amici RM, O'Gara JF, Hung C-C, Jones AA (1983) Macromolecules 16:1552
42. Roy AK, Jones AA, Ingelfield PT (1985) J Magn Reson 64:441

43. Poliks MD, Gullion T, Schaefer J (1990) Macromolecules 23:2678
44. Walton JH, Lizak MJ, Conradi MS, Gullion T, Schaefer J (1990) Macromolecules 23:416
45. Henrichs PM, Linder M, Hewitt JM, Massa D, Isaacson H (1984) Macromolecules 17:2412
46. Hutnik M, Argon AS, Suter UW (1991) Macromolecules 24:5956
47. Laskowski BC, Yoon DY, McLean D, Jaffe RL (1988) Macromolecules 21:1629
48. Bicerano J, Clark HA (1988) Macromolecules 21:585
49. Bicerano J, Clark HA (1988) Macromolecules 21:597
50. Theodorou DN, Suter UW (1985) Macromolecules 18:1467
51. Hutnik M, Argon AS, Suter UW (1991) Macromolecules 24:5970
52. Shih JH, Chen CL (1995) Macromolecules 28:4509
53. Petrie SEB, Moore RS, Flick JR (1972) J Appl Phys 43:4318
54. Wyzgoski MG, Yeh GSY (1973) Polymer 4:29
55. Jones AA (1985) Macromolecules 18:902
56. Schaefer J, Stejskal EO, Perchak D, Skolnick J, Yaris R (1985) Macromolecules 18:368
57. Beaume F, Lauprêtre F, Monnerie L, Maxwell A, Davies GR (2000) Polymer 41:2677
58. Beaume F, Lauprêtre F, Monnerie L (2000) Polymer 41:2989
59. Garin N, Hirschinger J, Beaume F, Lauprêtre F (2000) Polymer 41:4281
60. Beaume F, Brulé B, Halary J-L, Lauprêtre F, Monnerie L (2000) Polymer 41:5451
61. Choe S, Brulé B, Bisconti L, Halary J-L, Monnerie L (1999) J Polym Sci B Polym Phys 37:113
62. Cukierman S, Halary J-L, Monnerie L (1991) Polym Eng Sci 31:1476
63. Heux L, Halary J-L, Lauprêtre F, Monnerie L (1997) Polymer 38:1767
64. Merritt ME, Heux L, Halary J-L, Schaefer J (1997) Macromolecules 30:6760
65. Ochi M, Okasaki M, Shimbo M (1982) J Polym Sci B Polym Phys 20:689
66. Garroway AN, Ritchey MR, Moniz WB (1982) Macromolecules 15:1051
67. Shi JF, Ingelfield PT, Jones AA, Meadows MD (1996) Macromolecules 29:605
68. Halary J-L, Bauchiere D, Lee PL, Monnerie L (1997) Polymery (Poland) 42:86
69. Daly J, Britten A, Garton A (1984) J Appl Polym Sci 29:1403
70. Heux L, Lauprêtre F, Halary J-L, Monnerie L (1998) Polymer 39:1269
71. Merritt ME, Goetz JM, Whitney D, Chang-Po PC, Heux L, Halary J-L, Schaefer J (1998) Macromolecules 31:1214
72. Muzeau E, Perez J, Johari GP (1991) Macromolecules 24:4713
73. Heijboer J (1960) Makromol Chem 35A:86
74. Heijboer J (1977) Int J Polym Mater 6:11
75. Bordes B (1999) PhD Thesis, Université Pierre et Marie Curie, Paris
76. Gomez Ribelles JL, Diaz Calleja R (1985) J Polym Sci Polym Phys Ed 23:1297
77. Schmidt-Rohr K, Kulik AS, Beckham HW, Ohlemacher A, Pawelzik U, Boeffel C, Spiess HW (1994) Macromolecules 27:4733
78. Lousteau B (2001) PhD Thesis, Université Pierre et Marie Curie, Paris
79. Tordjeman P, Teze L, Halary J-L, Monnerie L (1997) Polym Eng Sci 37:1621
80. Teze L, Halary J-L, Monnerie L, Canova L (1999) Polymer 40:971

Adv Polym Sci (2005) 187: 215–364
DOI 10.1007/b136957
© Springer-Verlag Berlin Heidelberg 2005
Published online: 13 October 2005

Deformation, Yield and Fracture of Amorphous Polymers: Relation to the Secondary Transitions

Lucien Monnerie[1] (✉) · Jean Louis Halary[1] · Hans-Henning Kausch[2]

[1]Laboratoire PCSM (UMR 7615), Ecole Supérieure de Physique et de Chimie Industrielles de la Ville de Paris, 10 rue Vauquelin, 75231 Paris cedex 05, France
l.monnerie@noos.fr, jean-louis.halary@espci.fr

[2]Ecole Polytechnique Fédérale de Lausanne, Station 6, 1015 Lausanne, Switzerland
kausch.cully@bluewin.ch

Abstract This paper deals with the mechanical properties (plastic deformation, micromechanism of deformation, fracture) of various amorphous polymers: poly(methyl
methacrylate) and its maleimide and glutarimide copolymers, bisphenol A (and/or
tetramethyl bisphenol A) polycarbonate, and aryl-aliphatic copolyamides. First, the required background on molecular characteristics, micromechanisms of deformation and
fracture characterisation is recalled. Then, the results are discussed for each polymer series, considering information obtained on the motions involved in secondary transitions
(mostly β transitions) analysed in a previous paper. The importance of the cooperative motions occurring in the high temperature part of the transition is unambiguously pointed out. Furthermore, in glutarimide copolymers, as well as in aryl-aliphatic
copolyamides, it is shown that the dependence of toughness on the entanglement density fails and only the consideration of cooperative β transition motions can consistently
account for the results. Finally, concerning the change of strain softening with increasing
temperature, two opposite behaviours are observed for polymers in which β transitions
result, on the one hand from side-group motions (strain softening decreases) and on
the other hand from certain main-chain units (strain softening increases). Two different mechanisms have been proposed, based on a softening of the polymer medium by
β transition motions associated with either an intramolecular cooperativity (in the first

case), or an intermolecular cooperativity (in the second case) of these motions. Thus, combination of the results of the analysis of the relation of chemical structure to β transition motions with the present conclusions yields a molecular description of the whole set of behaviours involved in the mechanical properties of amorphous polymers, till fracture.

Keywords Mechanical properties · Plastic deformation · Poly(methyl methacrylate) · Methyl methacrylate-co-N-cyclohexyl maleimide copolymer · Methyl methacrylate-co-N-methyl glutarimide copolymer · Bisphenol A polycarbonate · Tetramethyl bisphenol A polycarbonate · Aryl-aliphatic copolyamides

Abbreviations

$°C$	Degrees Celsius
B	Specimen thickness
C	Compliance
C_∞	Characteristic ratio
E	Young's modulus
E''	Mechanical loss modulus
E_a	Activation energy
G'	Shear modulus
G''	Mechanical loss shear modulus
G_{Ic}	Critical strain-energy release
$G_{N°}$	Rubbery plateau modulus in viscoelastic measurements
Hz	Hertz
I	Polydispersity index
K	Degrees Kelvin
K_I	Stress intensity factor in mode I
K_{Ic}	Critical stress intensity factor in mode I
K'_{Ic}	Critical stress intensity factor in mode I and plane strain conditions
K''_{Ic}	Critical stress intensity factor in mode I and plane stress conditions
M	Molar mass of a polymer chain
M_0	Molar mass of a polymer repeat unit
M_e	MW between entanglements
M_n	Number average MW
M_w	Weight average MW
N_A	Avogadro number
N_e	Number of bonds between entanglements
$N_{eq,e}$	Number of equivalent bonds between entanglements
P	Load
P_{max}	Maximum load
R	Perfect gas constant
T	Absolute temperature
T_{12}	Temperature of CSC–SDZ transition
T_g	Glass–rubber transition temperature
T_α	Temperature at which the mechanical loss E'' or G'' goes through a maximum
U	Energy required to break a covalent bond
V_0	Activation volume
V_p	Pressure activation volume
W	Specimen length

Y	Correction factor for configuration of specimen and crack geometry
a	Crack length
j_{eq}	Number of equivalent bonds per repeat unit
l_{eq}	Length of an equivalent bond
m_{eq}	Molar mass of an equivalent repeat unit
n_{eq}	Number of bonds of an equivalent chain
n_v	Number of bonds per polymer repeat unit
r	Polar coordinate distance
r_0	Specimen thickness over which occurs the transition from K_{Ic}'' to K_{Ic}'
$\langle r_0^2 \rangle$	Mean squared end-to-end distance of unperturbed chain
$\langle r_e^2 \rangle$	Mean squared end-to-end distance of a chain of molar mass M_e
v	Velocity of the void surface between the fibrils in a craze
Δ	Displacement
$\langle \Delta \phi_i \rangle$	Average change of the torsion angle of bond i
Γ	Surface energy within a craze
ε	Strain
$\dot{\varepsilon}$	Strain rate
ε_y	Yield strain
γ	Van der Waals surface energy
$\eta'(\omega)$	Newtonian viscosity measured at frequency ω
λ_{max}	Maximum extension ratio of a single chain of mass M_e
λ_{net}	Maximum extension ratio of a chain strand of molar mass M_e in an entanglement network
ν	Poisson ratio
ν_e	Entanglement density
θ	Polar coordinate angle
ρ	Polymer density
σ	Stress
σ_{CDC}	Characteristic stress associated with chain disentanglement craze
σ_{CSC}	Characteristic stress associated with chain scission craze
σ_{el}	Limit of purely elastic behaviour
σ_{pf}	Plastic flow stress
σ_{SDZ}	Characteristic stress associated with shear deformation zone
σ_y	Yield stress
ζ	Monomeric friction coefficient

Acronyms

Ar-Al-PA	Aryl-aliphatic copolyamide
BPA-PC	Bisphenol A polycarbonate
CDC	Chain disentanglement craze
CMI	N-Cyclohexyl maleimide unit
CMIMx	Methyl methacrylate-co-N-cyclohexyl maleimide copolymer containing x Molar percent of the latter monomer
CSC	Chain scission craze
CT	Compact-tension test
FTIR	Fourier transform infra-red spectroscopy
MGI	N-Methyl glutarimide unit
MGIMx	Methyl methacrylate-co-N-methyl glutarimide copolymer containing x molar percent of the latter monomer

MT$_y$I$_{1-y}$ Copolyamide whose repeat unit contains 1,5 diamino-2-methyl pentane, y and $1 - y$ *tere-* and *iso*-phthalic moieties
MW   Molecular weight of a polymer
nSSA   Normalised strain softening
PMMA   Poly(methyl methacrylate)
SDZ   Shear deformation zone
SSA   Strain softening
TEM   Transmission electronic microscopy
TMBPA-PC   Tetramethyl bisphenol A polycarbonate
xT$_y$I$_{1-y}$ Copolyamide whose repeat unit contains x lactam-12 sequences, y and $1 - y$ *tere-* and *iso*-phthalic moities, respectively, and a 3,3′-dimethyl cyclohexyl methane unit in the regular order

1
Introduction

In another paper in this issue [1], the molecular motions involved in secondary transitions of many amorphous polymers of quite different chemical structures have been analysed in detail by using a large set of experimental techniques (dynamic mechanical measurements, dielectric relaxation, ^{1}H, ^{2}H and ^{13}C solid state NMR), as well as atomistic modelling.

The purpose of this paper is to investigate the mechanical properties (plastic deformation, micromechanisms of deformation, fracture) of several amorphous polymers considered in [1], i.e. poly(methyl methacrylate) and its maleimide and glutarimide copolymers, bisphenol A polycarbonate, aryl-aliphatic copolyamides. Then to analyse, in each polymer series, the effect of chemical structure on mechanical properties and, finally, to relate the latter to the motions involved in the secondary transitions identified in [1] (in most cases, the β transition).

To achieve this goal, the developed approach is based on a gradual change of chemical structure in a given polymer series; an approach similar to that used in [1].

It is worth noting that in addition to the previously published results, some of which have been re-treated, unpublished results have also been included, leading in several systems to new interpretations.

2
Background

Before analysing the mechanical behaviour of amorphous polymers, it is useful to briefly give information on their molecular characteristics, the main descriptors used for plastic deformation and fracture, the micromechanisms of deformation, and some of the experimental procedures.

More details about these various features can be found in classic textbooks, such as [2–5].

2.1
Molecular Characteristics

When considering amorphous polymers, an essential characteristic for mechanical properties concerns the glass–rubber transition temperature. It is usual to denote as T_g the temperature of occurrence of the glass–rubber transition determined by techniques which operate on samples at rest (thermal measurements, spectroscopic techniques, dielectric relaxation). When dealing with techniques involving a mechanical solicitation, like dynamic mechanical measurements, the temperature corresponding to the glass–rubber transition is denoted as T_α. In the case of the dynamic mechanical technique, T_α is chosen as the temperature at which the mechanical loss, E'' or G'', goes through a maximum. Thus, T_α measured at 1 Hz is shifted upwards by 10 K relative to T_g measured by differential scanning calorimetry at a heating rate of 20 K min^{-1}.

Another important characteristic of a polymeric material deals with the molecular weight (MW), more precisely its number and weight averages, M_n and M_w, respectively and, in order to get an estimate of the broadness of the MW distribution, the polydispersity index $I = M_w/M_n$.

For many mechanical behaviours, like crazing or fracture, the MW, per se, is not the pertinent characteristic. Indeed, long-enough polymer chains in the melt or in solid state are entangled. The MW between entanglements, M_e, is quite an important molecular characteristic since any behaviour involving displacement of the whole chain will be controlled by the number of entanglements per chain M_w/M_e. The M_e value is strongly dependent on the chemical structure of the polymer chain and it varies from about 2000 g mol^{-1} for polyethylene to around 18 000 g mol^{-1} for polystyrene. Thus, for each considered polymer chemical structure, M_e has to be determined from melt viscoelasticity measurements performed as a function of temperature above T_α, at an angular frequency, ω, ranging usually from 0.01 to 100 rad s^{-1}. The rubbery plateau modulus, G_N^0, is taken to be the value of G' at the frequency corresponding to the minimum of tan δ in the plateau region. Value of M_e is derived from G_N^0 using:

$$G_N^0 = \rho R T M_e^{-1} \qquad (1)$$

where ρ is the polymer density at the absolute temperature T and R is the gas constant.

Instead of M_e, a frequently used characteristic is the entanglement density, ν_e, given by:

$$\nu_e = \rho N_A / M_e \qquad (2)$$

where N_A is the Avogadro number.

The number of bonds between entanglements in a given chain is easy to calculate for simple ethylenic chains like polyethylene, polystyrene, poly(methyl methacrylate). However, some difficulties arise for polymers containing, in their main chains, groups like conjugated amides, phenyl rings, maleimide, or glutarimide units. Indeed, it requires first a determination of the virtual bonds which will replace those groups and around which rotations of other bonds can be performed [6], as illustrated in Fig. 1 for polypeptide chains. Some other examples are treated in Sects. 3.2.2.2 and 3.3.2.2. Then, the number of bonds between entanglements of the virtual chain, also called the equivalent chain (it is equivalent to the real chain from the point of view of conformations), $n_{e,eq}$, is calculated.

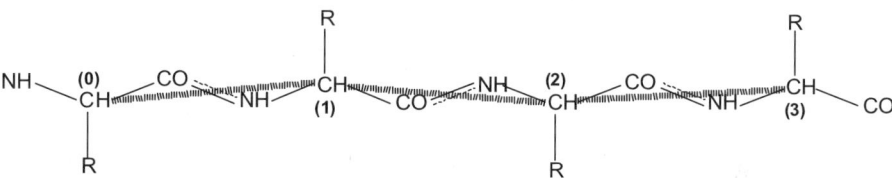

Fig. 1 Polypeptide chain. The partial double bond character of the amide bonds is indicated. Virtual bonds are shown by *dashed lines* connecting consecutive CHR groups (From [6])

Another important characteristic for describing the mean conformation of a polymer chain is the characteristic ratio, C_∞, defined as:

$$C_\infty = \left[\langle r_0^2 \rangle / n_{eq} l_{eq}^2 \right]_{n_{eq} \to \infty} \tag{3}$$

where $\langle r_0^2 \rangle$ is the mean squared end-to-end distance of the unperturbed real chain, and n_{eq} is the number of bonds of the equivalent chain, each of them getting a mean length l_{eq}. Experimental determination of C_∞ requires solution measurements of $\langle r_0^2 \rangle$ for narrow dispersed samples, which is not an easy task. To avoid such difficulties, an empirical relationship has been proposed [7]:

$$C_\infty = \left(n_{e,eq} M_e / 3 M_0 \right)^{1/2} \tag{4}$$

where M_0 is the molar mass of the polymer repeat unit. This relationship leads to satisfactory estimates.

In any circumstance where the whole displacement of a polymer chain is involved, it is important to characterise friction arising from the surroundings. This is achieved through the monomeric friction coefficient, ζ, which can be determined from melt viscoelasticity measurement of the Newtonian viscosity, corresponding to the low frequency plateau in $\eta'(\omega)$. ζ is calculated from:

$$\zeta = \eta_0 M_0^2 M_e^{5/2} / 0.0027 \rho j_{eq} l_{eq}^2 C_\infty M_w^{7/2} N_A \qquad (5)$$

where j_{eq} is the number of equivalent bonds per repeat unit [8].

2.2
Stress–Strain Curves

Typical true stress–strain curves for amorphous polymers are shown in Fig. 2.

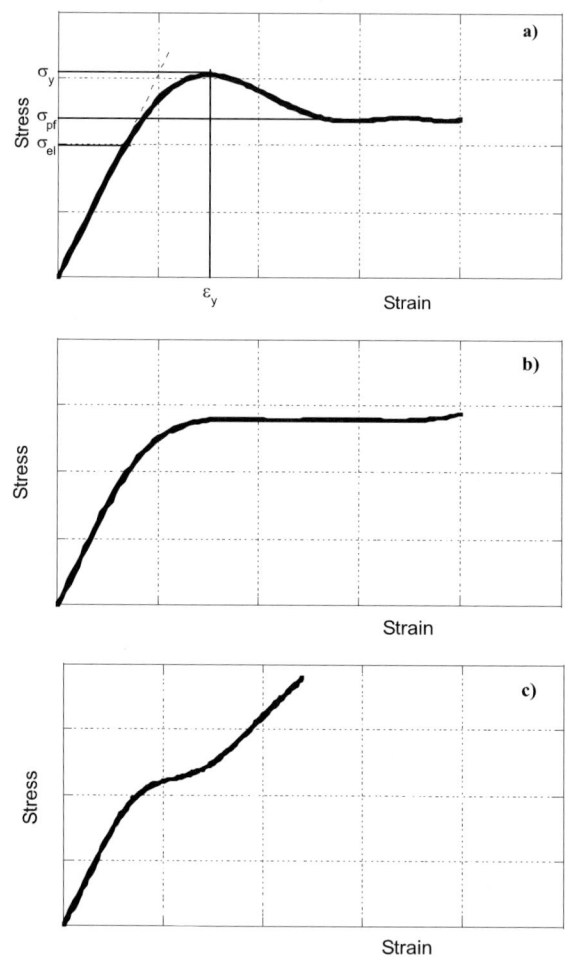

Fig. 2 Typical stress–strain curves for amorphous polymers. **a** Elastic, anelastic, strain softening, and plastic flow regions can be seen. **b** Plastic flow occurs at the same stress level as required for yielding so strain softening does not exist. **c** Strain hardening occurs very close to yielding, suppressing both strain softening and plastic flow behaviour

It is worth pointing out that the most reliable stress–strain curves are obtained by uniaxial compression or shear, in order to avoid the craze formation that can occur in tensile measurements.

Let us first examine the behaviour shown in Fig. 2a. The initial part is linear and perfectly reversible till a true stress value denoted as σ_{el}. It corresponds to purely elastic behaviour.

For stress value higher than σ_{el}, a slight departure from linearity is observed, which is more pronounced when increasing the stress level. In this stress range, when unloading the specimen, it goes back to zero strain, but along a curve slightly lower than the initial one, due to some energy dissipation. The behaviour in this stress regime is called pre-plastic or anelastic.

Further on, the true stress reaches a maximum, denoted as σ_y, corresponding to the yield point. At this stress value, the specimen no longer goes back to zero strain, a permanent deformation remains, which can be removed only by heating the deformed specimen above T_α. It is worth noting that the permanent deformation obtained when reaching σ_y is much smaller than the strain at the yield point, ε_y.

Straining the specimen beyond ε_y, a decrease of true stress is observed; this is the strain softening behaviour.

At a certain level, denoted as σ_{pf}, the true stress stabilises and a plateau is obtained. It corresponds to the plastic flow behaviour, associated with an increase of strain under a constant stress, the plastic flow stress, σ_{pf}.

At higher strain, a new increase of stress occurs, the strain hardening behaviour, which end in specimen fracture.

Depending on the material and deformation conditions (strain rate, temperature) other stress–strain curve shapes can be observed (Fig. 2b and c). In Fig. 2b, the plastic flow occurs at the same stress level as that required for the yielding so the strain softening does not exist. In the case shown in Fig. 2c, the strain hardening happens very close to yielding, suppressing both strain softening and plastic flow behaviour.

2.2.1
Yield Stress

2.2.1.1
Eyring Equation

Increasing the strain rate, $\dot{\varepsilon}$, leads to an increase of the yield stress, σ_y.

By considering that the applied stress induces molecular flow, which can be treated along the lines of the Eyring viscosity theory, the strain rate, $\dot{\varepsilon}$, and yield stress, σ_y, are related by the so-called Eyring equation:

$$\dot{\varepsilon} = \dot{\varepsilon}_0/2 \exp\left[-\left(\frac{E_a - V_0\sigma_y}{RT}\right)\right] \tag{6}$$

or, equivalently:

$$\sigma_y/T = E_a/V_0 T + R/V_0 \ln\left(2\dot{\varepsilon}/\dot{\varepsilon}_0\right) \tag{7}$$

where V_0 is the activation volume, which is considered to represent the polymer volume, and E_a is the activation energy.

Thus, plotting σ_y/T versus ln(strain rate) should give a straight line from which V_0 can be determined.

In a similar way, the yield stress increases under applied hydrostatic pressure, P, leading to a modified Eyring equation:

$$\dot{\varepsilon} = \dot{\varepsilon}_0/2 \exp\left[-\left(\frac{E_a - V_0\sigma_y + V_P T}{RT}\right)\right] \tag{8}$$

where V_p is the pressure activation volume.

2.2.1.2
Molecular Approach

Concerning the molecular process associated with polymer yielding, a particularly interesting approach [9, 10] considers that the applied stress induces conformation changes of main-chain bonds. In average, at the yield point, the statistical weight of the high-energy conformation is increased relative to its value in the glassy state (as no more conformation changes occur below T_g, the statistical weights of the main-chain conformation in the glassy state correspond to their values at T_g). Thus, under the applied σ_y stress, the main chain reaches an average conformational state which corresponds to that which would be obtained by increasing the temperature above T_g, without any applied stress.

Such a theoretical model has been quite successfully checked by Fourier transform infra-red spectroscopy (FTIR) on polystyrene [11, 12]. Indeed, the FTIR spectrum of polystyrene contains vibrational lines associated with out-of-plane modes of the phenyl ring, whose position depends on the *gauche* or *trans* conformation of the main-chain C – C bonds. By recording FTIR spectra during sample stretching, it is possible to determine the amount of *gauche* or *trans* conformation as a function of strain.

In the case of atactic polystyrene [11], the *gauche* conformation has an energy 9 kJ mol^{-1} higher than the *trans* conformation. So, according to the theoretical model, the yield point should be associated with an increase in the amount of *gauche* conformation. This is precisely what is observed in Fig. 3, in which the stress–strain curve recorded at the same time is plotted as the dotted line. The decrease observed beyond yielding corresponds to the beginning of stretching-induced orientation (favouring the *trans* conformation). At the yield point, the amount of *gauche* conformation is equivalent to that measured at $T_g + 30$ K, whereas the theory leads to an estimate of 20–50 K

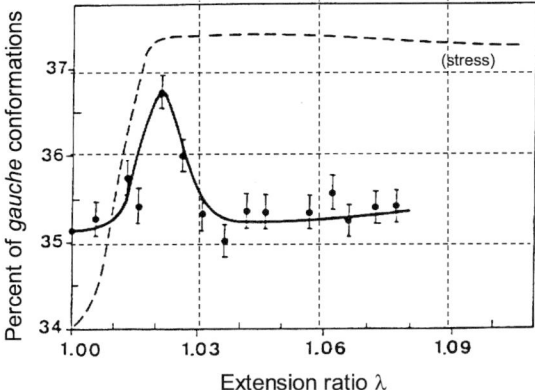

Fig. 3 Percentage of *gauche* conformations along the stress–strain curve of atactic poly-styrene. The *dashed curve* is the stress–strain curve (From [11])

above T_g. Such a result has also been found on plasticised polystyrene, as well as on poly(styrene-*co*-acrylonitrile) [12].

Quite interestingly, in the case of isotactic polystyrene, the *trans* conformation gets the higher energy, contrary to what happens for atactic polystyrene. The same type of FTIR experiment under stretching, performed on isotactic polystyrene [12], actually shows an increase in the amount of *trans* conformations.

Thus, the FTIR experiments on atactic and isotactic polystyrene unambiguously support the theoretical model, confirming that the occurrence of yielding is associated with an average increase of the higher-energy conformation, independently of whether it could be a *gauche* or a *trans* conformation.

These FTIR results under stretching are quite important for they represent the first experimental evidence that a precise molecular change is involved in the yielding process.

Using the molecular approach, the stress-induced conformational change of a single bond inside a polymer main chain raises a question. Indeed, changing from *gauche* to *trans*, or vice-versa, the conformation of a single C – C bond would imply rotating the remaining part of the chain. This is, of course, absolutely unlikely for a polymer chain in the glassy state, but also for a chain in solution. This feature has been recognised for quite a long time and a clear answer has been given by simulation of Brownian dynamics performed on a single polyethylene chain [13, 14]. It turns out that the single bond conformation change between *trans* and *gauche*, or vice-versa, is actually accommodated by the coordinated small-amplitude rearrangements of neighbouring units. This is illustrated in Fig. 4, taken from the results on a single polyethylene chain by using a different theoretical approach [15]. In Fig. 4a, the average change $\langle \Delta \phi_i \rangle$ of the torsion angles in response to 120°

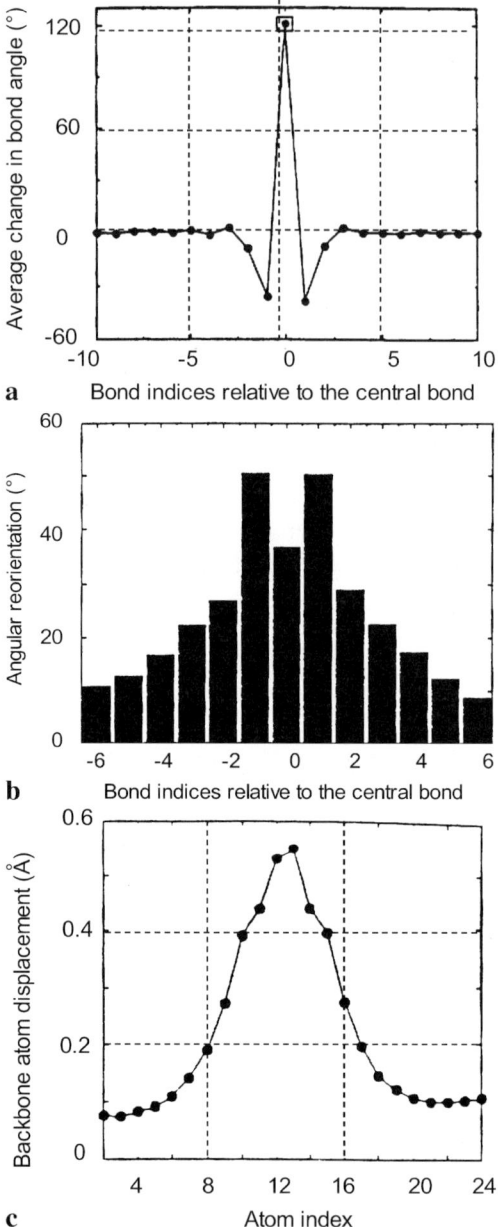

Fig. 4 Response to $120°$ rotation of the middle bond. **a** Average change, $\langle\Delta\phi_i\rangle$, in bond rotation angles as a function of location relative to the central bond. **b** Mean angular spatial reorientation of backbone bonds as a function of location relative to the central bond. **c** Mean spatial displacements of backbone atoms as a function of atom index along the chain for rotating chains of 25 bonds. The length of each bond is taken as 1 Å (From [15])

rotation of the middle bond as a function of bond location relative to the central bond, shows that counter-rotations of about 40° happen for the two first bonds on each side of the central rotating bond. The conformation change of the central bond also induces reorientations in space of the other neighbouring bonds (Fig. 4b), as well as spatial displacements of backbone atoms (Fig. 4c). From the latter two figures, it appears that the intramolecular rearrangements induced by a conformation change involve four bonds and four backbone atoms on each side of the concerned bond, with an amplitude gradually decreasing when bonds or atoms are situated farther away along the chain backbone.

2.2.2
Plastic Flow

In the plastic flow process, large displacements of the whole chain are performed to reach the involved strain values. Such displacements are analogous to those that happen during polymer flow above the glass–rubber transition (α transition) temperature. These motions are performed by segmental backbone conformation changes with intra- and intermolecular cooperativity specific of the α transition. They will be called α transition motions, in order distinguish them from the β transition motions analysed in [1].

Thus, for a given polymer chemical structure, the plastic flow corresponds to specific α transition motions, which are identical whatever the considered temperature. Consequently, the plastic flow, and the associated stress σ_{pf}, can be considered as constituting reference behaviour.

2.2.3
Strain Softening

The strain softening effect can be characterised by its amplitude, SSA, defined as:

$$SSA = \sigma_y - \sigma_{pf} \tag{9}$$

However, when considering temperature or strain dependencies of strain softening, the absolute stress difference ($\sigma_y - \sigma_{pf}$) does not allow one to discriminate between a change of the absolute values of σ_y and σ_{pf} and an intrinsic effect on the strain softening behaviour.

To overcome this ambiguity and focus on the intrinsic change of strain softening, another descriptor is more appropriate, consisting in normalising SSA by σ_{pf}:

$$nSSA = (\sigma_y - \sigma_{pf}) / \sigma_{pf} \tag{10}$$

Indeed, as above discussed, plastic flow stress serves as a reference for describing the temperature dependence observed with a given polymer.

Though SSA was used in our previous papers on this topic, hereafter the strain softening behaviour will be discussed in considering preferentially nSSA as descriptor.

2.3
Micromechanisms of Deformation

Depending on the polymer chemical structure and MW and on the deformation conditions (temperature and strain rate), two types of deformation heterogeneities are observed: *crazes* and *shear deformation zones*.

In spite of possible observations on bulk samples under strain, the most suitable technique for investigating the deformation mechanisms is the thin film technique [16]. Thin polymer films (thickness around 0.5 μm), prepared from a polymer solution spread on the surface of a water bath, are picked up on a well-annealed copper grid and bonded to the grid first by drying under vacuum at room temperature then by heating at $T_g + 15$ K for 30 s. The grid is strained in tension at the selected temperature and strain rate, monitoring the test under an optical microscope (sometimes with record on videotape). After straining, individual grid squares are observed by transmission electron microscopy (TEM).

2.3.1
Crazes

Many studies have been performed on crazes and their relation to fracture behaviour; most of them are reviewed in [17, 18].

2.3.1.1
Craze Morphology

Limited information will be given on the mechanisms involved.

Firstly, crazes always appear perpendicular to the stretching direction. A typical example of craze is shown in Fig. 5a. The craze is constituted of polymer fibrils of about 5–10 nm in diameter, separated by interfibrillar distances between 20 anA 60 nm. In the middle, a lighter zone is observed, called midrib.

The craze nucleation in bulk polymers results from nucleation of voids in the plane strain region of the sample to relieve the triaxial constraints.

The craze growth occurs by two different mechanisms [19, 20]:

1. At the craze tip, the advance mechanism would be by a Taylor meniscus instability leading to a series of void fingers occurring in the plastically deformed and strain-softened polymer formed at the craze tip. As the finger-like craze tip propagates, fibrils develop.

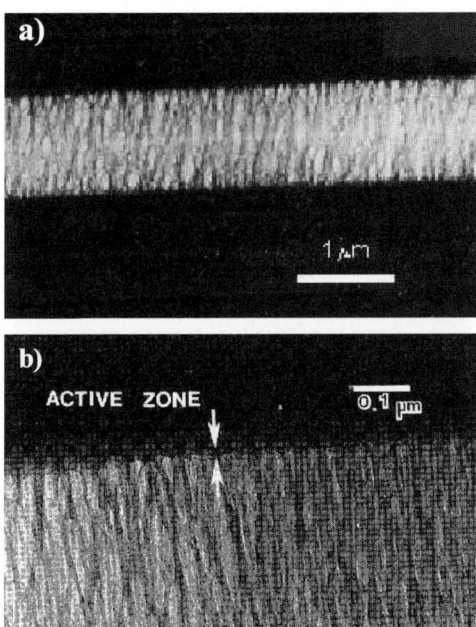

Fig. 5 TEM micrographs of thin films: **a** craze in MGIM76 at 0 °C (From [51]), **b** an active zone at the craze–bulk interface in polystyrene (From [21])

2. The craze thickening, associated with the craze growth, implies an increase of fibril length. This is achieved by pulling out polymer chains from the craze–bulk interface, according to a behaviour analogous to plastic flow within the active layer (5–10 nm thick), as shown in Fig. 5b [21].

When the craze propagates over a certain length, the fibril located in the central part (midrib) of the craze breaks, yielding a crack in the middle of the craze. Such a craze fibril breakdown also occurs in the craze ahead of a crack tip and results in a crack propagation. The broken down fibril parts retract on each crack surface and can be observed on fracture surfaces. The fibril breakdown mechanisms will be described later on in this section.

As regards fibril formation or growth, for polystyrene of $M_{\mathrm{w}} \sim 500\,000$ g mol^{-1}, a typical fibril spacing is in the range 10–20 nm, whereas the radius of gyration of a polymer chain is of the order of 50 nm, the distance between entanglements being around 5 nm. Furthermore, there are many different chains within a volume spanned by a radius of gyration. As a consequence, to generate the void-fibril structure of a craze, there is a geometrically necessary entanglement loss [19]. This loss can occur either by chain scission or by disentanglement. These two mechanisms lead to two types of crazes with similar morphologies but different dependencies on polymer characteristics. Each type of craze is considered hereafter.

2.3.1.2
Chain Scission Craze

Under conditions where chain mobility is very low, such as at low temperatures, the loss of entanglements occurs by chain scission; this is what happens for polystyrene air crazes. Of course, chain scission crazes (CSC) are MW-independent as soon as chains are long enough to be entangled (at too-low MWs, no craze can be formed, only cracks happen).

The characteristic stress associated with CSC, σ_{CSC}, may be quantitatively modelled [20], leading to:

$$\sigma_{CSC} \propto \left(\sigma_{pf}\Gamma\right)^{1/2} \tag{11}$$

where Γ is the surface energy within the craze.

For CSC, Γ takes the form:

$$\Gamma_{CSC} = \gamma + 0.25\nu_e\langle r_e^2\rangle^{1/2}U \tag{12}$$

where γ is the Van der Waals surface energy, U is the energy required to break a covalent bond in the polymer chain backbone (about 6×10^{-19} J), and $(\langle r_e^2\rangle)^{1/2}$ is the rms end-to-end distance of a chain whose molar mass equals the mass between entanglements M_e, i.e. $N_{eq,e}l_{eq}^2C_\infty$. Furthermore, γ is much smaller than the scission energy term, leading to the following relationship for the critical stress:

$$\sigma_{CSC} \propto \sigma_{pf}^{1/2}\nu_e^{1/2}\left(N_{eq,e}\right)^{1/4}l_{eq}^{1/2}U^{1/2} \tag{13}$$

Concerning the proportionality constant, it involves quantities difficult to determine experimentally, such as the thickness and rate of advance of the craze–bulk interface, the coefficients $\dot{\varepsilon}_f$ and n_e in the constitutive equation of non-Newtonian flow:

$$\sigma = \sigma_{pf}\left(\dot{\varepsilon}/\dot{\varepsilon}_f\right)^{1/n}e \tag{14}$$

where $\dot{\varepsilon}$ is the strain rate.

2.3.1.3
Chain Disentanglement Craze

Under conditions where chain mobility is high enough, typically at high temperature and low strain rate, the loss of entanglement in the active layer at the craze–bulk interface can occur by chain disentanglement, resulting in chain disentanglement craze (CDC).

As a direct consequence, the polymer MW enters in the chain disentanglement mechanism. More precisely, the relevant quantity is the ratio $C_\infty(M/M_e)$, which takes into account M_e of the considered polymer.

The characteristic stress associated with CDC, σ_{CDC}, is still modelled by:

$$\sigma_{CDC} \propto \left(\sigma_{pf}\Gamma\right)^{1/2} \tag{15}$$

However, in the case of CDC, the surface energy Γ takes the form:

$$\Gamma = \gamma + W(M, \zeta) \tag{16}$$

where $W(M, \zeta)$ is the energy required to disentangle the polymer chains in the active zone. $W(M, \zeta)$ mainly depends on the ratio (M/M_e) and the monomeric friction coefficient ζ.

For high entangled chains, two expressions have been proposed, that of Kramer and Berger [20]:

$$[W(M, \zeta)]_K = n_e \left(\langle r_e^2 \rangle\right)^{1/2} l_{eq} \nu \zeta M^2 / 48 M_e M_0 \tag{17}$$

and that of McLeish [22]:

$$[W(M, \zeta)]_{ML} = n_e \left(\langle r_e^2 \rangle\right)^{1/2} \left(\dot{\varepsilon}/\dot{\varepsilon}_p\right) \zeta \left(M/M_e\right)^{3/2} \tag{18}$$

where, in addition to the above defined quantities, ν is the interface velocity, i.e. the velocity of the void surface between the fibrils.

Regardless of the chosen model, the Van der Waals surface energy, γ, can be neglected compared with $W(M, \zeta)$, so that:

$$[\sigma_{CDC}]_K \propto \sigma_{pf}^{1/2} C_\infty^{1/4} \zeta^{1/2} M M_e^{-3/4} M_0^{-1/2} \tag{19}$$

or

$$[\sigma_{CDC}]_{ML} \propto \sigma_{pf}^{1/2} C_\infty^{1/4} \zeta^{1/2} M^{3/4} M_e^{-1} \tag{20}$$

It is worth noting that the proportionality constant for σ_{CDC} is the same as for σ_{CSC}, involving quantities difficult to determine experimentally.

2.3.1.4
Craze Fibril Breakdown

When considering fracture behaviour of polymers, an important feature, as mentioned, deals with craze fibril breakdown. Indeed, this latter mechanism leads to crack propagation and easier specimen fracture.

In the case of CSCs, under conditions where there is a low mobility of polymer chains within fibrils, it is likely [19] that under the applied surface stress σ_{CSC}, the polymer chain, elongated between entanglements, breaks down, thus decreasing the number of load-bearing chains in the fibril. Estimates performed on a blend of high MW ($M_w = 3 \times 10^5$ g mol^{-1}) and low MW ($M_w = 1.9 \times 10^4$ g mol^{-1}) polystyrenes [19] support such a mechanism.

At higher mobility of polymer chains within the fibrils, another mechanism for fibril breakdown can happen: chain slippage allowing chain disentanglement and fibril creep till rupture. Such a mechanism emphasises the time during which a polymer chain, entering the fibril at the craze tip, has been submitted to the applied surface stress before the fibril breaks down [23]. The longer this time, the higher fibril and craze stability is. Of course, the breakage time of a fibril increases with the MW of the polymer chains constituting the fibrils and, consequently, fibril stability.

It is worth noticing that this chain slippage mechanism applies both to CSCs and CDCs. The only difference concerns the MW of the fibril polymer chains. Indeed, in the case of CDCs, this is the bulk polymer MW since only disentanglement has occurred. In contrast, for CSCs, the fibril polymer MW is lower than that of the bulk polymer, nevertheless, the average MW of chain fragments in the fibril is higher when the bulk polymer MW is larger.

2.3.2
Shear Deformation Zone

When polymer thin films are deformed under certain conditions (usually low strain rates and temperatures higher than those at which only CSCs are ob-

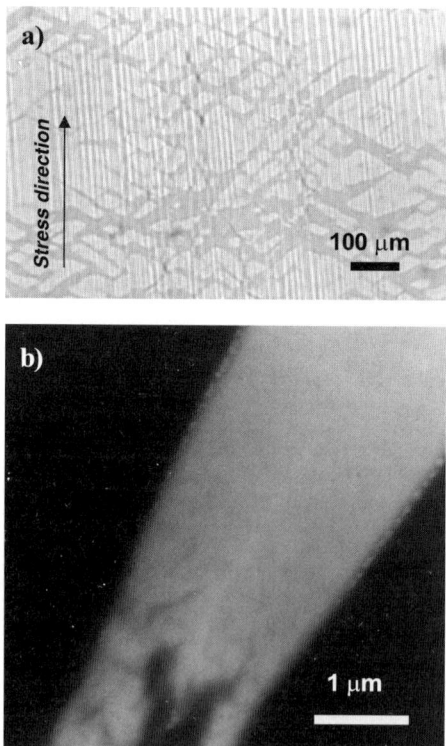

Fig. 6 SDZs at 25 °C. **a** Optical microscopy of localised SDZs in aryl-aliphatic copolyamide. **b** TEM photograph of diffuse SDZs in MGIM copolymer

served), another type of micromechanism occurs – shear deformation zones (SDZ).

SDZs do not contain any void. They can be either localised or diffuse, as shown in Fig. 6a and b, respectively. In the case of localised SDZs, they develop at about 45° of the tensile stress, which correspond to the maximum shear stress direction.

Localised SDZs are usually constituted of a set of micro-SDZs in which the main-chain deformation is quite significant and homogeneous. In contrast, in diffuse SDZs there is a gradual change of the main-chain deformation both along the length and the width of the SDZ. Furthermore, diffuse SDZs are considerably larger than localised SDZs [24].

The occurrence of localised or diffuse SDZs depends on deformation conditions $(T, \dot{\varepsilon})$, on the chemical structure of the considered polymer (this feature will be illustrated through the various polymers analysed in this paper), as well as on physical ageing.

As growth of SDZs is only due to shear yielding, without chain scission or disentanglement, the characteristic stress associated with SDZ, σ_{SDZ}, is considered to be:

$$\sigma_{SDZ} = \sigma_y \tag{21}$$

2.3.3
Influence of Molecular Entanglements in Crazing

TEM micrographs of crazes in thin films have been used to determine the extension ratio, λ, of the polymer chains along the craze [19]. At the crack tip a higher extension ratio is observed, which remains in the midrib. Otherwise, the measured λ is identical along the craze and depends on the polymer being considered .

An entangled polymer, with a characteristic M_e value, can be considered as a temporary network with polymer strands of molar mass M_e. The maximum extension ratio, λ_{net}, that can be achieved is:

$$\lambda_{net} = 3^{1/2} \lambda_{max} \propto M_e^{1/2} \tag{22}$$

where λ_{max} is the maximum extension ratio of a single chain with molar mass M_e.

The extension ratios obtained at room temperature on a large number of homopolymers and copolymers have been analysed [19]. Polymers with high λ_{max}, i.e. large M_e or low entanglement density, ν_e, lead to crazes. In contrast, polymers with low λ_{max}, and thus low M_e and large ν_e, show SDZs.

It has been concluded [19] that the highest entanglement density polymer (bisphenol A polycarbonate) forms SDZs readily and rarely crazes. The intermediate entanglement density polymers may exhibit both crazing and SDZs. The lower entanglement density ($\lambda_{max} > 3.5$) polymers only lead to crazes.

A few comments have to be made on these conclusions:

- PMMA does not obey this rule, for it crazes instead of deforming by SDZs and crazes
- The results refer to room temperature behaviour and it will be shown that the deformation micromechanism can change, depending on temperature and strain rate
- We will show that, whereas the entanglement density can be considered in many polymers as the relevant characteristic, there are polymer series where such a criterion is no longer valid

2.3.4
Craze-Shear Deformation Zone Transition

2.3.4.1
Qualitative Approach

Any polymer submitted to a given stress will develop the micromechanism (CSC, SDZ, CDC) that requires the lowest critical stress among $\sigma_{CSC}, \sigma_{SDZ} = \sigma_y$, and σ_{CDC}.

As these critical stresses have different temperature and strain rate dependencies, depending on conditions (T or $\dot{\varepsilon}$), one or another micromechanism is favoured.

At low temperatures, CSCs are formed, as mentioned, and σ_{CSC} scales as $\sigma_{pf}^{1/2}$, whereas σ_{SDZ} behave as σ_y (higher but close to σ_{pf}). With decreasing

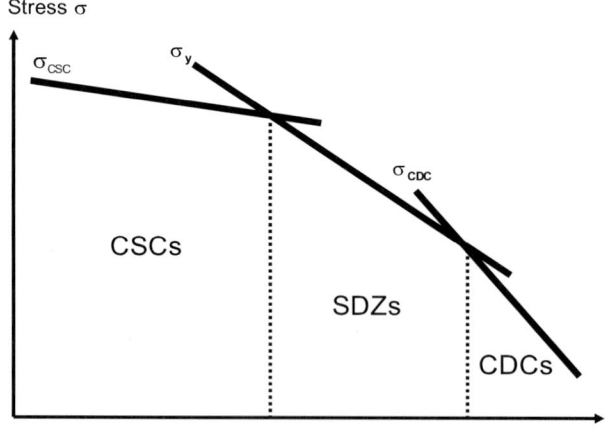

Fig. 7 Representation of the temperature or strain rate dependence of the critical stresses σ_{CSC}, σ_y and σ_{CDC}, and the effect on the favoured micromechanism

temperature, σ_y will always increase more rapidly than σ_{CSC}, in such a way that at a sufficiently low temperature, CSCs will be favoured relative to SDZs (Fig. 7).

At higher temperatures, due to the steeper decrease of σ_y, a temperature will be reached at which SDZs are favoured and a CSC–SDZ transition will occur.

At much higher temperatures, where CDCs are able to develop, one has a temperature dependence of σ_{CDC}, which involves both $\sigma_{pf}^{1/2}$ and the monomeric friction coefficient, ζ. In systems where the activation energy of ζ is high enough, its temperature dependence can lead to a steeper decrease of σ_{CDC} with increasing temperature than for σ_y. For such polymers, at temperatures not much lower than T_α, CDCs can be the favoured micromechanism and a SDZ–CDC transition will take place.

It is worth mentioning that there is a temperature-strain rate equivalence: increasing temperature at a fixed strain rate leads to the same behaviour as decreasing strain rate at a fixed temperature.

The favoured micromechanisms as a function of temperature or strain rate, for a given polymer, are shown in Fig. 7.

2.3.4.2
Quantitative Approach

In order to calculate, for a considered polymer under defined experimental conditions (constant strain rate or temperature), the occurrence of the micromechanism transition, one has to plot the temperature or strain rate dependence of the various critical stresses. Hereafter, we will assume that experiments are performed as a function of temperature at constant strain rate.

In the case of σ_y, the experimental values are available since $\sigma_y(T)$ is known.

The difficulty arises when considering σ_{CSC} and σ_{CDC}. One needs the absolute values of the critical stresses, thus the proportionality constant in the above relationships for σ_{CSC} and σ_{CDC} is required. Unfortunately, as indicated, they involve quantities which are not experimentally available. Thus, for a particular polymer, it is not possible to draw the corresponding critical stress-temperature diagram.

However, the situation is different when we are interested in a series in which the chemical structure of the polymer is gradually changed, as will be the case in this paper when dealing with poly(methyl methacrylate) and its maleimide and glutarimide copolymers (Sect. 3), or with the aryl-aliphatic copolyamides (Sect. 5). In these cases, one can take one of the polymers in the series as a reference and use the corresponding experimental transition temperature for determining the proportionality constant.

The CSC–SDZ transition, denoted as T_{12}, occurs at T_{12}^{ref} for the reference, such that:

$$\sigma_{\mathrm{CSC}}^{\mathrm{ref}}\left(T_{12}^{\mathrm{ref}}\right) = \sigma_y^{\mathrm{ref}}\left(T_{12}^{\mathrm{ref}}\right) \tag{23}$$

From this relationship, the proportionality constant, A, is given by:

$$A = \frac{\sigma_y^{\mathrm{ref}}\left(T_{12}^{\mathrm{ref}}\right)}{\left[\sigma_{\mathrm{pf}}^{\mathrm{ref}}\left(T_{12}^{\mathrm{ref}}\right)\right]^{1/2}\left(v_e^{\mathrm{ref}}\right)^{1/2}\left(C_\infty^{\mathrm{ref}}\right)^{1/4}\left(N_{\mathrm{eq,e}}^{\mathrm{ref}}\right)^{1/2}l_{\mathrm{eq}}^{\mathrm{ref}}U^{1/2}} \tag{24}$$

For the reference polymer at any temperature T:

$$\sigma_{\mathrm{CSC}}^{\mathrm{ref}}(T) = \sigma_y^{\mathrm{ref}}\left(T_{12}^{\mathrm{ref}}\right)\left[\frac{\sigma_{\mathrm{pf}}^{\mathrm{ref}}(T)}{\sigma_{\mathrm{pf}}^{\mathrm{ref}}\left(T_{12}^{\mathrm{ref}}\right)}\right]^{1/2} \tag{25}$$

For any other polymer of the series, designated as i:

$$\sigma_{\mathrm{CSC}}^{i}(T) = \kappa_{\mathrm{CSC}}^{i}(T)\sigma_{\mathrm{CSC}}^{\mathrm{ref}}(T) \tag{26}$$

where the proportionality factor x can be expressed by:

$$\kappa_{\mathrm{CSC}}^{i}(T) = \left[\frac{\sigma_{\mathrm{pf}}^{i}(T)}{\sigma_{\mathrm{pf}}^{\mathrm{ref}}(T)}\right]^{1/2}\left(\frac{v_e^{i}}{v_e^{\mathrm{ref}}}\right)^{1/2}\left(\frac{C_\infty^{i}}{C_\infty^{\mathrm{ref}}}\right)^{1/4}\left(\frac{N_{\mathrm{eq,e}}^{i}}{N_{\mathrm{eq,e}}^{\mathrm{ref}}}\right)^{1/4}\left(\frac{l_{\mathrm{eq}}^{i}}{l_{\mathrm{eq}}^{\mathrm{ref}}}\right)^{1/2} \tag{27}$$

Through these expressions, $\sigma_{\mathrm{CSC}}^{i}(T)$ can be calculated at each temperature and plotted as a function of temperature. The intercept of $\sigma_{\mathrm{CSC}}^{i}(T)$ and σ_y^{i} leads to the transition temperature, T_{12}^{i}, for the polymer i of the series.

The same approach can be applied for determining the transition temperature from SDZ to CDC, denoted T_{233}. It leads to:

$$\sigma_{\mathrm{CdC}}^{\mathrm{ref}}(T) = \sigma_y^{\mathrm{ref}}\left(T_{233}^{\mathrm{ref}}\right)\left[\frac{\sigma_{\mathrm{pf}}^{\mathrm{ref}}(T)}{\sigma_{\mathrm{pf}}^{\mathrm{ref}}\left(T_{233}^{\mathrm{ref}}\right)}\right]^{1/2}\left[\frac{\zeta^{\mathrm{ref}}(T)}{\zeta^{\mathrm{ref}}\left(T_{233}^{\mathrm{ref}}\right)}\right]^{1/2} \tag{28}$$

and for any other polymer i of the series to:

$$\sigma_{\mathrm{CDC}}^{i}(T) = \kappa_{\mathrm{CDC}}^{i}(T)\sigma_{\mathrm{CDC}}^{\mathrm{ref}}(T) \tag{29}$$

Two different expressions can be obtained for $\kappa_{\mathrm{CDC}}^{i}(T)$, depending on the model. From the Kramer model [20], we get:

$$\left[\kappa_{\mathrm{CDC}}^{i}(T)\right]_K = \left[\frac{\sigma_{\mathrm{pf}}^{i}(T)}{\sigma_{\mathrm{pf}}^{\mathrm{ref}}(T)}\right]^{1/2}\left(\frac{C_\infty^{i}}{C_\infty^{\mathrm{ref}}}\right)^{1/4}\left(\frac{\zeta^{i}(T)}{\zeta^{\mathrm{ref}}(T)}\right)^{1/2}$$
$$\times\left(\frac{M^{i}}{M^{\mathrm{ref}}}\right)\left(\frac{M_e^{i}}{M_e^{\mathrm{ref}}}\right)^{-3/4}\left(\frac{M_0^{i}}{M_0^{\mathrm{ref}}}\right)^{-1/2} \tag{30}$$

and for the McLeish model [22]:

$$\left[\kappa_{CDC}^{i}(T)\right]_{ML} = \left[\frac{\sigma_{pf}^{i}(T)}{\sigma_{pf}^{ref}(T)}\right]^{1/2} \left(\frac{C_{\infty}^{i}}{C_{\infty}^{ref}}\right)^{1/4} \left(\frac{\zeta^{i}(T)}{\zeta^{ref}(T)}\right)^{1/2} \left(\frac{M^{i}}{M^{ref}}\right)^{3/4} \left(\frac{M_{e}^{i}}{M_{e}^{ref}}\right)^{-1}$$

(31)

The transition temperature, T_{233}, for the polymer of the series under consideration is obtained from the intercept of $\sigma_{y}^{i}(T)$ and $\sigma_{CDC}^{i}(T)$.

2.4
Fracture Behaviour

Depending on the chemical structure of the polymer and on the experimental conditions (T and $\dot{\varepsilon}$), polymer solids can present a brittle behaviour, a ductile behaviour, or an intermediate fracture behaviour.

In order to get a quantitative characterisation of fracture, different concepts of fracture mechanics have been developed and we will focus on two of them (i.e. the K_{Ic} and G_{Ic} approaches) which will be defined and briefly considered. More information about polymer fracture is given in the first review of this volume and in [4, 5].

2.4.1
Fracture Characteristics

A crack at a notch tip in a solid sample can propagate according to three different loading modes. The propagation occurs in:

- Mode I, under tension
- Mode II, under shearing in the plane
- Mode III, under shearing out of the plane

The lowest resistance of the material corresponds to a tension solicitation (i.e. mode I) and it is for this reason that it is the more frequently considered mode for fracture characterisation.

Furthermore, the fracture characteristics will be defined in the frame of linear elastic fracture mechanics (LEFM), assuming a purely elastic response of the material (stress proportional to infinitesimal strains). However, LEFM can be extended to materials that exhibit inelastic deformation around the crack tip, provided that such deformations are confined to the immediate vicinity of the tip [25].

2.4.1.1
Energy Criterion

The first fracture criterion was defined by Griffith, supposing that fracture occurs when sufficient energy is released (from the stress field) by growth of the crack. Thus, it provides a measure of the energy required to extend a crack over unit area; this is termed the fracture energy or critical strain-energy release and is denoted as G_{Ic} and is expressed in kJ m^2.

Thus, for a body of thickness B, with a crack of length a, subjected to a load P leading to a displacement Δ, the compliance C of the body is defined as:

$$C = \Delta/P \tag{32}$$

If P_c represents the load at the onset of crack propagation, G_{Ic} is expressed as:

$$G_{Ic} = \frac{P_c^2}{2B} \frac{\partial C}{\partial a} \tag{33}$$

By experimentally determining C as a function of a, then $\partial C/\partial a$ may be found; so, if P_c is measured, the value of G_{Ic} may be deduced.

2.4.1.2
Stress Intensity Factor

Another fracture criterion has been introduced by Irwin, based on the stress intensity existing at a point of polar coordinates, r and θ from the tip of a sharp crack of length $2a$ in a body uniformly stressed by an applied stress, σ_0. For regions close to the crack tip, the components of the stress tensor at the considered point take the form:

$$\sigma_{ij} = \sigma_0 (a/2r)^{1/2} f_{ij}(\theta) \tag{34}$$

or:

$$\sigma_{ij} = \frac{K}{(2\pi r)^{1/2}} f_{ij}(\theta) \tag{35}$$

The parameter K is the stress intensity factor, whose level defines the stress field around the crack tip. In the case of a mode I loading, it is denoted as K_I.

Irwin postulated that the condition, $K_I \geq K_{Ic}$, represented a fracture criterion. K_{Ic} is a critical value for crack growth in the materials; it is frequently termed the fracture toughness. The units for K_{Ic} are MPa m$^{1/2}$.

It is interesting to notice that K_{Ic} can be expressed as:

$$K_{Ic} = \sigma_c Y a^{1/2} \tag{36}$$

where σ_c is the applied stress at the onset of crack growth, a is the total notch and crack length, and Y is a correction factor that takes into account the particular configuration of specimen and crack geometry.

Experimentally, K_{Ic} is determined by the relationship:

$$K_{Ic} = f(a/W) P_{max}/BW^{1/2} \tag{37}$$

where B and W are the thickness and length of the specimen, and P_{max} is the maximum load achieved during the fracture test.

2.4.2
Plastic Zone

The σ_{ij} expression indicated above shows that at the crack tip ($r = 0$), the stress tensor components become infinite. Actually, for a material able to undergo plastic deformation, above some stress level yielding occurs and limits the stress to the corresponding value, σ_y. Thus, around the crack tip a zone exists in which the material is plastically deformed. Such a zone is called the plastic zone, and it is represented in Fig. 8 in the case of a crack across a plate thickness.

It is important to note that within the plastic zone there is an energy dissipation originating from the deformation micromechanisms (SDZs, CSCs, or CDCs) occurring. The amount of energy thus dissipated represents almost the whole energy involved in crack propagation.

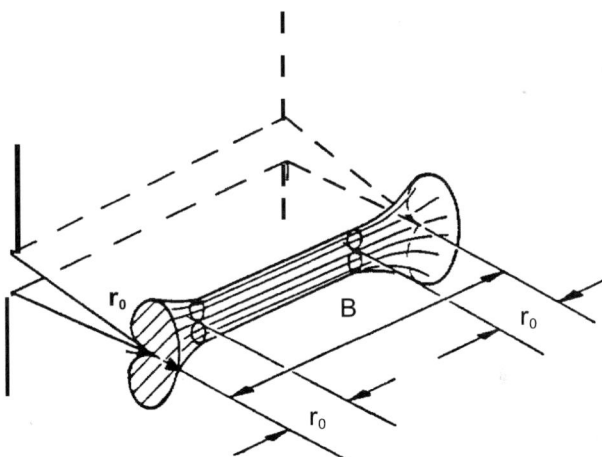

Fig. 8 Plastic zone shape across a plate thickness

2.4.3
Influence of Sample Thickness

When analysing the stress and strain tensors at various positions through the thickness of a plate, it turns out that:

- At the surface, the stress is planar and the strain is three-dimensional; such a situation is referred to as plane stress
- In the inner part, the stress is three-dimensional and the strain is planar; it is referred to as plane strain
- Going away from the surface, there is a gradual change from plane stress to plane strain

The shape of the plastic zone through a plate thickness, shown in Fig. 8, is a direct manifestation of the stress and strain tensor change. It is also reflected in the change of K_{Ic} through the thickness. Indeed, at and near the surface, the plastic zone is large, more energy is dissipated during crack propagation and, so, the local K_{Ic} value is high (K_{Ic}'' refers to planar stress conditions). In contrast, in the central part, the plastic zone is small and, consequently, K_{Ic} value is low (K_{Ic}' refers to plane strain conditions). The smooth transition from K_{Ic}'' to K_{Ic}' extends over a thickness r_0, which is inversely proportional to σ_y^2. The easier the material is plastically deformed, the larger r_0.

For characterising the fracture behaviour of a bulk material, the interesting K_{Ic} value is that corresponding to plane strain, K_{Ic}'. Thus, in order to measure K_{Ic}', it is necessary to get a large enough sample thickness, B. Actually, one has to use $B \geq 50r_0$ or $B \leq 2.5(K_{Ic}'/\sigma_y)^2$.

2.4.4
Relation of K_{Ic} and G_{Ic}

In the frame of LEFM a simple relationship exists between K_{Ic} and G_{Ic}. Thus, for mode I:

$$G_{Ic} = \left(K_{Ic}^2/E\right) \text{ for plane stress} \tag{38}$$

$$G_{Ic} = \left(K_{Ic}^2/E\right)\left(1 - \nu^2\right) \text{ for plane strain} \tag{39}$$

where E is the elastic modulus and ν the Poisson ratio.

2.4.5
Crack Propagation

G_{Ic} determines the applied load, P_c, at the onset of crack propagation. Then the crack propagation itself takes place. Depending on the material and the experimental conditions ($T, \dot{\varepsilon}$), two different types of crack propagation may occur:

1. Unstable propagation: once the crack starts to grow, it propagates spontaneously, without requiring any further load input (see Kausch/Michler in this volume).
2. Stable propagation: once the crack starts to grow, it is necessary to keep an applied load to get a crack propagation. The propagation stops when the specimen is unloaded.

It is worth noting that most of the fracture characteristics reported in this paper correspond to stable crack propagation.

2.4.6
Specimen Geometries

Several kinds of specimen geometries have been used to determine fracture characteristics. Only two are considered here and are schematically shown in Fig. 9; information on others can be found in [4].

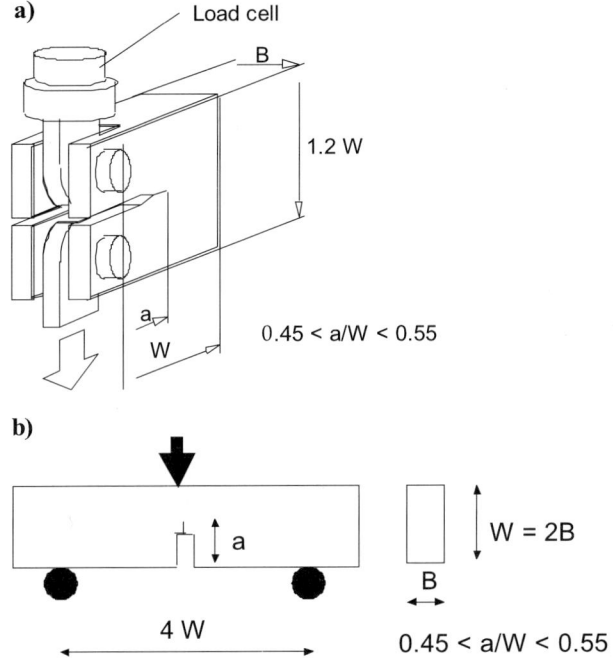

Fig. 9 Specimen geometries for fracture measurements: **a** compact-tension, **b** three-point bending

2.4.6.1
Compact-Tension

Specimen geometry for the compact-tension (CT) test is shown in Fig. 9a.

The correction factor, $f\left(\frac{a}{W}\right)$, for determining K_{Ic}, from Eq. 37, is expressed as a function of specimen dimensions by:

$$f\left(\frac{a}{W}\right) = \frac{2 + \frac{a}{W}}{\left(1 - \frac{a}{W}\right)^{3/2}} \tag{40}$$
$$\times \left[0.826 + 4.64\left(\frac{a}{W}\right) - 13.32\left(\frac{a}{W}\right)^2 + 14.72\left(\frac{a}{W}\right)^3 - 5.6\left(\frac{a}{W}\right)^4\right]$$

The sample thickness is defined in Sect. 2.4.5. The initial total length a has to be in the range $0.45 < \frac{a}{W} < 0.55$.

2.4.6.2
Three-Point Bending

Specimen geometry for the three-point bending (TPB) test is shown in Fig. 9b.

The corresponding correction factor, $f\left(\frac{a}{W}\right)$, is expressed as:

$$f\left(\frac{a}{W}\right) = 6\left(\frac{a}{W}\right)^{1/2} \frac{1.99 - \frac{a}{W}\left(1 - \frac{a}{W}\right)\left[2.15 - 3.93\frac{a}{W} + 2.7\left(\frac{a}{W}\right)^{2}\right]}{\left(1 + 2\frac{a}{W}\right)\left(1 - \frac{a}{W}\right)^{3/2}} \tag{41}$$

The requirements for B and the other dimensions, as well as for a, are identical to those indicated for CT specimens.

2.4.6.3
Notch and Pre-crack

In order to get relevant G_{Ic} and K_{Ic} measurements, it is compulsory to get a very sharp pre-crack tip. Indeed, any blunting of the crack tip will result in higher G_{Ic} and K_{Ic} values, misleading conclusions about the actual material toughness.

For most of the reported results, a machined notch was made, then a sharp pre-crack formed with a fresh razor blade, using a falling weight apparatus specially designed to operate the blades in reproducible conditions. Quite a satisfactory sharp pre-crack is obtained, as shown in Fig. 10.

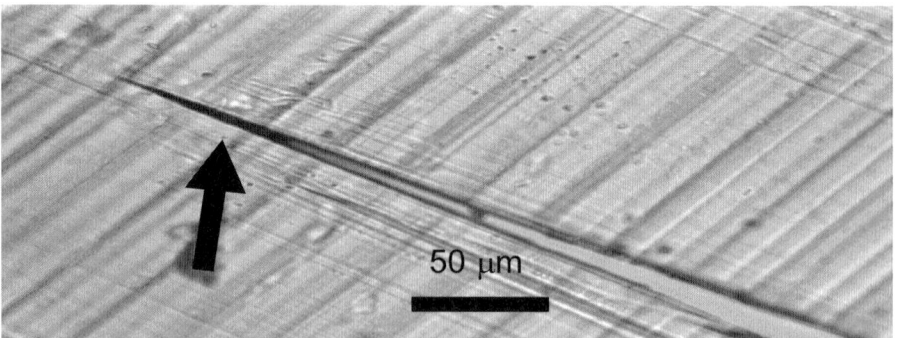

Fig. 10 Optical micrograph showing a pre-crack in poly(methyl methacrylate-*co-N*-methyl glutarimide)

2.4.7
Effect of Polymer Chemical Structure on Fracture

Understanding the effect of the chemical structure of polymers on their fracture behaviour has been a challenging question, which has been studied for quite a long time.

As mentioned, the molecular weight between entanglements, M_e, or equivalently the entanglement density, ν_e, is involved in crazing, in craze fibril stability and, thus, in crack formation and propagation.

So, it is not surprising that a relationship between K_{Ic} and ν_e has been considered. Empirical approaches [26] as well as theoretical models [27–29] lead to an increase of K_{Ic}, with increasing ν_e. Another approach [30, 31] yields a dependence of K_{Ic} with $\nu_e^{1/2}$. Figure 11 shows, for a large variety of polymer

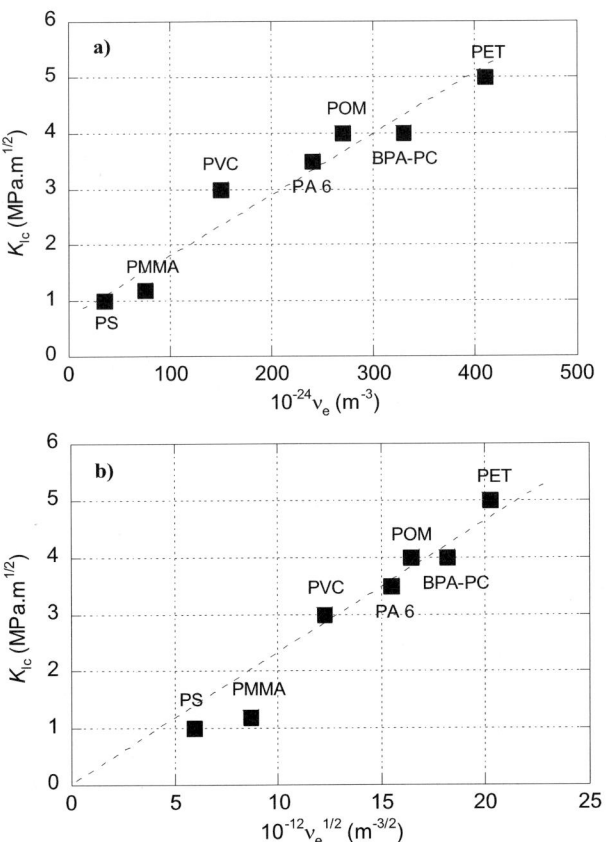

Fig. 11 K_{Ic} for different amorphous or semi-crystalline polymers, plotted versus **a** ν_e and **b** $\nu_e^{1/2}$

structures either amorphous or semi-crystalline, K_{Ic} as a function of v_e or $v_e^{1/2}$. It is clear that a general trend exists when considering the quite different chemical structures of the polymers.

Nevertheless, it will be shown in this paper that, by changing the chemical structure within a given polymer series, such a relationship fails and other molecular effects have to be taken into account.

3
Poly(methyl methacrylate)
and its Maleimide and Glutarimide Copolymers

The mechanical properties of poly(methyl methacrylate), PMMA, have been studied for quite a long time and, in addition to its industrial interest, PMMA constitutes a kind of reference material. Indeed, among the amorphous linear polymers it represents an intermediate between the very brittle polystyrene and the tough bisphenol A polycarbonate considered in Sect. 4. Furthermore, as shown in [1] (Sect. 8.1), the molecular motions responsible for its large β transition are precisely identified, as well as the nature of the cooperativity that develops in the high temperature range of the β transition.

Introducing comonomers like maleimide or glutarimide units in the PMMA chain backbone strongly affects the cooperativity of the molecular motions involved in the β transition, as described in [1], Sects. 8.2 and 8.3, respectively.

For these reasons, PMMA and its maleimide and glutarimide copolymers represent very suitable materials for investigating the effect of the chemical structure and of the solid state molecular motions on the plastic deformation, the occurrence of the various micro-mechanisms of deformation (chain scission crazes, shear deformation zones, chain disentanglement crazes), as well as the fracture behaviour.

In this part, PMMA and the two series of copolymers are successively considered and, for each of them, the various mechanical features are analysed in order to reach a better understanding of the influence of the chemical structure and, in particular, of the β transition motions and their cooperativity.

3.1
Poly(methyl methacrylate)

The plastic deformation, the micro-mechanisms of deformation, and the stable fracture are successively analysed. In each case the relation to the β transition motions is emphasised.

3.1.1
Plastic Deformation of PMMA

First, it is important to notice that the stress–strain curves and, consequently, the derived characteristics (yield stress, σ_y, plastic flow stress, σ_{pf}, and strain softening) have been studied in a temperature range extending to, typically, $T_\alpha - 20$ K. Indeed, for temperatures closer to T_α, the experimental results are less reliable, as some creep behaviour can occur.

3.1.1.1
Physical Ageing Effect

The investigation of the β transition of PMMA ([1], Sect. 8.1.3) has revealed that physical ageing has an influence on the dynamic mechanical response in the β–α crossover temperature region. As a consequence, it is worth checking whether it leads to some changes in the plastic deformation behaviour [32].

The compression stress–strain curves obtained at various temperatures for quenched PMMA are shown in Fig. 12 and for physically aged PMMA (sample cooled down at 5 K h^{-1} to 87 °C, then maintained at this temperature for 80 h) in Fig. 13. It clearly appears that the physical ageing induces quite a significant development (around 10 MPa) of the strain softening at all considered temperatures. This change in strain softening amplitude comes from an increase of the yield stress in the physically aged PMMA, whereas the plastic flow stress is unchanged, as shown in Fig. 14.

These results show the dependence of the plastic deformation of PMMA on the thermal history of the sample. The relation with the cooperativity of the β transition motions is addressed later in Sect. 3.1.1.6.

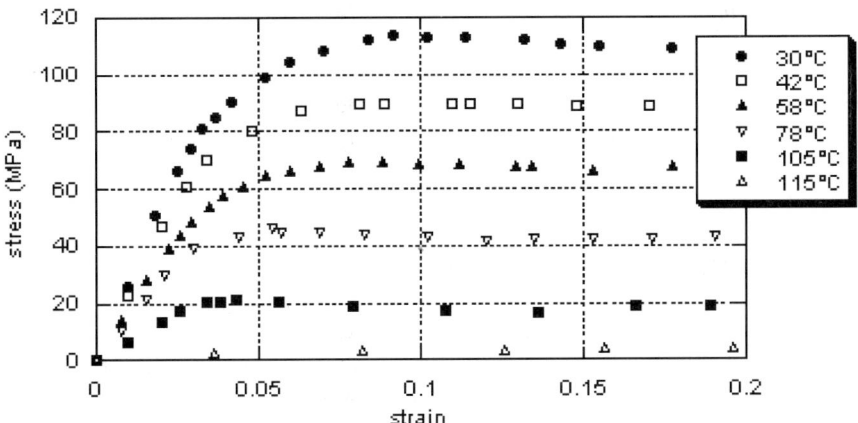

Fig. 12 Stress–strain compression curves of quenched PMMA obtained at a strain rate of 2×10^{-3} s^{-1} at various temperatures (From [32])

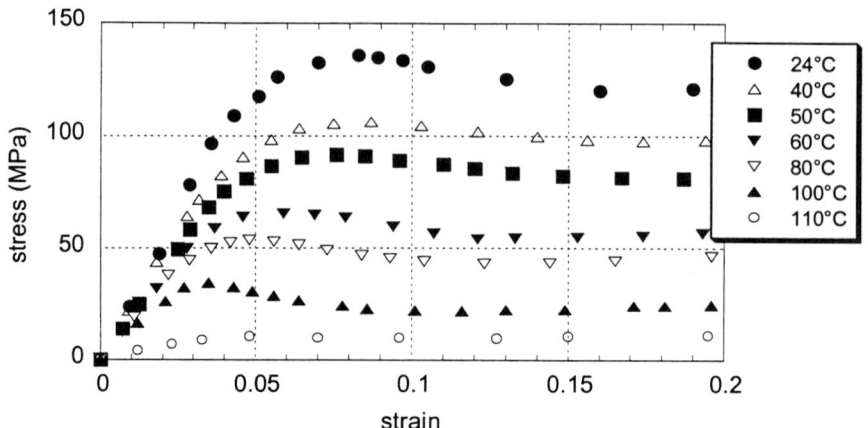

Fig. 13 Stress–strain compression curves of physically aged PMMA obtained at a strain rate of 2×10^{-3} s^{-1} at various temperatures (From [32])

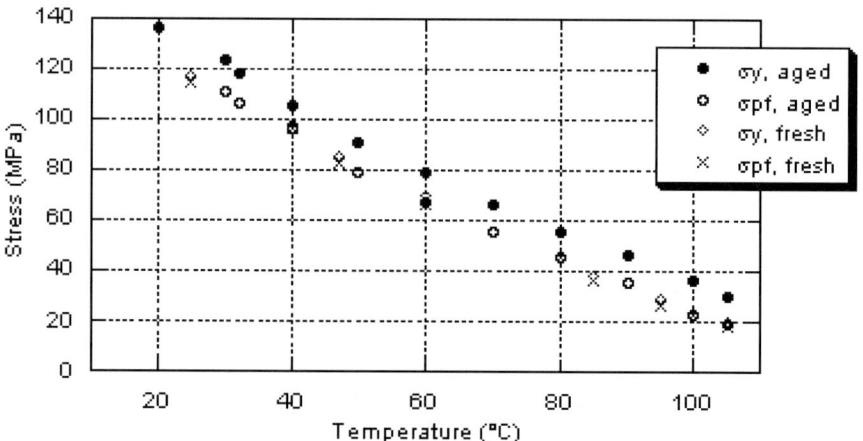

Fig. 14 Temperature dependence of yield stress, σ_y, and plastic flow stress, σ_{pf}, for quenched and physically aged PMMA. Strain rate is 2×10^{-3} s^{-1} (From [32])

It is important to notice that all the results presented hereafter deal with quenched PMMA.

3.1.1.2
Stress–Strain Curves

The compression stress–strain curves obtained over quite a broad temperature range are shown in Fig. 15 [33]. It appears that the strain softening, weak and slightly temperature-dependent in the temperature range from 40 to 100 °C, increases when lower temperatures are considered.

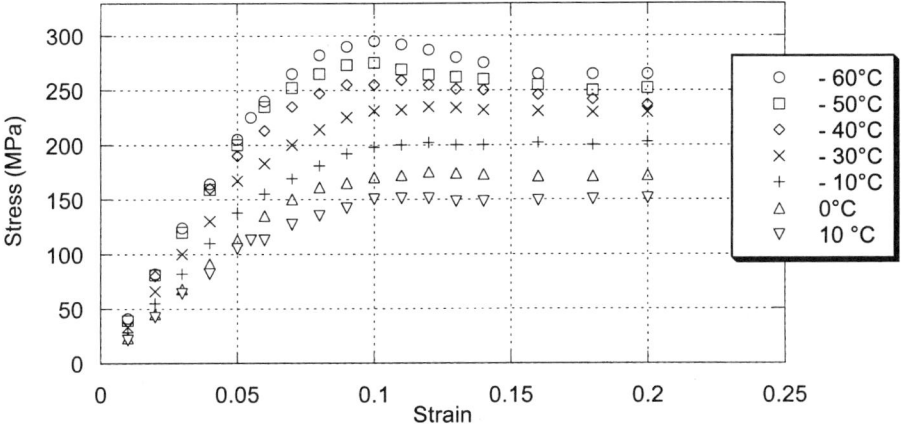

Fig. 15 Temperature dependence of nominal stress–strain curves under uni-axial compression for PMMA at a strain rate of $2 \times 10^{-3} \text{ s}^{-1}$ (From [33])

The effect of strain rate, $\dot{\varepsilon}$, at $-50\,°C$ and $50\,°C$ is shown in Fig. 16. Qualitatively, increasing the strain rate is analogous to decreasing temperature (the temperature and strain rate dependencies of the stress–strain curve shapes are summarised in Fig. 17). However, it is worth noting that an equivalence temperature-strain rate does not apply over the whole stress–strain curve.

3.1.1.3
Yield Stress

The temperature dependence of the yield stress, σ_y, of PMMA obtained at a strain rate, $\dot{\varepsilon} = 2 \times 10^{-3} \text{ s}^{-1}$, is shown in Fig. 18. A sigmoidal curve is observed, which looks like the temperature dependence of the Young's modulus, E. When increasing the strain rate a similar behaviour is observed.

In order to check whether the temperature dependence of σ_y would reflect the change of modulus only, the ratio σ_y/E is plotted in Fig. 19. It is clear that the modulus does not normalise the yield stress behaviour, the latter decreasing more than the modulus when temperature increases.

The strain rate dependence of the yield stress is shown at various temperatures in Fig. 20. To go further in the analysis, it is interesting to use the Eyring approach presented in Sect. 2.2.1.1. For this purpose, the ratio σ_y/T, K is plotted versus $\log(\dot{\varepsilon}, \text{s}^{-1})$ at various temperatures in Fig. 21. A linear dependence is observed at each temperature, in agreement with the Eyring expression. However, the slopes show two different temperature regimes at low and high temperatures. Of course, the activation volume, V_0, directly related to the slope, reflects the change in behaviour, as shown in Fig. 22. At low temperature, the activation volume is small (around 0.1 nm^3) and independent of temperature, whereas it increases rapidly above room temperature

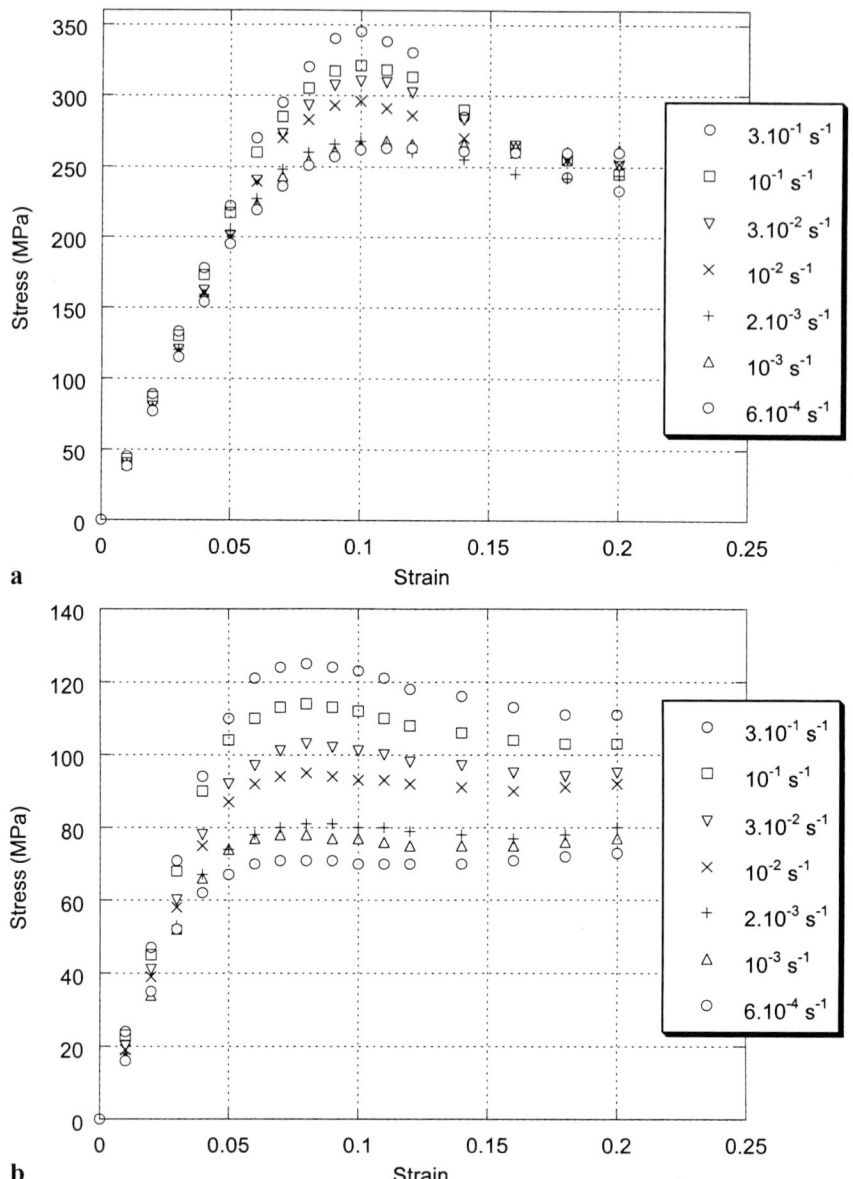

Fig. 16 Effect of strain rate on uni-axial compression stress–strain curves of PMMA:
a at $-50\,°C$ and **b** at $50\,°C$ (From [33])

reaching 2 nm^3 at 80 °C. A similar result has been obtained in [34]. It clearly indicates that two different mechanisms are involved, with a change around 0–20 °C.

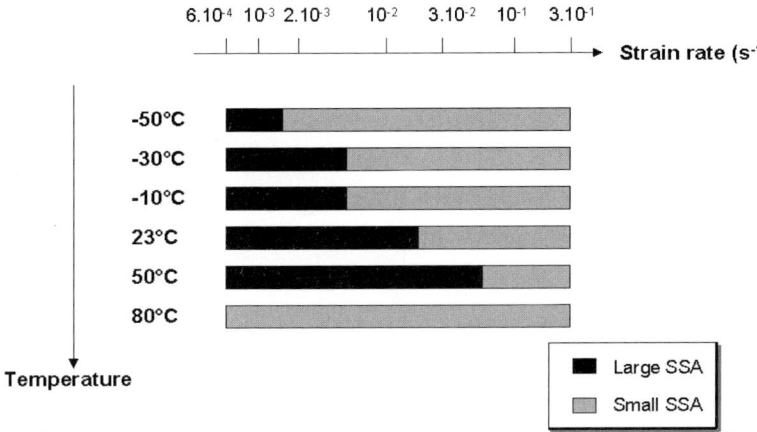

Fig. 17 Temperature and strain rate dependencies of the stress–strain curve shapes of PMMA: small (less than 10 MPa) and larger strain softening (From [33])

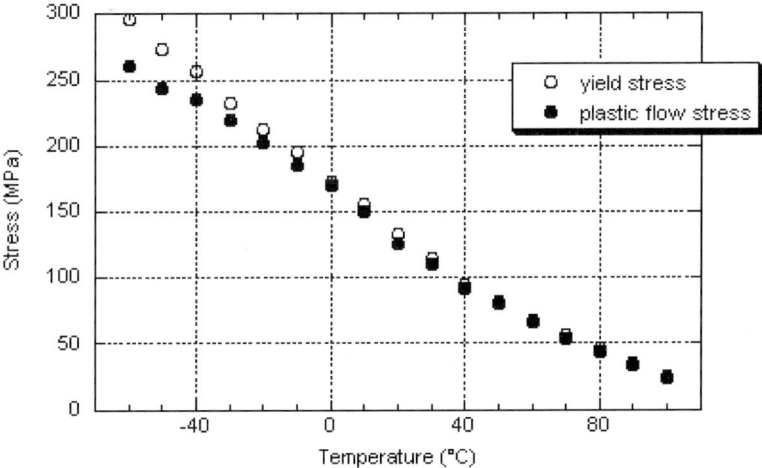

Fig. 18 Temperature dependence of yield stress and plastic flow stress of PMMA, obtained under uni-axial compression at a strain rate of 2×10^{-3} s^{-1}

3.1.1.4
Plastic Flow Stress

The temperature dependence of the plastic flow stress, σ_{pf}, of PMMA is shown in Fig. 18. At high temperatures ($T \geq 50\,°C$) the σ_{pf} curve is parallel to that of σ_y, but at lower temperatures it is gradually lower and lower, the decrease being particularly pronounced below 0 °C.

The strain rate dependence of the plastic flow, σ_{pf}, is shown at various temperatures in Fig. 20. It is clear that σ_{pf} is lightly dependent on the strain rate,

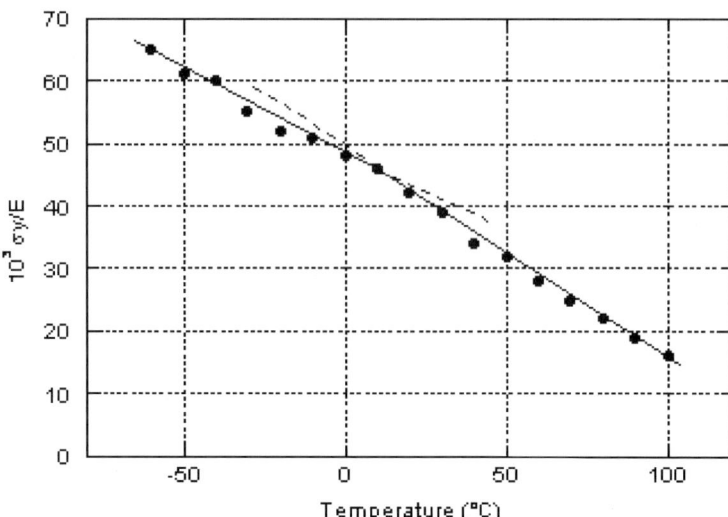

Fig. 19 Temperature dependence of the ratio σ_Y/E for PMMA

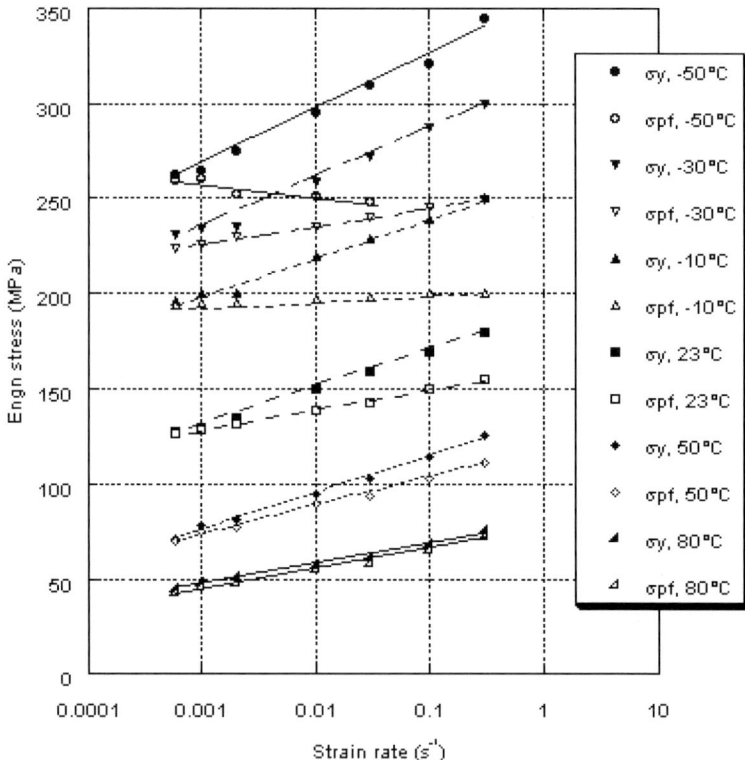

Fig. 20 Strain rate dependence of yield stress, σ_y, and plastic flow stress, σ_{pf}, of PMMA at the indicated temperatures (From [33])

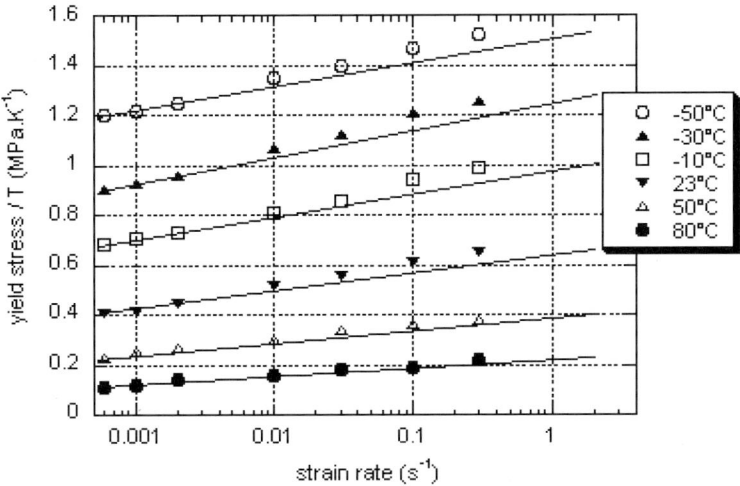

Fig. 21 Strain rate dependence of σ_y/T for PMMA at various temperatures (From [33])

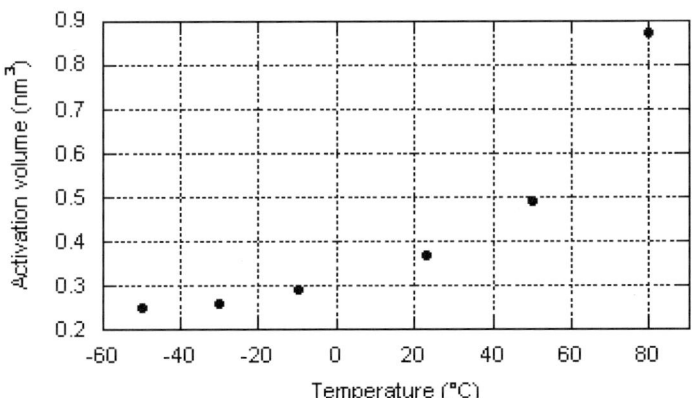

Fig. 22 Temperature dependence of activation volume for PMMA (From [33])

whatever the considered temperature. If an Eyring treatment is applied to σ_{pf}, it leads to an activation volume around $2\,nm^3$ for the whole temperature range.

At this stage, the difference in temperature and strain rate dependencies of σ_y and σ_{pf} explains why a strain rate and temperature equivalence cannot be achieved over the whole stress–strain curve.

3.1.1.5
Strain Softening

As above described, depending on temperature and strain rate, the stress–strain curves present a strain softening of variable amplitude.

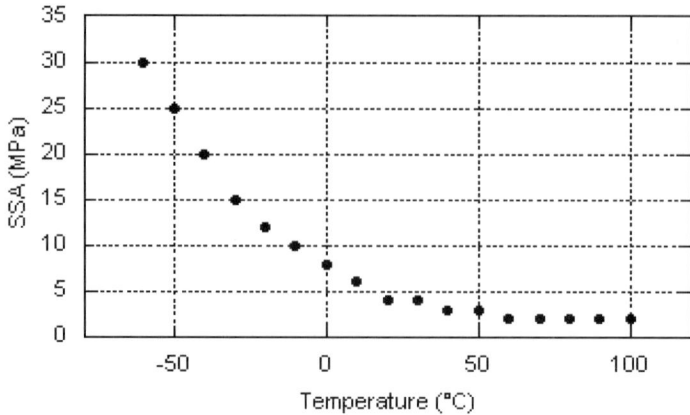

Fig. 23 Temperature dependence of strain softening amplitude (SSA) of PMMA at a strain rate of 2×10^{-3} s^{-1}

The strain softening amplitude, SSA, used in our previous papers, is shown for PMMA in Fig. 23. It is clear that SSA values are quite small and constant above 50 °C and sharply increase at temperatures lower than 0 °C. This feature was already pointed out when comparing σ_y and σ_{pf} versus temperature (Fig. 18).

It is interesting to consider also the normalised SSA, nSSA, defined in Sect. 2.2.3. The result is shown in Fig. 24 for PMMA. At first, over quite a broad temperature range (0–80 °C), SSA corresponds to a constant percent (around 4%) of σ_{pf}. At lower temperatures, nSSA reflects the different behaviours of σ_y and σ_{pf}. Finally, an additional increase of nSSA is observed above 80 °C, owing to the sharp decrease of σ_{pf} in this temperature range closer to T_α.

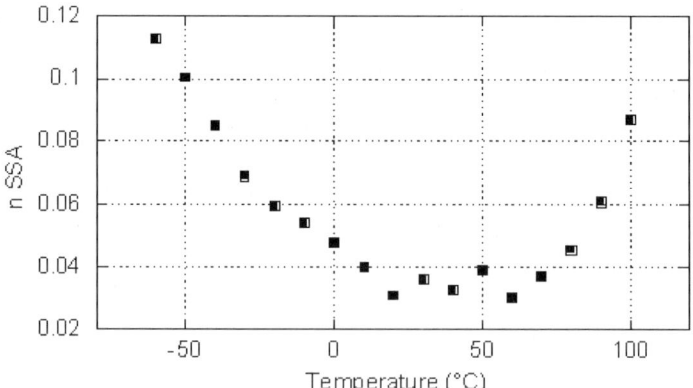

Fig. 24 Temperature dependence of normalised strain softening amplitude (nSSA) of PMMA at a strain rate of 2×10^{-3} s^{-1}

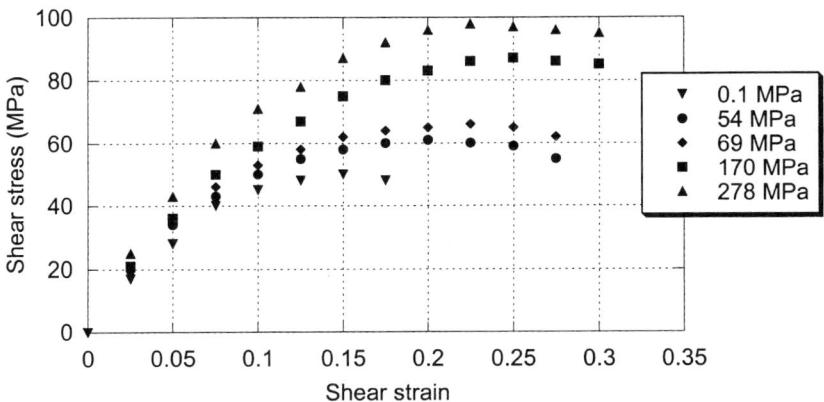

Fig. 25 Pressure dependence of yield stress and strain softening of PMMA (From [36])

An interesting feature deals with the effect of pressure on the yield stress and the strain softening amplitude [36], as shown in Fig. 25. Thus, under shear, increasing pressure leads to higher yield stress, as expected from the higher molecular packing. More surprisingly is the decrease of the strain softening obtained by going from about 10 MPa under 500 bars to 3 MPa under 1.7 kbar. Increasing pressure gives the same trend as increasing temperature.

3.1.1.6
Molecular Analysis of Plastic Deformation and Relation to the β Transition

First, let us examine the temperature dependence of the plastic flow, σ_{pf}, shown in Fig. 18. As already mentioned in Sect. 2.2.2, the plastic flow requires whole chain motions like those occurring above the α transition, whatever the considered temperature. This feature is reflected in the large activation volume associated with σ_{pf} and in the independence to temperature. Consequently, one can consider the plastic flow stress as a reference behaviour in the molecular analysis of plastic deformation.

Several results on the plastic deformation of PMMA as a function of temperature (activation volume, strain softening) clearly show a change of behaviour occurring around 0 °C. Looking at the temperature dependence of the dynamic mechanical loss modulus, E'', Fig. 26 shows that such a change occurs at the temperature corresponding to the maximum of the β peak at 1 Hz and concerns the high temperature part of the β transition. This latter statement is even more valid if one takes into account the downward shift of the β peak by about 20 °C, owing to the low strain rate (2×10^{-3} s^{-1}) used in the experiments. Furthermore, physical ageing both increases the yield stress, σ_y, and changes the loss modulus, E'', in the β–α crossover temperature range ([1], Sect. 8.1.3).

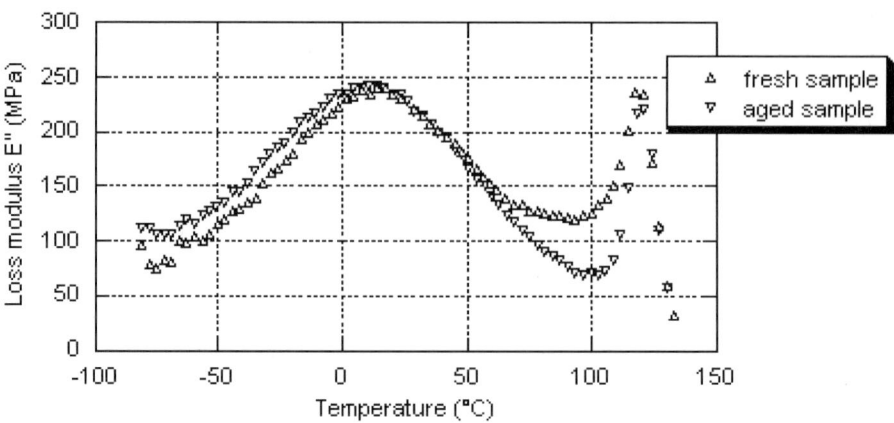

Fig. 26 Temperature dependence of mechanical loss modulus, E'', at 1 Hz for PMMA (From [47])

These observations lead one to wonder how the motions involved in the β transition could affect the plastic deformation behaviour of PMMA.

First, as described in [1] (Sect. 8.1 and summarised in Sect. 8.1.6), the β transition of PMMA, illustrated in Fig. 26, originates from ester group π-flips occurring either isolated in the low temperature part of the β transition, or coupled with a change of the internal rotation angles of the main chain in the high temperature part. In the latter case, it corresponds to the development of an intramolecular cooperativity. In addition, in the $\beta-\alpha$ crossover temperature range, the extent of cooperativity with the main chain is such that these cooperative motions can be considered as the precursors of the larger segmental motions that appear at the glass–rubber transition temperature.

As mentioned in Sect. 2.2.1.2, the yield point originates from a conformational change of the main chain, leading to an increase of conformations corresponding to what would happen at a temperature above the glass–rubber transition temperature.

On these bases, one can examine how the β transition motions could be involved in the yielding.

At temperatures below the onset of the β transition, occurring around $-70\,^\circ\mathrm{C}$ (or $-90\,^\circ\mathrm{C}$ at $2 \times 10^{-3}\,\mathrm{s}^{-1}$), conformational changes occur in sites of lower local packing due to the existing heterogeneity of local density, under a sufficient applied stress. It is worth mentioning that for polymer chains bearing side groups, like PMMA, the local conformational change of the main chain implies a displacement of the side group. Consequently, the whole moving unit (main chain segment and side group) induces quite a significant adaptation of its local surrounding.

The first conformational changes occur at the end of the purely elastic behaviour, corresponding to a stress value σ_{el}. By applying higher and higher stresses, in the stress range for pre-plastic behaviour, conformational changes occur in sites with higher and higher density. At a stress level equal to σ_y, numerous-enough sites are been involved to initiate a plastic deformation (supporting this picture is the fact that yielding is not an instantaneous process, but requires some time for appearing). At this stage, the numerous induced local packing adaptations and the numerous conformational changes allow whole chain motions under a stress, σ_{pf}, which is lower than σ_y. This leads to a strain softening effect.

In the temperature range where side group motions occur, leading to a secondary transition like the β transition of PMMA, these motions soften the surrounding of the moving units and, consequently, allow local conformational changes of the main chain under a lower stress than in the low temperature case. In the case of the β transition of PMMA, on the low temperature side, the ester group π-flips are localised and the mechanism above described operates.

On the high temperature side, the changes of internal rotation angles of the main chain, associated with the ester group motion and intramolecular cooperativity, favour occurrence of the conformational changes involved both in yielding and in plastic flow, in such a way that they occur at stresses σ_y and σ_{pf} close to each other. This is actually what is observed above $0–10\,°C$ in the strain softening described either by SSA (Fig. 23) or by nSSA (Fig. 24). This change in mechanism between low and high temperature ranges of the β transition is also reflected in the activation volume related to the yield stress (Fig. 22), which increases above $0\,°C$.

In the $\beta–\alpha$ crossover region, typically above $50\,°C$, the intramolecular cooperativity of the β motions is such that they can be considered as precursors of the α motions. A direct consequence is that yielding and plastic flow can occur under almost identical stresses, as shown in the strain softening amplitude (Fig. 23).

Regarding the effect of physical ageing, the associated densification hinders the intramolecular $\beta–\alpha$ cooperativity (as discussed in [1], Sect. 8.1.3). This makes the conformational changes more difficult and, so, require a larger stress. The resulting increase of σ_y corresponds to the situation encountered at a temperature shifted downward by $20\,°C$. The physical ageing densification does not influence the plastic flow stress. The reason is that the plastic flow happens in a material that has been strained, and it is known that strain induces a disappearance of the physical ageing effects, this is called "rejuvenation".

It is worth pointing out that hindering of the $\beta–\alpha$ cooperativity due to physical ageing leads to quite a different temperature dependence of nSSA, as shown in Fig. 27, compared to the case of quenched PMMA shown in Fig. 24. Instead of a plateau, a large increase of nSSA is observed in the temperature range $30–80\,°C$.

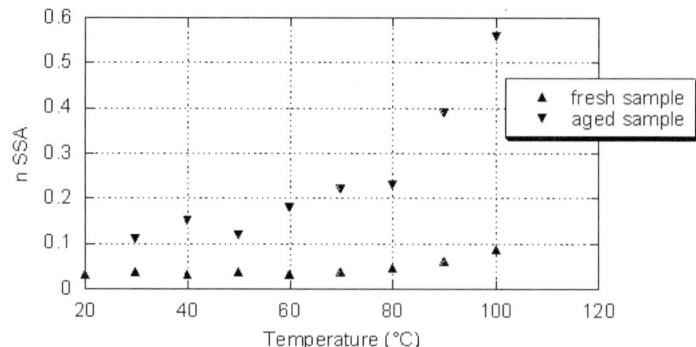

Fig. 27 Temperature dependence of nSSA of physically aged PMMA at a strain rate of $2 \times 10^{-3} \ s^{-1}$

Finally, it is important to emphasise that in the proposed molecular mechanism, the absence or development of intramolecular cooperativity between the ester side group motion and the change of internal rotation angles of the main chain, is the determinant feature for the temperature dependence of PMMA plastic behaviour, particularly of strain softening amplitude. This is supported by the results obtained on PMMA under pressure. Applying pressure leads to an increase in molecular packing, as does a decrease of temperature. Thus, if intermolecular effects were responsible for strain softening amplitude, one should find under pressure a behaviour similar to that observed at low temperature. Actually, the opposite result is obtained, i.e. increasing pressure is equivalent to increasing temperature. This proves that the large strain softening occurring at low temperatures does not originate from intermolecular effects but, as proposed here, from an absence of intramolecular cooperativity between the ester side group motion and chain distortion.

3.1.2
Micromechanisms of Deformation of PMMA

In order to understand fracture behaviour, it is important to analyse the types of deformation micromechanisms undergone under strain: chain scission craze (CSC), shear deformation zone (SDZ), chain disentanglement craze (CDC) and the temperature range over which each one occurs. Furthermore, it is worth wondering whether these micromechanisms are related to β transition motions.

3.1.2.1
Thin Film Investigation

The thin film technique described in Sect. 2 has been applied to PMMA [37, 38] from room temperature to the glass–rubber transition temperature.

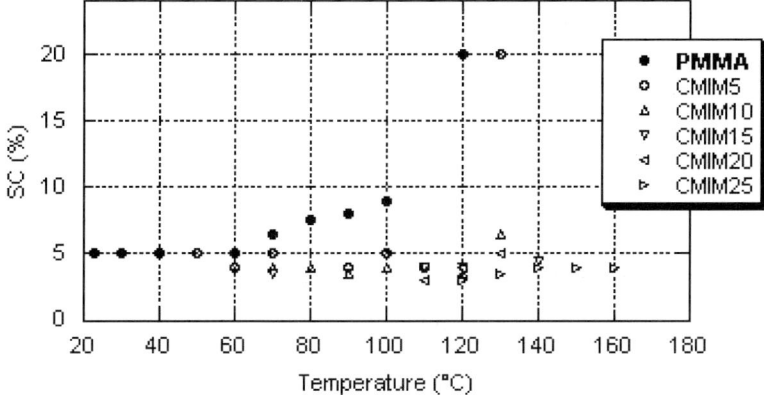

Fig. 28 Temperature dependence of the strain to craze for PMMA thin films at a strain rate of 2×10^{-3} s^{-1}

In the whole temperature range to $(T_\alpha - 20$ K), crazes are observed in PMMA films. However, the strain to craze shown in Fig. 28 exhibits two quite different regimes.

From room temperature to 60 °C, crazes are generated at a strain value around 5%, independently of temperature. At higher temperature, a much larger strain is required, which increases with temperature. Furthermore, the number of crazes significantly decreases above 60 °C. Such a break in strain to craze is interpreted as a change from CSCs to CDCs. Indeed, by investigating, with the same technique, PMMA either monodisperse ($M_n = 100\,250$ g mol^{-1}) or blended with 5% of low MW PMMA ($M_n = 3000$ g mol^{-1}), it has been found [39] that a change from CSC to CDC occurs at 65 and 50 °C, respectively.

In addition to the observed crazes, at 50 °C and above, diffuse SDZs appear at the craze tip, at an angle of about 45° towards the strain axis, as shown in Fig. 29.

Finally, close to T_g (at about 20 °C), the films deform homogeneously.

3.1.2.2
Relation of Micromechanisms of Deformation to β Transition Motions

As above mentioned, SDZs appear in PMMA above 50 °C. Furthermore, it is also in this temperature range that the crazing mechanism shifts from CSC to CDC.

Both behaviours can be accounted for by considering the effect of the co-operativity of the β transition motions. Indeed, it is precisely in this temperature range that α precursor motions develop (as described in [1], Sect. 8.1.6). Thus, the occurrence of such cooperative motions is able to generate large chain motions, analogous to those involved in the glass–rubber transition, in

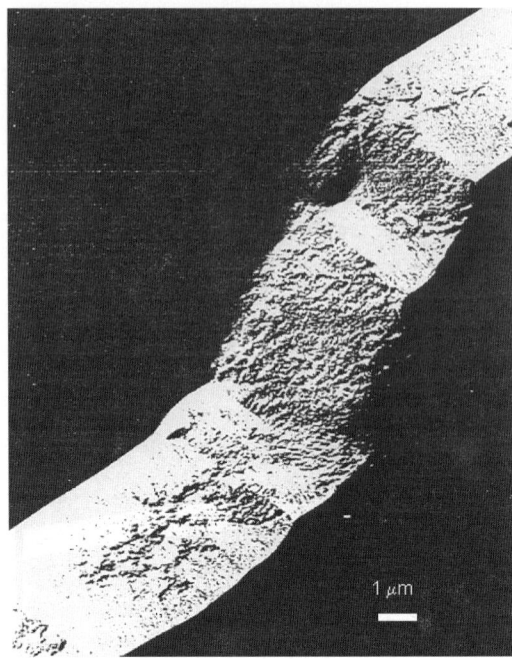

Fig. 29 Typical example of shear deformation zone. Thin film of PMMA at 53 °C and a strain rate of 2×10^{-3} s^{-1}

such a way that chain scissions are less favoured than chain slippage involved in SDZs or in CDCs.

3.1.3
Fracture Behaviour of PMMA

3.1.3.1
Fracture Characteristic, K_{Ic}

Many studies [40–43] have been performed on the fracture behaviour of PMMA as a function of temperature or cross-head speed. As an illustration, Fig. 30 shows the critical stress intensity factor, K_{Ic}, in a log–log plot as a function of temperature for various crack speeds [40]. The temperature range is limited to + 80 °C in order to avoid ductile tearing. In the stable crack growth regime of interest here, whatever the crack speed, K_{Ic} decreases with increasing temperature.

At each temperature, K_{Ic} increases with crack speed and a linear log–log dependence is observed. However, when considering the slopes, two regimes clearly appear, depending on temperature, with a transition between − 20 °C and 0 °C. In the low temperature range a lower sensitivity to crack speed is

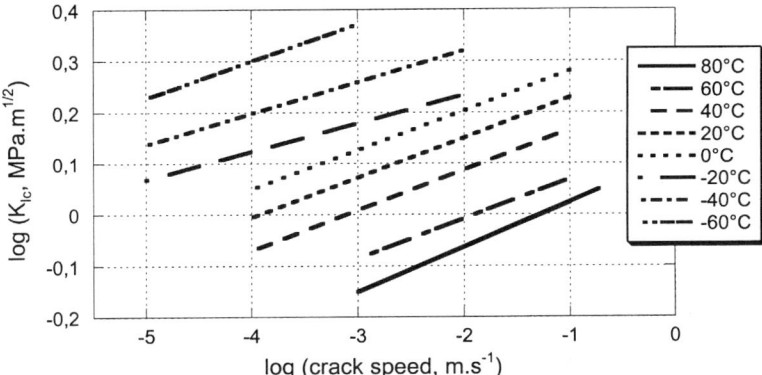

Fig. 30 Fracture toughness, K_{Ic}, as a function of crack speed at various temperatures for stable crack growth (From [40])

observed. Furthermore, as the temperature decreases, the lines are displaced along the log(crack speed) axis in accordance with an Arrhenius-type equation. It leads to an activation energy of 82 kJ mol⁻¹ for the high temperature range and 66 kJ mol⁻¹ for the low temperature range. These values are close to the activation energy of the β transition of PMMA.

3.1.3.2
Micromechanics of Fracture

Investigation of the crack tip of PMMA after a stable crack growth reveals that, at room temperature, a single craze is present.

A convenient technique for studying the crack tip craze propagation in amorphous polymers deals with optical interferometry. It has been applied to the examination of PMMA behaviour at various temperatures and crack speeds under conditions of stable propagation [44, 45].

Interestingly, under static loading conditions and at constant velocity (1 μm s⁻¹) it appears [46] that below – 20 °C multiple crazes are formed at the crack tip, whereas above this temperature a single craze propagates. Thus, the sharp increase of K_{Ic} below – 20 °C, as seen in Fig. 31, originates from the higher energy dissipation due to multiple crazing.

In the higher temperature range, where a single craze develops at the crack tip, it has been shown [46] that a breaking time τ of the craze fibrils under a stress, σ_c, can be defined from the craze length S and the crack-craze velocity v_c by:

$$\tau\left(\sigma_c\right) = S/v_c \tag{42}$$

Furthermore, this breaking time is thermally activated. The corresponding activation energy, 96 kJ mol⁻¹, completely agrees with the activation energy of

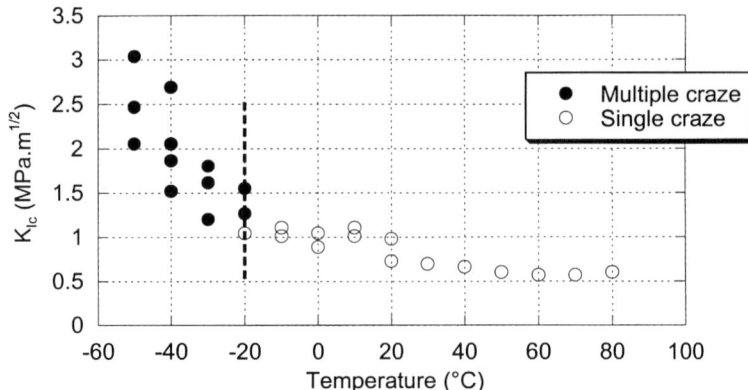

Fig. 31 Fracture toughness, K_{Ic}, at about $10\,\mu m\,s^{-1}$ for PMMA as a function of temperature (From [46])

the β transition of PMMA. So, the breakage of the microscopic craze fibril itself is controlled by β processes. This means that in this temperature range, the craze fibrils break by chain slippage or disentanglement rather than by chain scission. Such a break should happen in the oldest part of the fibril (i.e. the midrib) as is actually observed. Consequently, $\tau(\sigma_c)$ can be considered as a disentanglement time.

Besides, analysing the effect of MW of PMMA on the increase in craze dimensions with time [44], leads to the conclusion that the molecules in the craze fibrils are not fully stretched, retaining folds and thus keeping entangled. So, it supports the consideration of $\tau(\sigma_c)$ as a disentanglement time.

Finally, it is worth mentioning that, looking at the material characteristics undergoing a significant change of their temperature dependence in the

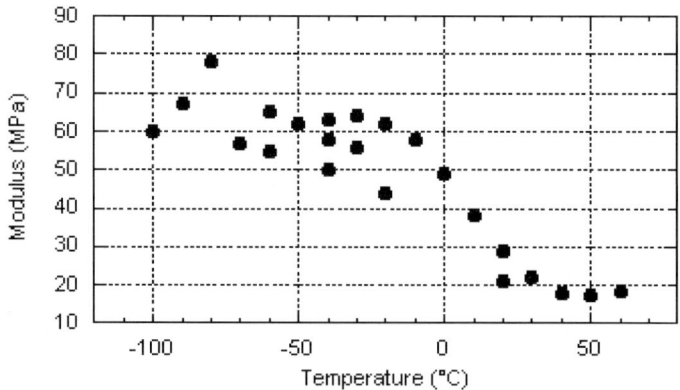

Fig. 32 Young's modulus of craze versus temperature at zero strain (From [47])

temperature range around $-20\,^{\circ}\text{C}$ [47], it turns out that only the Young's modulus of the craze shows such a change, as shown in Fig. 32. Thus, it has been proposed [47] that the occurrence of multiple crazes at temperatures below $-20\,^{\circ}\text{C}$, would be due to the increase in the stiffness of the craze, which would hinder the craze to open enough to cancel out the stress singularity at the crack tip. Then the material would create another type of crack tip when the crack propagates: multiple crazes.

3.1.3.3
Relation of Fracture Behaviour to β Transition Motions

First, let us examine the effect of the β transition on the micromechanics. Considering on the one hand the temperature dependence of the dynamic mechanical loss of PMMA presented in Fig. 26, and on the other hand the equivalent frequency range (10^{-1} to 10^{-2} Hz) corresponding to the optical interferometry investigation, it appears that the temperature ($-20\,^{\circ}\text{C}$) at which multiple crazes develop at the crack tip corresponds either to the low temperature side or to the maximum of the β peak. In this temperature range (as described in [1], Sect. 7.1), the β transition motions are isolated π-flips of the ester groups, without cooperativity with the main chain. Under these conditions, the fibrils can no longer break by creep in the midrib and some other mechanism takes place [45]: craze breaking in a rather disordered way or multiple crazes.

In contrast, in the high temperature range of the β transition, the cooperative motions (implying π-flips of ester groups coupled with changes of the internal rotation angles of the main chain) allow chain slippage to occur, leading to a single craze development with a fibril break by creep.

Besides, the lower sensitivity of K_{Ic} to the crack speed, observed in Fig. 30 at temperatures below $-20\,^{\circ}\text{C}$, has to be related to the occurrence of the multiple craze mechanism (it does not require any slippage time, but only a time for fibrillation) and, so, to the isolated β transition motions, as described above.

Another feature dealing with the β transition concerns the occurrence of unstable crack propagation. Indeed, other data [40] show that, in the very low temperature range (below $-90\,^{\circ}\text{C}$) unstable crack propagation happens even at very low crack speed. This temperature range is located lower than the onset of the β transition. To explain the unstable crack propagation, it has been proposed [44] that the fibrillation rate at the craze tip is decelerated, leading to a craze shortening. Thereby, the craze stress also increases leading to an acceleration of fibril failure, which generates an adiabatic heating inducing a softening of the fibril material and, thus, favouring fibril failure. This self-accelerating process leads to the unstable crack propagation.

So, it is quite clear that β transition motions are involved in the fracture behaviour of PMMA. The continuous smooth decrease of K_{Ic} with increasing temperature above $-20\,°C$ could result from the increase of mobility, which allows chain fragments in the CSC fibrils to disentangle. Finally, investigation of the micromechanisms of deformation of thin films of PMMA (Sect. 3.1.2) has shown that SDZs appear at $50\,°C$. One could expect that the interaction of crazes and SDZs would increase the toughness of PMMA above $50\,°C$; this is not observed, contrary to what happens with other polymers, e.g. the methyl methacrylate-*co*-*N*-methyl glutarimide copolymers considered in Sect. 3.3.

3.2
Methyl Methacrylate-*co*-*N*-cyclohexyl Maleimide Random Copolymers

The structure of the methyl methacrylate-*co*-*N*-cyclohexyl maleimide copolymers is shown in Fig. 33. The maleimide cycle is a rigid structure whose $C-C$ bond contributes to the chain backbone.

   These copolymers will be coded CMIM*x*, where *x* indicates the molar % of the CMI units in the copolymer. Their T_gs as well as their T_αs at 1 Hz are gathered in Table 1. It is worth noting the strong dependence of T_g and T_α with the CMI content.

   Prepared by radical polymerisation of MMA and maleimide, the copolymers have a random distribution of the maleimide units.

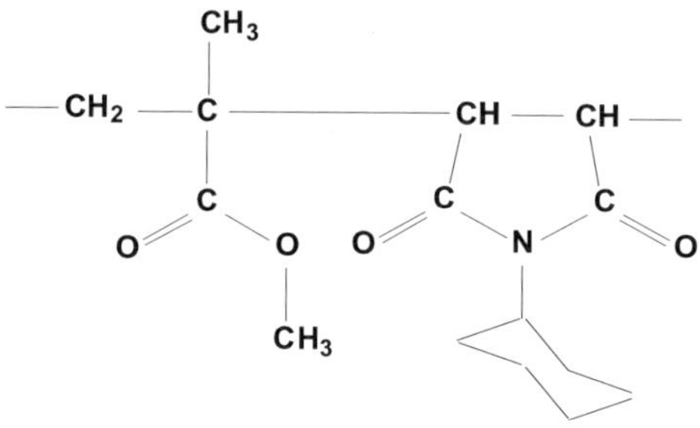

Fig. 33 Chemical structure of CMIM*x* copolymers

Table 1 Characteristics of PMMA and CMIM*x* copolymers

Polymer code	M_w g mol^{-1}	M_n g mol^{-1}	T_g °C	T_α °C	M_e g mol^{-1}	ν_e 10^{24} m^{-3}	m_{eq} g mol^{-1}	n_{eq}	l_{eq}	N_e	$C_{\infty eq}$
PMMA	118 500	66 000	116	126	9200	77	100	2	1.54	184	7.8
CMIM 5	118 500	65 000	122	132	10 000	69	109	2.05	1.65	189	7.9
CMIM 10	126 000	69 000	129	139	13 000	53	120	2.11	1.77	229	8.7
CMIM 15	151 500	73 500	135	145	20 000	35	132	2.18	1.89	331	10.5
CMIM 20	134 000	67 500	143	153	28 000	25	145	2.25	2.02	434	12
CMIM 25	156 000	71 000	151	161	40 000	17	160	2.33	2.16	583	13.9

3.2.1
Plastic Deformation of CMIMx Copolymers

3.2.1.1
Stress–Strain Curves

First, it is important to point out that the stress–strain curves and the derived characteristics have been studied in a temperature range extending to $T_\alpha - 20\,°\text{C}$, in order to avoid any creep behaviour.

The stress–strain curves obtained [32] for the whole series of CMIMx as well as for PMMA, as a reference, are shown in Fig. 34. It is clear that all the CMIMx exhibit a strain softening which is more and more pronounced when the CMI content increases. Quite a similar behaviour is observed at any temperature in the range 25 to $(T_\alpha - 20)\,°\text{C}$. It is worth noting that at lower temperatures, the CMIMx copolymers are brittle and, so, for CMIM25 the lowest possible temperature is $50\,°\text{C}$.

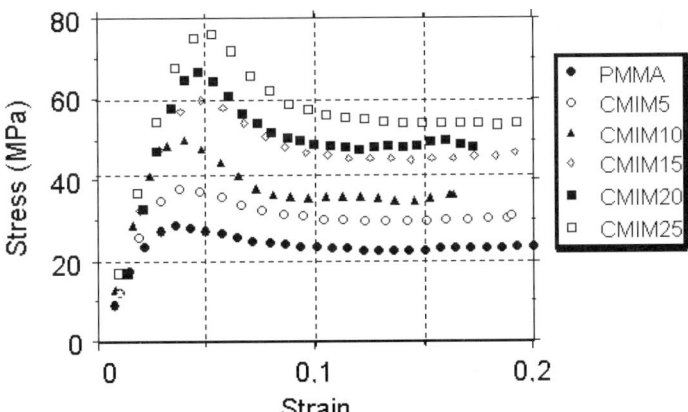

Fig. 34 Stress–strain curves at $100\,°\text{C}$ for PMMA and CMIM copolymers (strain rate of $2 \times 10^{-3}\,\text{s}^{-1}$) (From [32])

3.2.1.2
Yield Stress

The temperature dependence of the yield stress, σ_y, for the various CMIMx is shown in Fig. 35. For comparison, Fig. 36 gathers the σ_y values versus temperature for PMMA and all the CMIMx copolymers. Whereas all the curves converge at room temperature, σ_y increases with increasing CMI content at higher temperatures.

As for PMMA, it is interesting to check to what extent σ_y is controlled by the Young's modulus, E. The temperature dependence of the ratio σ_y/E is

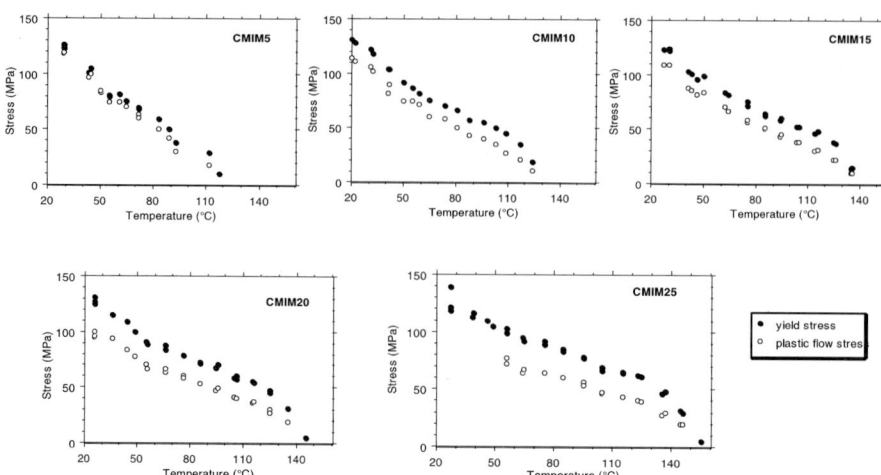

Fig. 35 Temperature dependence of the yield stress, σ_y, and plastic flow stress, σ_{pf}, for CMIM5, CMIM10, CMIM15, CMIM20 and CMIM25 (strain rate of 2×10^{-3} s^{-1}) (From [32])

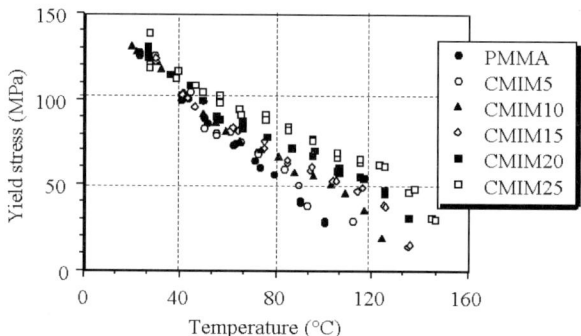

Fig. 36 Yield stress of PMMA and various CMIMx copolymers versus temperature

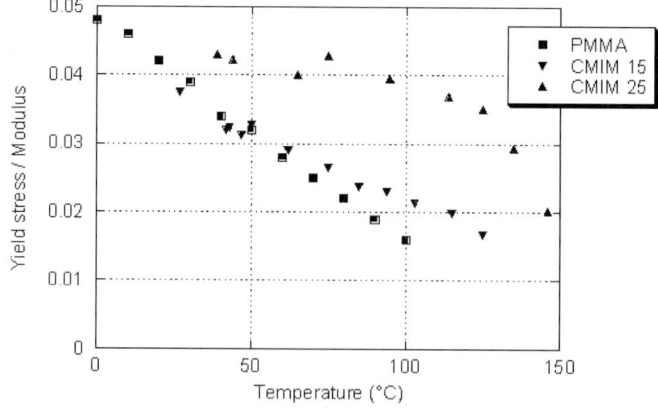

Fig. 37 Temperature dependence of σ_y/E for PMMA and various CMIMx copolymers

shown in Fig. 37 for PMMA and various CMIMx. It appears that in the investigated range (20 to $(T_\alpha - 20)$ °C), the ratio σ_y/E increases relatively to the values for PMMA, this effect being larger and larger when temperature increases.

3.2.1.3
Plastic Flow Stress

The temperature dependence of the plastic flow stress, σ_{pf}, is shown in Fig. 35 and Fig. 38. In the latter, very close σ_{pf} values are observed for PMMA and the various CMIMx up to 40 °C. At higher temperatures an increase of σ_{pf} with increasing CMI content is observed, similar to that seen for σ_y.

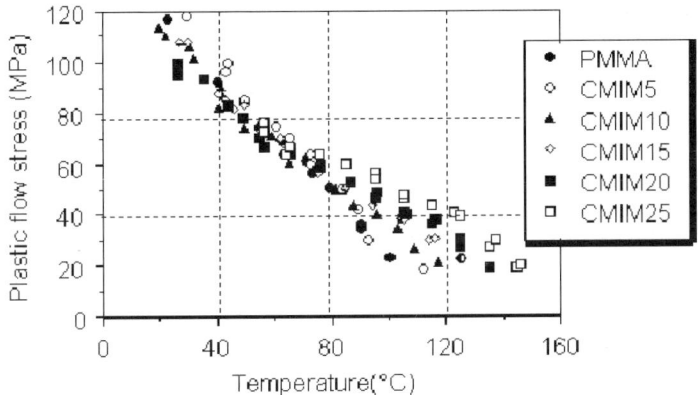

Fig. 38 Plastic flow stress of PMMA and various CMIMx copolymers versus temperature

3.2.1.4
Strain Softening

As shown in Fig. 35, the strain softening is more and more pronounced with increasing CMI content, whatever the temperature.

The strain softening amplitude, SSA, is plotted versus temperature in Fig. 39. The SSA values appear almost constant to 15 mol %, though at higher amounts a small decrease happens with increasing temperature.

SSA normalised by σ_{pf} (nSSA) is shown in Fig. 40 with those of quenched and physically aged PMMA. For comparing the temperature dependence, due to the large differences in T_α, it is interesting to plot nSSA versus $(T - T_\alpha)$, as shown in Fig. 41. In addition to the higher nSSA values obtained with increasing the CMI content, already manifested in SSA (Fig. 39), it is worth pointing out that, whereas nSSA is almost constant over a broad temperature range for PMMA, in the case of CMIMx copolymers a continuous increase occurs with

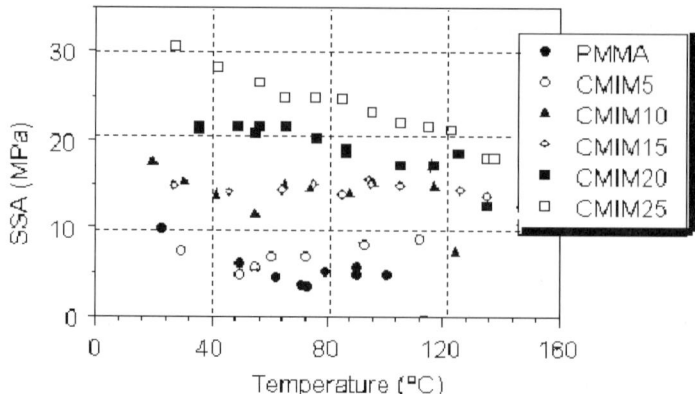

Fig. 39 SSA of PMMA and various CMIM*x* copolymers versus temperature

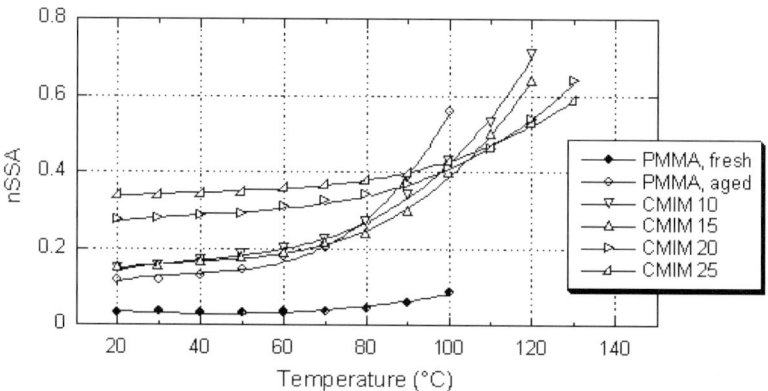

Fig. 40 Temperature dependence of nSSA of quenched and physically aged PMMA and various CMIM*x* copolymers

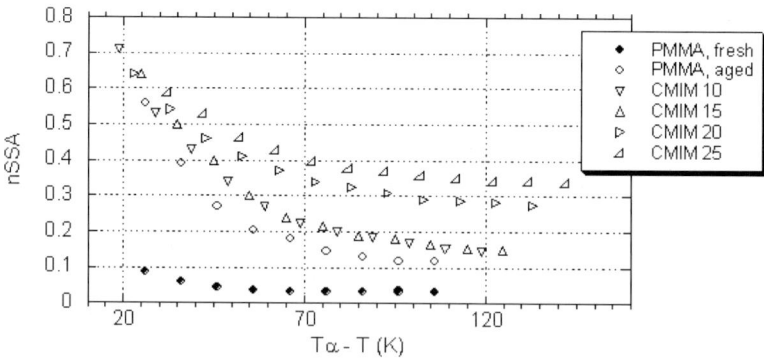

Fig. 41 nSSA versus ($T_\alpha - T$) for quenched and physically aged PMMA and various CMIM*x* copolymers

increasing temperature. Such a behaviour looks like that observed for aged PMMA.

3.2.1.5
Relation of Plastic Deformation and β Transition Motions

In the same way as for PMMA, the whole chain motions responsible for the plastic flow of CMIMx copolymers are identical to those occurring at the α transition temperature, whatever the considered temperature at which the plastic flow happens. Thus, plastic flow and the corresponding stress can be considered as a reference, as for PMMA.

Before considering the effect of β transition motions on the plastic deformation of CMIMx copolymers, it is useful to recall the information gathered on the β transition of these copolymers.

As described in [1] (Sect. 8.2 and summarised in Sect. 8.2.5) the introduction of the rigid CMI units within the PMMA chain backbone hinders the coupling between the ester group π-flips and the main chain adjustment occurring in the high temperature part of the β transition of pure PMMA. Whereas such a coupling still exists in long-enough MMA sequences, the probability of these sequences sharply decreases when the CMI content increases and, consequently, the cooperativity of the β transition motions disappears. In particular, a few mol% of CMI units lead to a behaviour in the β–α crossover temperature region which is equivalent to that is obtained by physical ageing of PMMA.

On another hand, dynamic computer modelling performed on isolated CMIMx chains [48] shows that conformational changes of MMA units are more difficult in the neighbourhood of CMI units. This result is confirmed by the ^{13}C NMR measurements in solution [49] (reported in [1], Sect. 8.2).

The molecular mechanism, described in Sect. 3.1.1.6, for the yielding and strain softening of PMMA accounts quite well for the plastic deformation behaviour of CMIMx copolymers.

Indeed, in the high temperature range of the β transition, which corresponds to the temperatures where the plastic deformation of CMIMx can be performed, the hindering of the ester group cooperativity, when the CMI content increases, leads to a gradual change from intramolecular cooperative to isolated motions. According to the considered mechanism, the local conformational changes of the main chain involved in the yielding process have to occur in a medium where only localised motions are present and, consequently, which has not been softened by the intramolecular cooperative β motions. In addition, the conformational changes of the main chain, implied in the yielding process, require the displacement of the rigid bulky CMI unit. These two effects lead to a larger difference between σ_{pf} and σ_y, i.e. to larger strain softening. The effect increases with the CMI content (Fig. 39). Furthermore, another effect of the CMI content, leading to an increase in the

σ_y values, deals with the fact that conformational changes of the MMA units close to CMI groups are more difficult. Such an occurrence increases with the CMI content.

It is worth noting that the same trend observed for the influence of physical ageing of PMMA, and for introduction of CMI units, on the decrease of α precursor motions is well reflected in the development of strain softening.

3.2.2
Micromechanisms of Deformation of CMIMx Copolymers

3.2.2.1
Thin Film Investigation

A series of CMIMx copolymers has been studied [38] by using the thin film technique in the temperature range 20 °C to T_α.

First, the high brittleness of the CMIMx copolymers with CMI content above 10 mol % is clearly manifested by the difficulty in avoiding film breakage before straining.

Except for CMIM5, which shows an homogeneous deformation close to T_g, similar to that of PMMA, all the other CMIMx copolymers develop crazes over the whole investigated temperature range, as indicated in Table 2. Furthermore, these crazes are more and more numerous when the CMI content increases. However, as shown in Fig. 28, with increasing temperature (except for CMIM5) no change in the strain to craze, similar to that observed for PMMA, indicates the occurrence of chain disentanglement crazes.

Shear deformation zones appear in all the CMIMx copolymers, except CMIM25, above a given temperature, denoted T_{12}. These SDZs are similar to

Table 2 Temperature dependence of the deformation mechanisms in thin films of PMMA and CMIMx copolymers

Polymer code	Investigated temperature range	Crazing temperature range	Shear deformation temperature range	Observed T_{12}	Calculated T_{12}	$T_\alpha - T_{12}$
	°C	°C	°C	°C	°C	°C
PMMA	20–126	20–110	50–110	50	Reference	76
CMIM 5	20–132	20–120	60–115	60	55–57	72
CMIM 10	20–139	20–139	100–139	100	65–75	39
CMIM 15	20–145	20–145	130–145	130	85–100	15
CMIM 20	20–153	20–153	140–153	140	120–125	13
CMIM 25	20–161	20–161	none	–	130	–

those observed in PMMA and shown in Fig. 29. The temperature T_{12} increases with the CMI content, as reported in Table 2.

3.2.2.2
Quantitative Approach of the Occurrence of SDZs

As described in Sects. 2.3.1.2 and 2.3.4.2, the critical stress, σ_{csc}, associated with CSCs, can be calculated from the molecular characteristics of the considered polymers and values of the plastic flow stress, σ_{pf}. The temperature at which SDZs appear corresponds to the intersect of σ_{csc} with the yield stress, $\sigma_y(T)$. However, as mentioned in Sect. 2.3.4.2, a reference compound is required to determine the proportionality constant involved.

In the case of CMIMx copolymers, the choice of PMMA as a reference is quite suitable. Indeed, the temperature, T_{12}, at which SDZs appear in PMMA is around 50 °C, the values of $\sigma_{pf}(50)$ and $\sigma_y(50)$ being 91 and 96 MPa, respectively.

The required molecular characteristics are:

M_e MW between entanglements
N_e Number of bonds corresponding to M_e
l Bond length
C_∞ Characteristic ratio

C_∞ can be estimated [50] from the relationship:

$$C_\infty = (N_e/3)^{1/2} \tag{43}$$

The other required characteristics are $\sigma_{pf}(T)$ and $\sigma_y(T)$ for the reference PMMA and the polymers under consideration.

Whereas the determination of N_e is straightforward for PMMA from the experimental value of M_e, all bonds having the same length (i.e. 1.54 Å) such determinations are more complex for the CMIMx copolymers. Indeed, due to the presence in the same chain of MMA repeat units and CMI units, each with a different number of bonds of different lengths, a new set of repeat units has to be defined, as well as an equivalent virtual chain.

First, let us split the CMIMx chain into the following units:

• MMA:

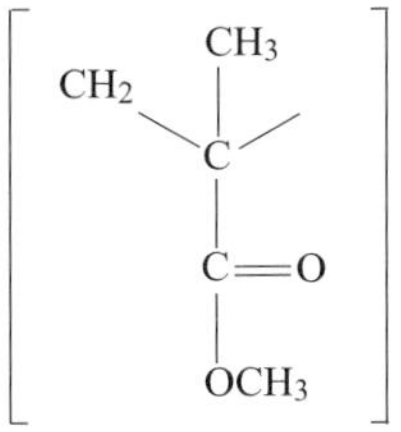

Scheme 1

The molar mass is $100\,\mathrm{g\,mol^{-1}}$ and it contains two rotating bonds with a length of $1.54\,\text{Å}$ each.

- MMA-CMI:

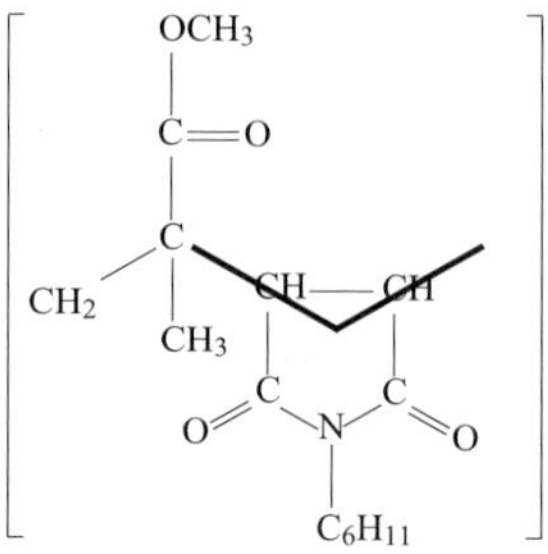

Scheme 2

The molar mass is $279\,\mathrm{g\,mol^{-1}}$ and it contains only three rotating bonds for no rotation can occur around the maleimide CH – CH bond. Among the three rotating bonds, one is a real one, corresponding to the included MMA unit, with a length of $1.54\,\text{Å}$, the two other ones are virtual bonds, (shown by bold lines) with a length of $3.70\,\text{Å}$ and an angle of $109°$ between them.

The equivalent chain corresponding to a CMIMx copolymer chain has the following characteristics:

1. A repeat unit of mass:

$$m_{\mathrm{eq}}(x) = (10\,000 + 79x)/(100 - x) \tag{44}$$

2. A number of bonds per repeat unit:

$$n_{\mathrm{eq}}(x) = (200 - x)/(100 - x) \tag{45}$$

3. A length of bond:

$$l_{\mathrm{eq}}(x) = (308 + 2.78x)/(200 - x) \tag{46}$$

4. A number of bonds between entanglements:

$$N_{\mathrm{e}}(x) = \left(M_{\mathrm{e}}(x)/m_{\mathrm{eq}}(x)\right) n_{\mathrm{eq}}(x) \tag{47}$$

The corresponding values, as well as the experimental values of M_{e} and the calculated C_∞ are gathered in Table 1 for PMMA and the CMIMx copolymers.

By applying the quantitative approach, described in Sect. 2.3.4.2, and the experimental results on $\sigma_{\mathrm{pf}}(T)$ and $\sigma_{\mathrm{y}}(T)$, the temperature, T_{12}, at which the SDZs appear can be determined. The ranges of calculated T_{12} values are reported in Table 1.

The quantitative approach accounts well for the observed increase of T_{12} when the CMI content increases. Furthermore, considering the uncertainties of experimental determination of both T_{12} and M_{e} (which controls N_{e} and C_∞ of the equivalent chain) the agreement seems reasonable.

3.2.2.3
Relation to the Molecular Structure

As already mentioned, the introduction of rigid CMI cycles within the PMMA chain hinders the cooperativity of the β transition motions. This effect plays against the chain slippage and, thus, the occurrence of SDZs.

However, in the case of CMIMx copolymers, the development of crazes with increasing CMI content can also originate from the large increase of M_e and, consequently, from the decrease of the entanglement density, ν_e, (Table 1). Indeed, small ν_e values favour the development of crazes [19], independently of the chemical structure, as described in Sect. 2.3.3.

3.2.3
Fracture of CMIMx Copolymers

As above mentioned, the presence of CMI units within the PMMA chain backbone makes the sample considerably more brittle. Such an effect is already reflected in the type of deformation micromechanisms: more and more crazes when the CMI content increases.

In spite of such difficulties, fracture characteristics of three CMIMx copolymers (i.e. CMIM10, CMIM15 and CMIM25) have been determined at room temperature (Mariot, private communication), using three-point bending samples. The corresponding K_{Ic} values, as well as that of PMMA, are reported in Table 3. It clearly appears that increasing CMI content quite significantly decreases the toughness of the CMIMx copolymers.

When considering the entanglement density, ν_e, reported in Table 1, one finds that K_{Ic} increases with ν_e, as shown in Fig. 42a and expected from [26]. However, the fitting straight line does not pass through the origin. Figure 42b shows the dependence of K_{Ic} with $\nu_e^{1/2}$, according to [30, 31]. In spite of a fitting line passing closer to the origin, it is not possible to decide what relationship is the more appropriate.

Table 3 K_{Ic} values obtained at room temperature for CMIMx copolymers

Polymer code	K_{Ic} [MPa m$^{1/2}$]
PMMA	1.2
CMIM 10	1.0
CMIM 15	0.8
CMIM 25	0.7

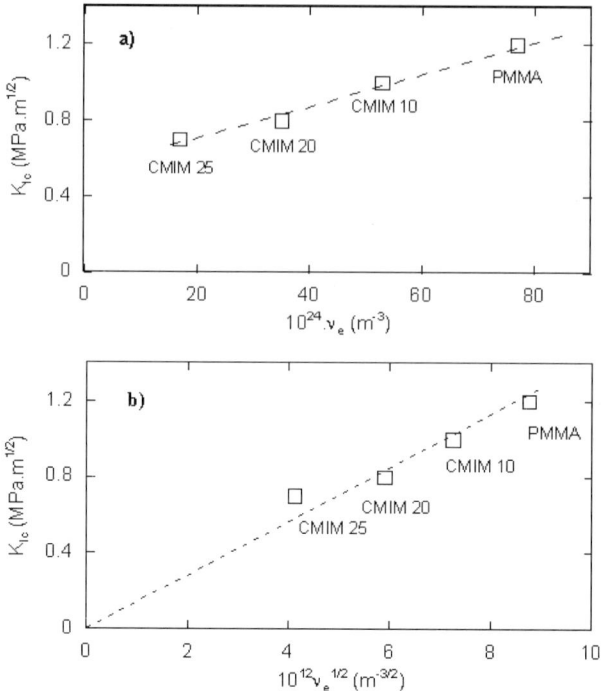

Fig. 42 Dependence of K_{Ic}, determined at room temperature on **a** ν_e, **b** $\nu_e^{1/2}$

3.3
Methyl Methacrylate-*co*-*N*-methyl Glutarimide Random Copolymers

Another series of methyl methacrylate copolymers is constituted by the methyl methacrylate-*co*-*N*-methyl glutarimide copolymers, whose the chemical structure is shown in Fig. 43.

These copolymers will be coded MGIMx, where x indicates the molar % of the glutarimide units in the copolymer. Their T_gs and T_αs at 1 Hz are given in Table 4. It is worth noting that T_gs and T_αs increase much less with the MGI content than with the CMI unit one (Table 1). Indeed, the CMI cycle is more rigid than the MGI one and, furthermore, the cyclohexyl group of CMI is bulkier than the methyl one of MGI.

An interesting feature of this series is that, contrary to the case of CMIMx copolymers, the entanglement characteristics, M_e and ν_e, are slightly dependent on the composition, as it can be seen in Table 4.

The copolymers are obtained by the reaction, in solution, of methylamine with PMMA, leading to glutarimide cycles randomly distributed along the original PMMA chain backbone.

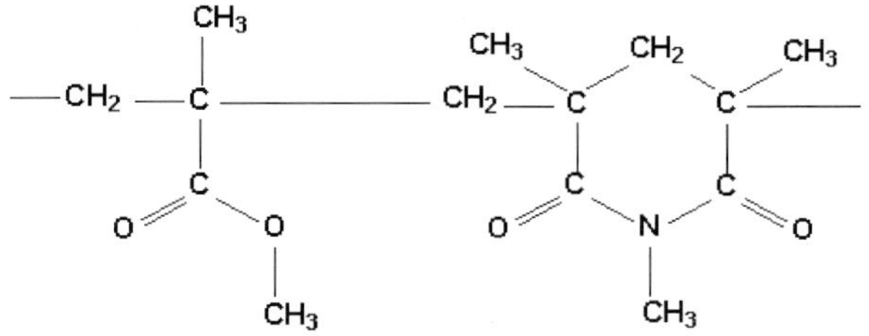

Fig. 43 Chemical structure of MGIM*x* copolymers

Table 4 Characteristics of PMMA and MGIM*x* copolymers

Polymer code	M_w g mol^{-1}	M_n g mol^{-1}	T_g °C	T_α °C	M_e g mol^{-1}	ν_e 10^{24} m^{-3}	m_{eq} g mol^{-1}	n_{eq}	l_{eq}	N_e	$C_{\infty eq}$
PMMA	119 000	66 000	112	119	9200	77	100	2	1.54	184	7.8
MGIM 4	75 000	37 000	115	120							
MGIM 21	80 000	37 000	121	127	11 500	60	114	2	1.89	201	8.2
MGIM 36	76 000	37 000	127	134	11 900	58	124	2	2.15	192	8.0
MGIM 58	110 000	50 000	140	147	12 400	56	139	2	2.52	178	7.7
MGIM 76	106 000	49 000	151	158	12 400	56	151	2	2.83	164	7.4

3.3.1
Plastic Deformation of MGIM*x* Copolymers

As for PMMA and CMIM*x* copolymers, the measurements have been performed to $(T_\alpha - 20 \text{ K})$ to avoid any creep effect.

3.3.1.1
Yield Stress

The temperature dependence of the yield stress and the plastic flow stress for PMMA and a large series of MGIM*x* copolymers [35] is shown in Fig. 44. For comparison purpose, the yield stress of PMMA and various MGIM*x* are presented together in Fig. 45.

It clearly appears that only slight changes of σ_y values are induced by the MGI content, even for MGI amounts as large as 76 mol%. A similar behaviour is obtained by considering the ratio σ_y/E.

3.3.1.2
Plastic Flow Stress

Figure 46 presents the temperature dependence of the plastic flow stress for PMMA and the various MGIM*x* copolymers. As it could be expected from Fig. 44, the behaviour of σ_{pf} is quite similar to that of σ_y.

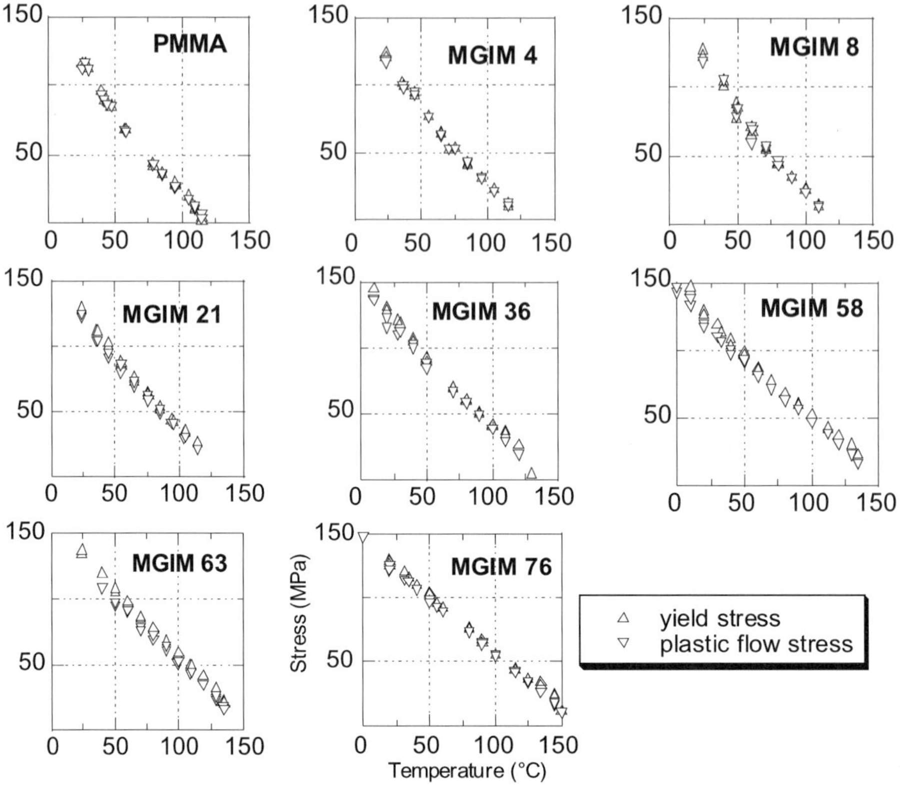

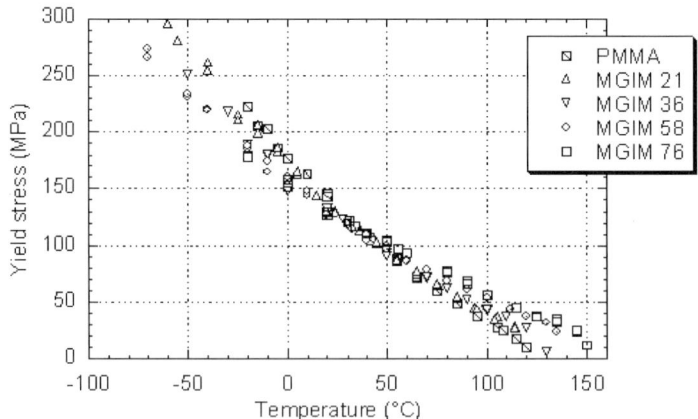

Fig. 44 Yield stress and plastic flow stress versus temperature (strain rate of 2×10^{-3} s^{-1}) for PMMA and various MGIMx copolymers (From [35])

Fig. 45 Temperature dependence of the yield stress for PMMA and the indicated MGIMx copolymers (From [52])

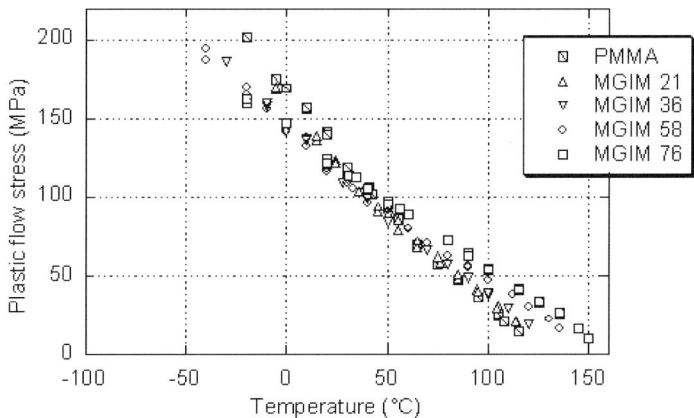

Fig. 46 Temperature dependence of the plastic flow stress for PMMA and the indicated MGIM*x* copolymers (From [52])

3.3.1.3
Strain Softening

The temperature dependence of the strain softening amplitude, SSA, is shown in Fig. 47. First, it is worth noticing the similarity between the behaviours of the four considered polymers, (i.e. PMMA, MGIM36, MGIM58 and MGIM76).

At low temperatures, typically below room temperature, an increase of the SSA values occurs with decreasing temperature, but the effect of the comonomer content tends to vanish.

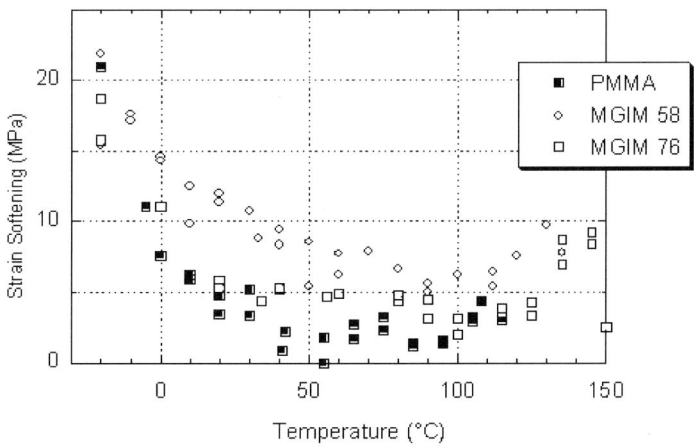

Fig. 47 Temperature dependence of SSA for PMMA and the indicated MGIM*x* copolymers (From [52])

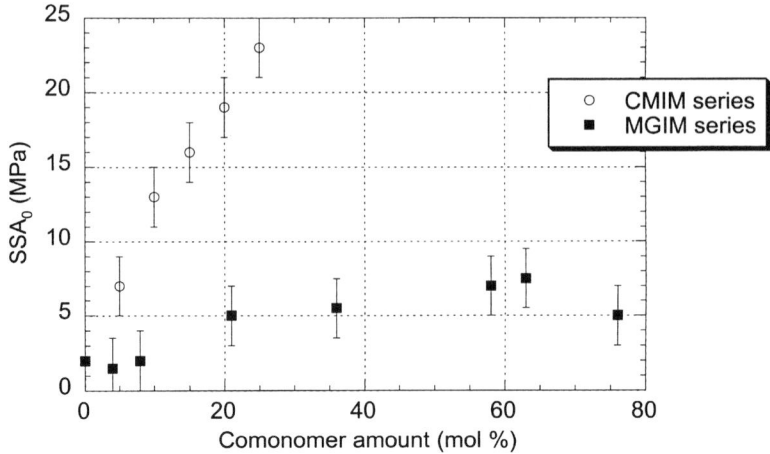

Fig. 48 Plateau value of SSA, SSA_0, versus copolymer composition in MGIMx copolymers (From [35])

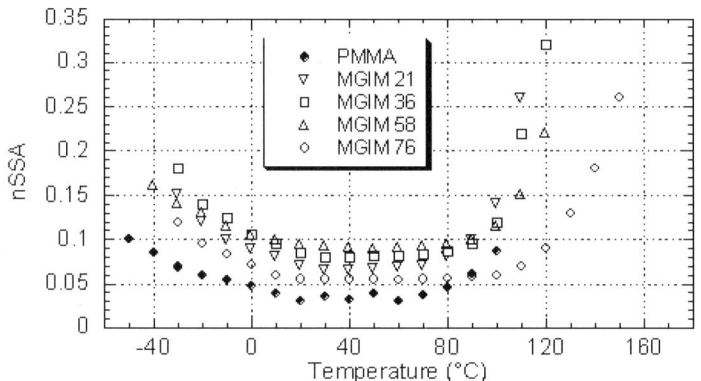

Fig. 49 Temperature dependence of nSSA for PMMA and the indicated MGIMx copolymers

At temperatures above room temperature, a leveling of the SSA values is observed which extends to $(T_\alpha - 20\,K)$. Interestingly, the corresponding plateau value, SSA_0, depends on MGI content, as shown in Fig. 48. SSA_0 values pass through a maximum for a MGI content around 60 mol%, then significantly drops for MGIM76.

The normalised SSA, nSSA, is shown in Fig. 49. The similarity of the curves observed for SSA is maintained with nSSA. In all the considered polymers, the decrease of nSSA occurring at low temperatures is followed by a plateau extending over a broad temperature range; the change in behaviour happens around $0 \pm 10\,K$. As regards the plateau value, $nSSA_0$, it depends on the MGI content, as shown in Fig. 50, and passes through a maximum around 60 mol%.

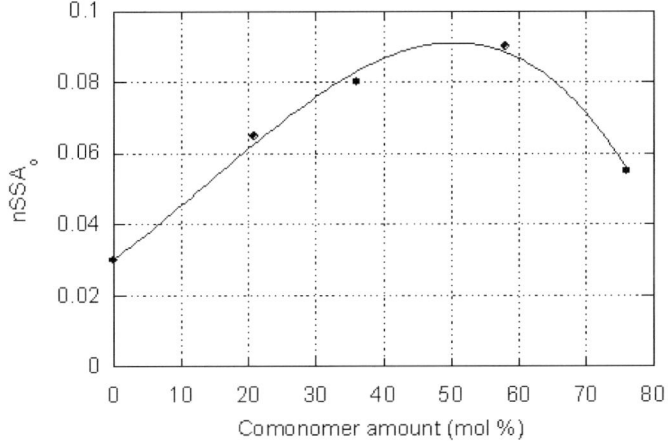

Fig. 50 Plateau value of nSSA, $nSSA_0$, versus copolymer composition in MGIMx copolymers

3.3.1.4
Comparison of Plastic Behaviours of CMIMx and MGIMx Copolymers

It is interesting to analyse the effect of comonomer nature on the plastic behaviour of the copolymers.

First, let us compare two copolymers with the same T_α, CMIM25 and MGIM76 [35] The temperature dependence of the corresponding yield stresses is shown in Fig. 51. It clearly appears that in the high temperature range, the nature of the comonomer greatly influences the yield stress.

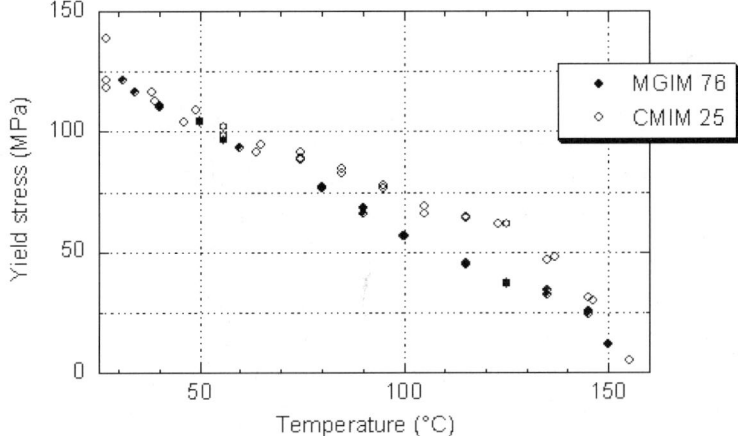

Fig. 51 Comparison of the yield behaviour of MGIM76 and CMIM25 versus temperature (From [35])

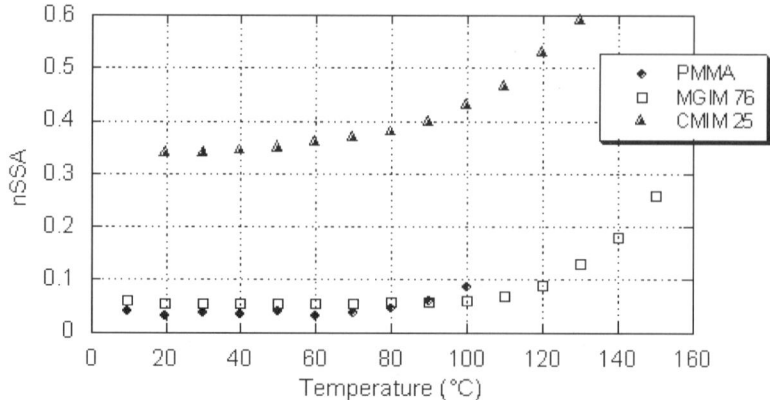

Fig. 52 Temperature dependence of nSSA for PMMA, MGIM76 and CMIM25

It is interesting to look at the temperature dependence of nSSA. The results are shown in Fig. 52, in which PMMA data are also plotted. Whereas PMMA and MGIM76 have plateau values, $nSSA_0$, that are rather close, there is a huge increase with CMIM25. Furthermore, in spite of the same T_α values, the plateau ends up at about 80 °C for CMIM25 instead of 115 °C for MGIM76.

Another interesting feature deals with the composition dependence of the strain amplitude, as shown in Fig. 48. Indeed, over the restricted range of maleimide compositions available, the effect of CMI units is considerably larger than that of MGI units.

3.3.1.5
Relation of Plastic Deformation of MGIM*x* Copolymers and β Transition Motions

The above reported results show that the plastic deformation behaviours of MGIM*x* copolymers are much closer to those of PMMA than those observed for CMIM*x* copolymers.

Concerning the β transition of MGIM*x* copolymers (described in [1] Sect. 8.3), the dynamic mechanical loss, E'', at 1 Hz is shown in Fig. 53. The investigations lead to the following main conclusions:

- In the low temperature part of the β transition, MMA ester group π-flips coupled with MGI distortions happen in sites of lower local density
- In the high temperature part, the MGI cycle takes part in the chain backbone distortions associated with the ester group π-flips and leading to cooperative β motions
- For MGI-rich copolymers, MGI–MGI cooperativity accompanies reorientation of the glutarimide cycle around the local chain axis

Another important point, revealed by ^{13}C NMR studies performed on MGIM*x* in solution [49] and confirmed by dynamic molecular modelling [48]

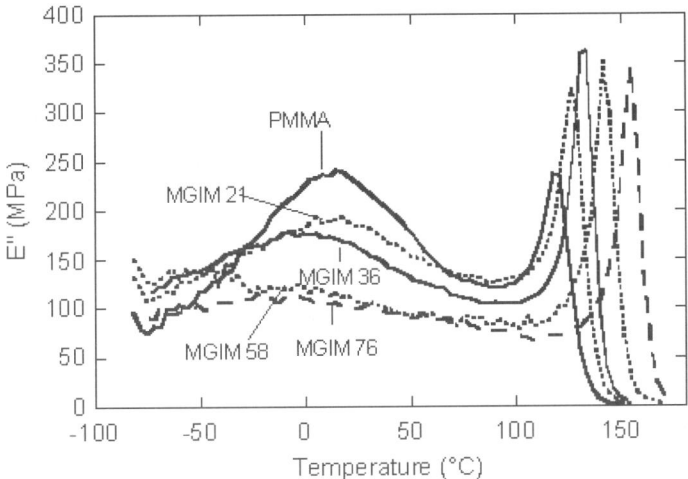

Fig. 53 Temperature dependence of the dynamic mechanical loss modulus, E'', at 1 Hz, for PMMA and various MGIMx copolymers (From [52])

([1], Sect. 8.3.4.2), deals with the fact that the conformational changes of MMA units, within isolated MGIMx copolymer chains, keep the same dynamics whatever the MGI content, in contrast to what happens with CMIMx copolymers ([1], Sect. 8.2.3.2). However, these MMA conformational changes have a faster dynamics than the reorientation of the MGI units.

The plastic deformation results of MGIMx copolymers mainly concern the temperature range above $-10\,°$C, corresponding to the high temperature part of the β transition of PMMA and MGIMx.

Before entering the discussion, it is worth noting that, as mentioned for PMMA, plastic flow implies main-chain motions such as those occurring above the α transition, whatever the considered temperature. Thus, the plastic flow constitutes a reference.

The relation of the plastic deformation of MGIMx copolymers with the corresponding β transition motions is based on the molecular mechanism of yielding developed for PMMA in Sect. 3.1.1.6.

In the MGIMx case, even if the conformational changes of the MMA unit are not affected by the MGI units, the conformational changes of the main chain, implied in the yielding process, require the displacement of the MGI unit. In spite of the allowed distortions of the $C-CH_2-C$ sequence of the MGI unit, the glutarimide cycle motion sweeps out a larger volume than the ester group of the MMA unit. As a consequence, for the same extent of cooperativity of the β motions, a MGIMx copolymer will require a larger value of the yield stress than PMMA needs.

Concerning the strain softening and its molecular origin according to the proposed mechanism, in the temperature range experimentally investigated, it is mostly controlled by the intramolecular cooperativity existing in the MGIMx polymer chain. It has been shown ([1], Sect. 8.3), that the introduction of MGI units decreases the extent of intramolecular cooperativity of the MMA ester group motions until an MGI content of between 36 and 58 mol %. At higher amounts, a new type of intramolecular cooperativity takes place, associated with MGI cycle reorientation. These changes in intramolecular cooperativity, according to the proposed mechanism for yielding and plastic flow, are directly reflected in the strain softening: larger strain softening when MMA intramolecular cooperativity decreases, then the opposite when MGI–MGI intramolecular cooperativity develops. This is what is observed for the MGI content dependence of strain softening, expressed either through SSA or nSSA (Fig. 48 and Fig. 50, respectively), which goes through a maximum around 60–70 mol %.

3.3.2
Micromechanisms of Deformation of MGIMx Copolymers

In contrast to the case of CMIMx copolymers, where M_e rapidly increases with the CMI content (Table 1), the series of MGIMx copolymers corresponds to a set of materials for which M_e is almost constant from 21 to 76 mol% of MGI content (Table 4). For this reason, it is particularly interesting to analyse how the micromechanisms of deformation are affected by the chemical structure.

3.3.2.1
Thin Film Investigation

The thin film technique has been applied to MGIMx copolymers in a temperature range from $-20\,°C$ to T_α [51].

Due to the low MW of MGIM21 and MGIM36, the former was too brittle to be studied by this technique and the latter one was brittle below room temperature and underwent homogeneous deformation above $110\,°C$, which limits the investigated temperature range.

The morphologies observed with MGIMx copolymers are much more complex that in the case of CMIMx compounds. Five different regimes of behaviour have been observed in thin films, as shown schematically in Fig. 54:

1. At low temperature (typically $T < 10\,°C$), long, high-aspect-ratio crazes are observed; an example is shown in Fig. 55a for MGIM76. They correspond to CSCs.
2. At intermediate temperature (between 10 and $80\,°C$ for all the copolymers), two types of deformation are observed to coexist, i.e. crazes and SDZs. At temperatures just above the transition temperature, T_{12}, from CSCs to mixed deformation, the crazes are of a high-aspect ratio, and the SDZs are generally restricted to the craze–bulk interfaces (Fig. 55a), al-

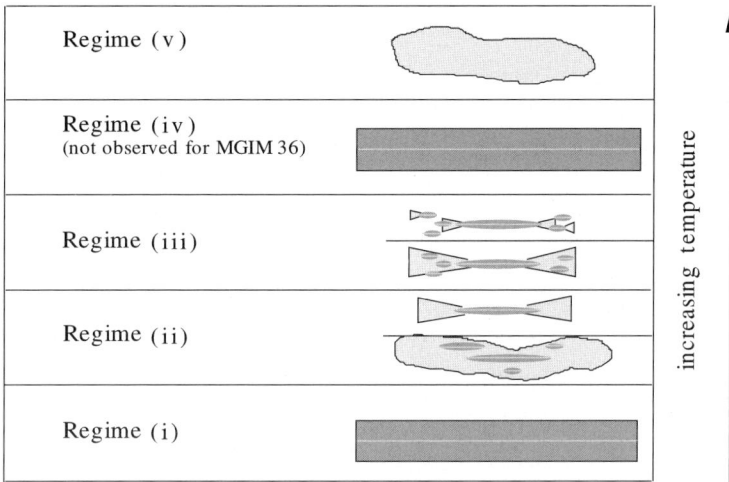

Fig. 54 The various regimes of behaviour of thin films of MGIM*x* copolymers (From [51])

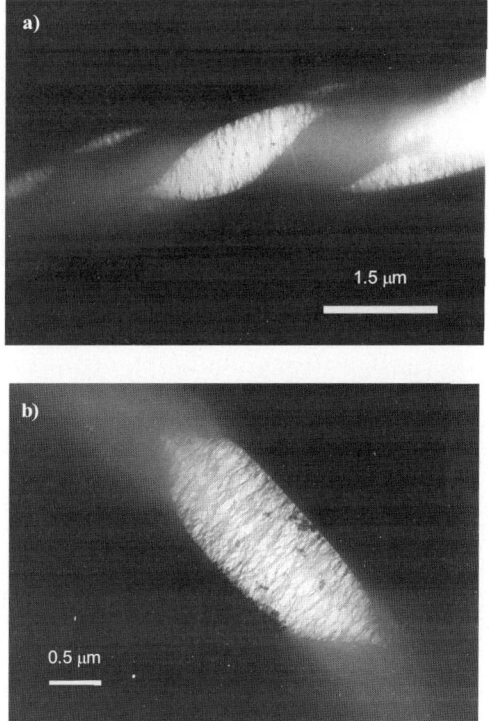

Fig. 55 Deformation behaviour at intermediate temperatures: **a** craze with diffuse edges in MGIM76 at 21 °C; **b** shear blunting at craze tips inMGIM76 at 50 °C

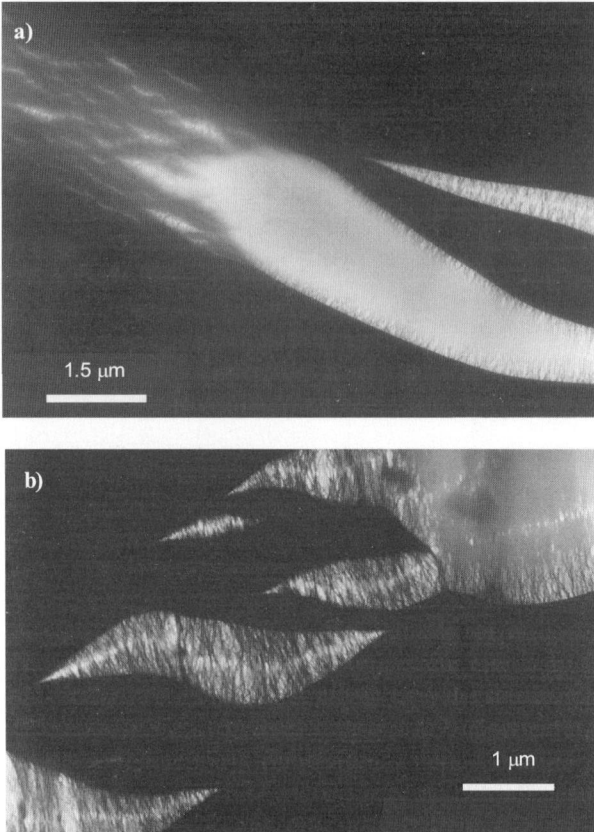

Fig. 56 Multiple crazing at a craze tip in MGIM76 at: **a** 90 °C; **b** 130 °C

though some large isolated SDZs are seen (Fig. 55b). As the temperature is raised further, the SDZs become more widespread and are also observed at craze tips. The crazes take on a relatively low-aspect ratio (Fig. 55c).

3. Above 70–80 °C, the deformation remains mixed, but crazing is dominant and corresponds to CDCs, as revealed by the behaviour of the strain to craze and the strain rate dependence; this temperature of occurrence of CDCs is denoted T_{223}. Furthermore, the SDZs become less widespread and tend to be accompanied at the tips of the main crazes by regions of multiple crazing (Fig. 56)

4. Above ca. 125 and 120 °C for MGIM76 and MGIM58, respectively, long, high-aspect-ratio crazes are observed, and the DSZs are no longer visible. This regime is absent in the case of MGIM36. This temperature at which only CDCs are present is denoted T_{233}.

5. Close to T_g, the films deform homogeneously.

Table 5 Temperature dependence of the deformation mechanisms in thin films of PMMA and MGIMx copolymers

Polymer code	Investigated temperature range °C	Scission crazing temperature range °C	Shear deformation temperature range °C	Disentanglement crazing temperature range °C	Homogeneous deformation temperature range °C	Obs T_{12} °C	Calc T_{12} °C	Obs T_{223} °C	$T_\alpha - T_{233}$ °C	Calc T_{223} °C	Calc $T_\alpha - T_{223}$ °C
PMMA MGIM 21	20–120	20–60	50–110	60–110	110–120	50	Reference 40–45	60	59	–	–
MGIM 36	20–134	20–70	20–110	70–110	110–134	20	35–40	70	64	–	–
MGIM 58	–20–147	–20–70	10–120	70–130	130–147	10	30–35	70	77	120	27
MGIM 76	–20–158	–20–80	10–125	80–145	145–158	10	25–30	80	78	125	33

The temperature ranges corresponding to CSCs, SDZs and CDCs, as well as the characteristic temperatures, T_{12}, T_{223} and T_{233}, of the various MGIMx copolymers are indicated in Table 5.

3.3.2.2
Quantitative Approach of the Occurrence of SDZs

It is tempting to apply the approach described in Sect. 2.3.4.2 and used in the case of CMIMx copolymers (Sect. 3.2.2.2) to analyse the effect of MGI units on the temperature, T_{12}, of occurrence of SDZs.

The choice of PMMA as a reference material is also appropriate for MGIMx copolymers.

For determining the equivalent virtual chain, the following units are considered:

- MMA

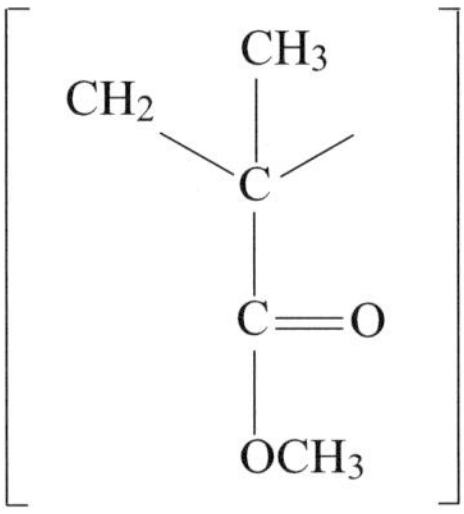

Scheme 3

with a molar mass of $100\,\text{g mol}^{-1}$, and two rotating bonds of length $1.54\,\text{Å}$ each.

- MGI$^+$

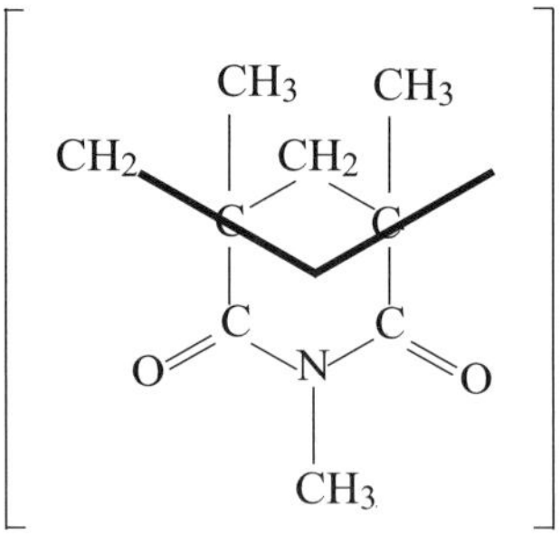

Scheme 4

with a molar mass of $167\,\text{g mol}^{-1}$, two virtual bonds, (shown in bold lines) of $3.23\,\text{Å}$ each and an angle of $109°$ between them.

The characteristics of an equivalent virtual chain corresponding to a MGIMx copolymer are the following:

- A repeat unit of mass:

$$m_{eq}(x) = 0.67x + 100 \tag{48}$$

- A number of bonds per repeat unit:

$$n_{eq}(x) = 2 \tag{49}$$

- A length of bond:

$$l_{eq}(x) = (308 + 3.38x)/(200) \tag{50}$$
$$N_{e,eq}(x) = M_e(x)/(0.34x + 50) \tag{51}$$
$$C_{\infty,eq}(x) = (N_{e,eq}(x)/3)^{1/2} \tag{52}$$

The corresponding values, as well as the experimental values of M_e, are reported in Table 4.

By using the experimental results on $\sigma_y(T)$ and $\sigma_{pf}(T)$, the temperature, T_{12}, at which the SDZs appear can be determined. The ranges of calculated T_{12} are reported in Table 5. It is worth noting that the calculated values of T_{12} depends on $\sigma_y(T)$ and on the CSC stress, $\sigma_{csc}(T)$. In the case of MGIMx copolymers these two quantities vary with the MGI content. In the temperature range where $\sigma_{csc}(T)$ intercepts $\sigma_y(T)$, as shown in Fig. 57, both increase with the MGI content.

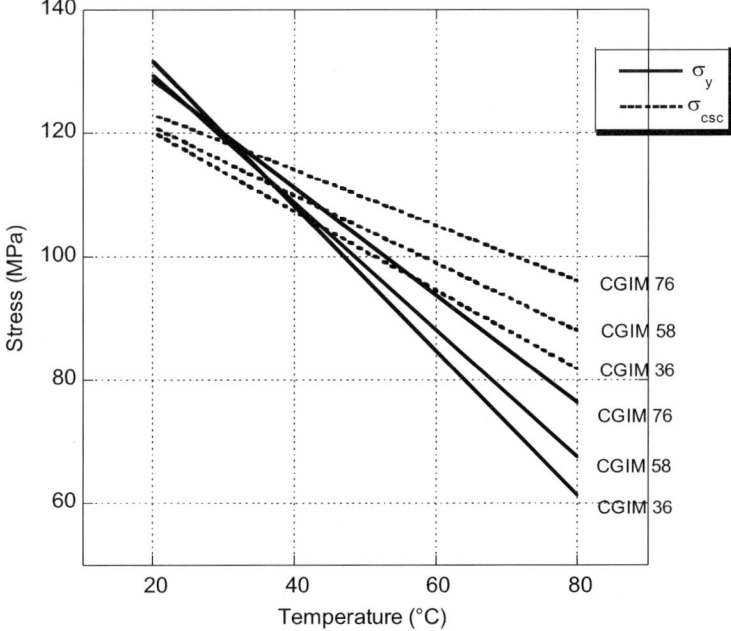

Fig. 57 Temperature dependence of σ_{CSC} and σ_y for the indicated MGIMx copolymers

The calculated T_{12} values decrease, relatively to those of PMMA, when the MGI content increases, in agreement with the observations. However, the predicted effect is smaller than that experimentally determined, maybe due to the experimental uncertainties on M_e and the estimation of $C_{\infty,eq}$. Indeed, as in the case of MGIMx copolymers, most of the ratios involved in the calculations, i.e. $(M_e(x)/M_e^{ref})^{-1/2}$, $(C_{\infty,eq}(x)/C_\infty^{ref})^{1/4}$, and $(N_{e,eq}(x)/N_e^{ref})^{1/4}$, are close to 1, so the final result is quite sensitive to M_e.

3.3.2.3
Relation of Micromechanisms of Deformation of MGIMx Copolymers to the β Transition Motions

From the experimental data reported in Table 5, two main effects of the introduction of MGI units within a PMMA chain are clearly evidenced.

First is the large decrease of the temperature, T_{12}, at which SDZs appear. Indeed, T_{12} drops from $50\,°C$ for PMMA to $10\,°C$ for MGIM58 and MGIM76, and is around $20\,°C$ for MGIM36. It is worth pointing out that this decrease has nothing to do with the effect of M_e, since the increase of M_e from PMMA to MGIM76 should be accompanied by an increase of T_{12}, as observed with CMIMx copolymers (Sect. 3.2.2). In the case of MGIM58 and MGIM76, the easier occurrence of SDZs could be attributed to the α-precursor motions, with a high cooperativity, existing in MGI-rich copolymers in the high temperature part of the β transition (as described in [1], Sect. 8.3). However, this argument cannot explain the quite significant decrease of T_{12} observed with MGIM36, for which no particular cooperativity has been evidenced.

The second effect concerns the occurrence of CDCs over the widest temperature range $(T_{233} - T_a)$ in the polymers with the highest T_α values, i.e. those with the highest MGI content. The theoretical approach developed for CDC and described in Sects. 2.3.1.3 and 2.3.4.2, involves the friction coefficient of the considered polymer chain. As this characteristic is not available for MGIMx copolymers, it is not possible to check whether one could account for the MGI content dependence of T_{233}. In spite of this impossibility, some qualitative reasoning can be attempted. Indeed, CDCs involve chain mobility related to T_α and the fact that the required motions can happen further away from T_α in the MGI-rich copolymers, compared to the case of PMMA and compounds of low MGI content, could originate from the higher cooperativity existing in the β motions of these copolymers, which get an α-precursor character (as discussed in [1], Sect. 8.3).

3.3.3
Fracture of MGIMx Copolymers

As mentioned for micromechanisms studies in Sect. 3.3.2, the MGIM36 copolymer has a low MW, which makes it brittle and limits the temperature range where stable fracture can occur.

The two MGIMx copolymers most extensively investigated [52] are MGIM36 and MGIM76.

3.3.3.1
Temperature Dependence of Toughness

Depending on the temperature, MGIM36 and MGIM76 show two quite different fracture behaviours:

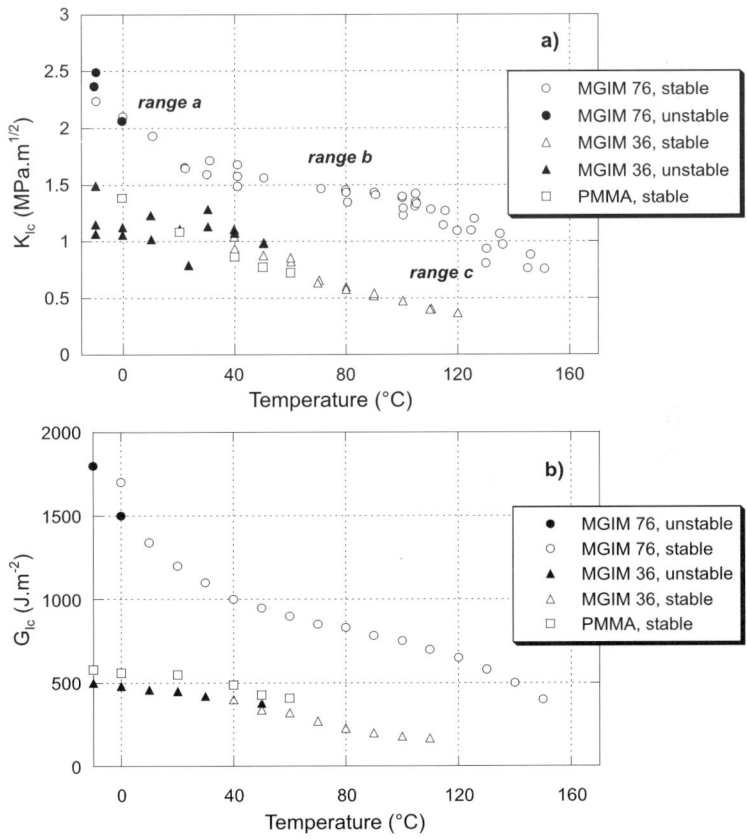

Fig. 58 Temperature dependence of **a** K_{Ic} and **b** G_{Ic} for PMMA, MGIM36 and MGIM76 copolymers; CT samples, cross-head speed of 10^{-4} m s^{-1} (From [52])

Table 6 Effect of composition on K_{Ic} values of MGIMx copolymers

Polymer code	K_{Ic} [MPa m$^{1/2}$]
PMMA	1.22
MGIM 36	1.34
MGIM 58	1.80
MGIM 76	2.04

CT specimens at room temperature with a cross-head speed of 10^{-4} m s^{-1}

1. An unstable fracture regime, observed at low temperature and extending to + 40 °C for MGIM36 and around − 20 °C for MGIM76
2. A stable fracture regime at higher temperatures

Only results obtained for stable fractures will be considered here. The K_{Ic} and G_{Ic} values as a function of temperature are shown in Fig. 58 for MGIM36, MGIM76 and PMMA.

It clearly appears that the toughness of MGIM76 is much higher than that of MGIM36 and PMMA. The latter two showing quite similar behaviours.

The K_{Ic} and G_{Ic} data obtained at room temperature for PMMA and the various MGIMx copolymers are reported in Table 6. They show a gradual increase of toughness with increasing MGI content.

3.3.3.2
Analysis of Fracture Morphology

In order to go further in the understanding of the temperature and composition dependence of toughness, it is useful to examine the morphology in front of the crack. Such a study has been carried out [52] in the case of MGIM58 copolymer. The applied technique concerns the stable fracture regime; optical and electron microscopies are combined to lead to a precise analysis.

Figure 59 shows a schematic representation of a crack and its accompanying instabilities. The lettered zones (a–g) correspond to the different pictures shown in Figs. 60–63.

The optical microscopy pictures of Fig. 60 show that many instabilities have developed along the crack and at its tip during crack propagation. Furthermore, electron microscopy picture of the crack tip, d, shown in Fig. 61, reveals three instabilities referred to as I_1, I_2 and I_3.

The central instability I_2, referred as zone e, is along the crack direction. It has a fibrillar structure, clearly seen in Fig. 62, which is characteristic of a craze.

The instabilities I_1 and I_3 are oriented at 45–60° from the crack propagating direction (Fig. 62 and Fig. 59). Furthermore, they end up by a craze

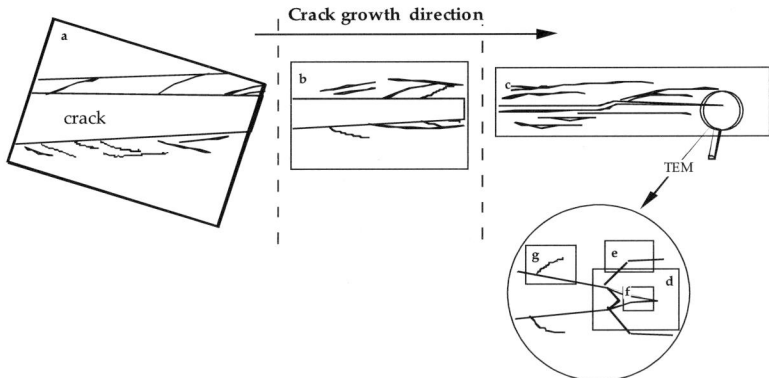

Fig. 59 Schematic representation of a crack and its accompanying instabilities during a stable fracture of MGIM58 at room temperature and a cross-head speed of 10^{-4} m s^{-1}. The letters $a -- g$ correspond to the zones whose pictures are presented in the following figures (From [11])

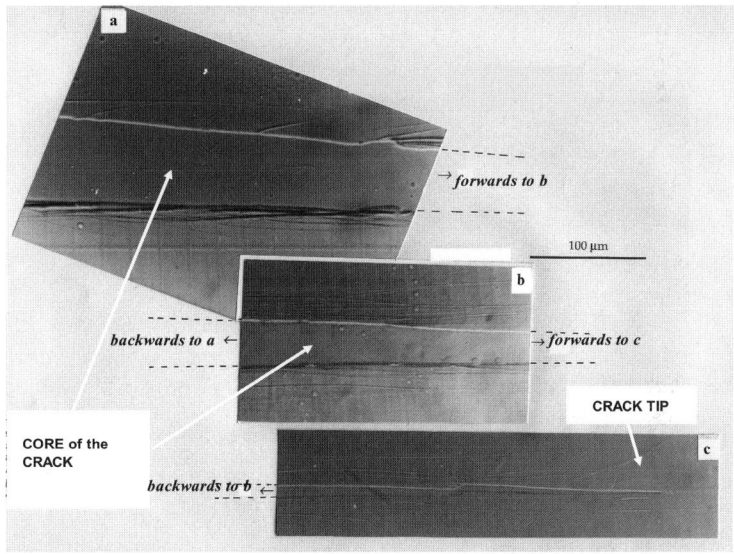

Fig. 60 Optical microscopy pictures of zones a, b and c indicated in Fig. 59 (From [52])

oriented perpendicular to the stretching direction, whose fibrillar structure is observed in the picture of zone f in Fig. 61. As can be seen in the optical microscopy picture of zone a in Fig. 60, this type of instability, which starts obliquely from the crack edges then turns abruptly to end up parallel to the crack, occurs in many places along the crack. The electron micrograph of the initial part of one of these instabilities, shown in Fig. 63 corresponding to zone g, proves that it deals with a shear band. Further away from the

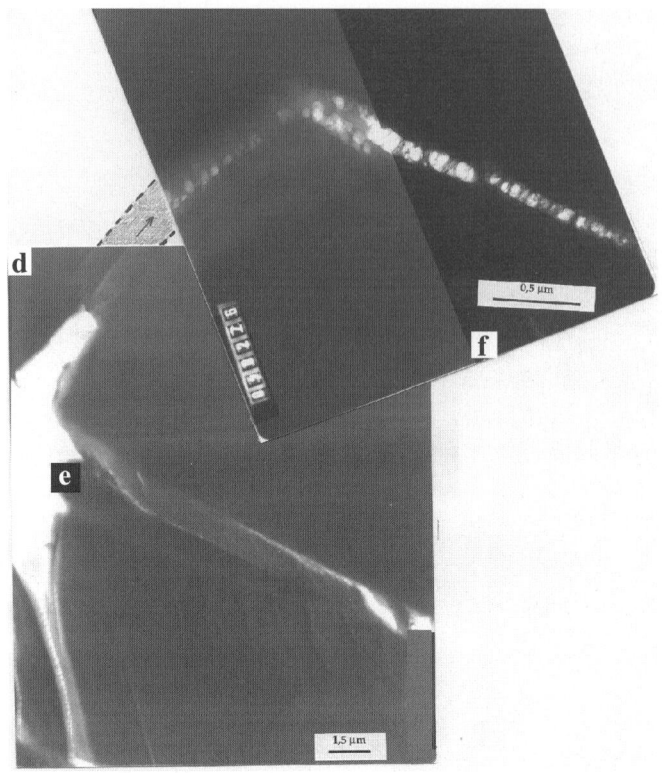

Fig. 61 TEM pictures of zones *d* and *f* indicated in Fig. 59 (From [52])

crack edge, the occurrence of a craze ends the shear band development, probably because the triaxiality of the stresses, which exists in the bulk material, favours craze formation.

This analysis of the fracture morphology of MGIM58 is quite important for two reasons:

1. It shows that shear bands are formed during crack propagation at room temperature, in agreement with the observations from thin film deformation
2. It illustrates, in the case of a bulk fracture, the way crazes and shear deformation zones interact

3.3.3.3
Mechanisms Involved in the Temperature Dependence of Toughness

Considering the temperature dependence of toughness through K_{Ic}, in the case of MGIM76 three temperature ranges (*a, b* and *c* in Fig. 58) can be dis-

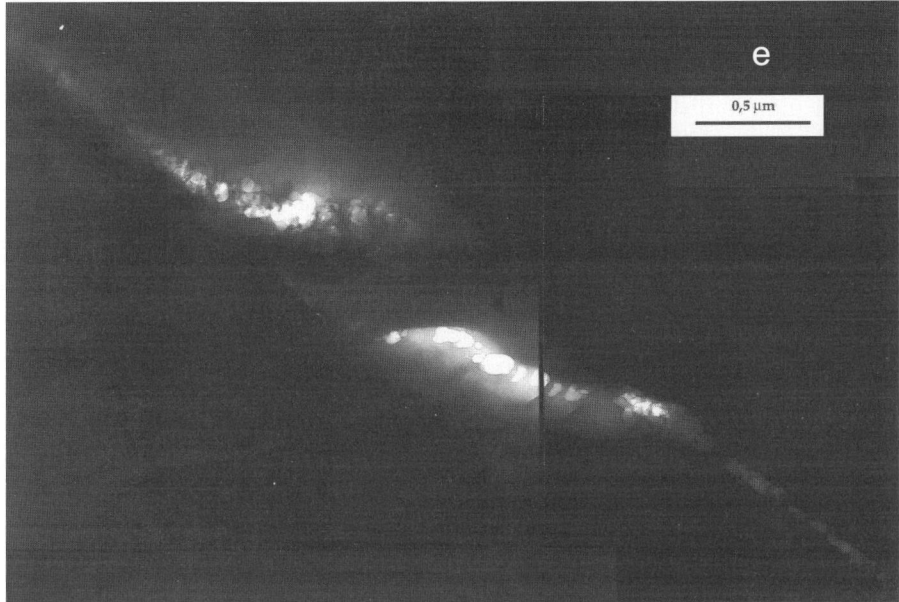

Fig. 62 TEM pictures of zone *e* indicated in Fig. 59 (From [52])

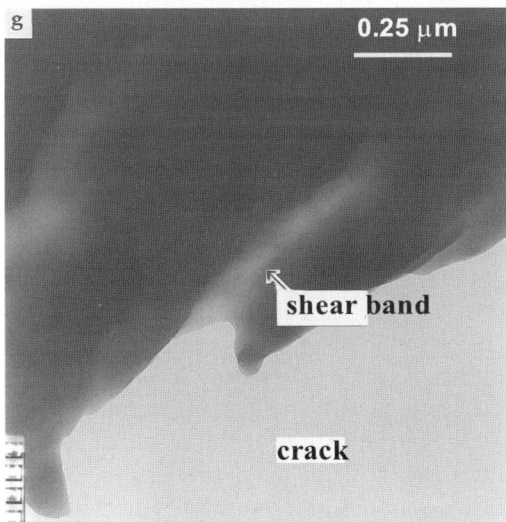

Fig. 63 TEM pictures of zone *g* indicated in Fig. 59 (From [52])

tinguished. In the first one, *a*, from – 20 to about 20 °C, one observes a rapid decrease of K_{Ic} when the temperature rises. In the temperature range *b*, from 20 to about 110 °C, K_{Ic} only slightly decreases when temperature increases. Finally, in the higher temperature range c, again a rapid decrease of K_{Ic} occurs. It is interesting to observe that similar regimes also appear for G_{Ic}.

In order to identify the processes involved in each temperature range and, in particular, to distinguish between CSCs and CDCs, it would be very useful to compare the MW dependence of $K_{Ic}(T)$, as has been done for aryl-aliphatic polyamides, xT_yI_{1-y} (Sect. 5.3). Unfortunately, various MWs were not available for MGIMx copolymers.

However, one can take advantage of the results on the micromechanisms of deformation obtained on thin films, as described in Sect. 3.3.2 and summarised in Table 5.

Thus, for MGIM76, in the temperature range corresponding to a, only CSCs are present. The temperature range c covers the domain where only CDCs are observed. For temperature range b, there is a mixed regime of SDZs with either CSCs (to 80 °C) or CDCs (at higher temperatures).

3.3.3.4
Effect of the Chemical Structure

As described above, the mechanisms of deformation depend on the temperature range, so, one must consider separately the low temperature range and the higher temperatures. In both cases, the stable fracture of MGIMx, as well as PMMA, occurs through breakdown of the fibrils forming the craze at the crack tip, thus leading to crack propagation.

Low Temperature Range
In this temperature range (to about 20 °C), data are available only for MGIM76 and PMMA.

Considering the plastic behaviour at these temperatures (described in Sect. 3.3.1 and Fig. 45) it is clear that the yield stress, σ_y, for PMMA is larger than that for MGIM76 at − 20 °C, but the σ_y values become closer and closer as temperature increases. Of course, the lower value of σ_y (MGIM76) could explain the increased toughness. However, at 20 °C, where almost identical σ_y values are observed, there is still the same difference of toughness between MGIM76 and PMMA. So, the origin of the effect of MGI content cannot be found in the σ_y values.

The entanglement density, ν_e, reported in Table 4 goes in the opposite direction to that observed, since the higher ν_e value (7.7×10^{25} m^{-3}) obtained for PMMA, compared to 5.6×10^{25} m^{-3} for MGIM76, should lead to a higher toughness for PMMA than for MGIM76 (Sect. 2.4.7).

The much larger capability of MGIM76 to lead to SDZs, as manifested by the occurrence of such a mechanism in thin film deformation at 10 °C for MGIM76 instead of 50 °C for PMMA, would go in the appropriate direction. However, it would mean that MGIM76 is able to develop SDZs under crack propagation (in a way similar to that found for MGIM58) at temperatures lower than in thin films, since the increased toughness of MGIM76, compared to PMMA, is also found at $-20\,°C$.

Another point concerns the β transition and the associated motions. Indeed, in PMMA the maximum of the β transition peak (Fig. 26) occurs around $10\,°C$, in such a way that the low temperature range corresponds to the low temperature part of the β transition, where the ester group motions are isolated, without any cooperativity with the main chain, and leading to multiple crazing, as described in Sect. 3.1.3.2. In contrast, in the case of MGIM76, the maximum of the β peak is found around $-20\,°C$ (Fig. 53), in such a way that the considered temperature range a is located in the high temperature part of the β transition, in which MGI–MGI cooperativity exists, as shown in [1] (Sect. 8.3).

Finally, the decrease of toughness with temperature should be related to the decrease of stability of fibrils within the present CSCs, as shown for PMMA in Sect. 3.1.3.3.

Higher Temperatures

In the temperature range b observed for MGIM76, there is an interaction between the SDZs and the CSCs in the lower temperature part, then with the CDCs in the higher temperature part. Such an interaction should be similar to that described for MGIM58 fractured at room temperature. With increasing temperature, the occurrence of SDZs is favoured and, consequently, they interact more, leading to a high energy dissipation within the plastic zone ahead of the crack. This increase of dissipation counterbalances the decrease of fibril stability arising from the easier chain mobility. It results in a lower decrease of toughness relatively to that expected from the craze stability decrease alone.

In the temperature range c, the molecular process is the creep of the fibrils in the only present CDCs, due to the easier chain slippage occurring when approaching T_α.

Owing to the important role played by fibril stability in the behaviour at high temperatures, it is more suitable to consider the toughness as a function of $(T_\alpha - T)$. Thus, K_{Ic} and G_{Ic} are plotted versus $(T_\alpha - T)$ in Fig. 64.

It is interesting to observe that in the same $(T_\alpha - T)$ range, corresponding to regime c, the decrease of MGIM36 toughness appears less pronounced than that of MGIM76, in spite of the low MW of MGIM36, which should lead to an opposite effect. The large MGI–MGI cooperativity of motions existing in this temperature range, which leads to α-precursor motions, should favour chain slippage within the fibrils. However, in MGIM36 there is a lower cooperativity, as described in [1] (Sect. 8.3.1.1).

The absence in MGIM36 of the expected favourable contribution of SDZs to toughness (at 40–70 °C, or for $(T_\alpha - T)$ at 100–50 K) is due to the low MW

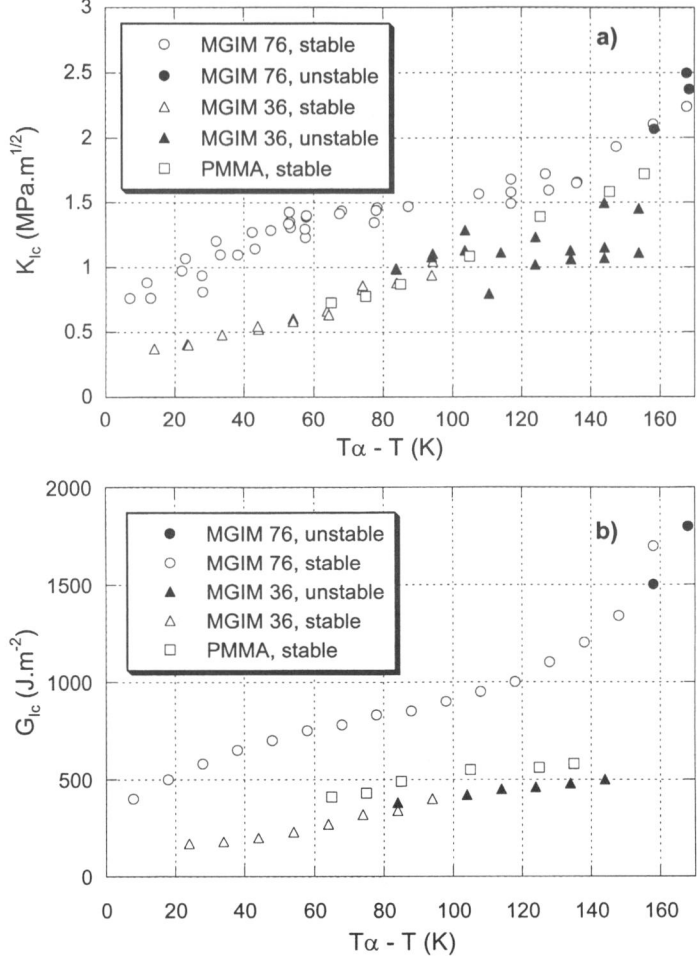

Fig. 64 Fracture results: **a** K_{Ic} and **b** G_{Ic} for PMMA, MGIM36 and MGIM76 copolymers, versus $(T_\alpha - T)$; CT samples, cross-head speed of 10^{-4} m s^{-1} (From [52])

of the copolymer, which leads to quite unstable craze fibrils and hinders the possible interaction between SDZs and CDCs.

As regards the increased toughness of MGIM76 in this high temperature range, relative to PMMA, consideration of ν_e leads to the opposite expectation. The yield stress, σ_y, plotted as a function of $(T_\alpha - T)$ (as in Fig. 65) shows that σ_y decreases when the MGI content increases, which is in qualitative agreement with the corresponding increase of toughness. Nevertheless, the greater facility for MGIM76 to lead to SDZs ($T_{12} = 10$ instead of $50\,°C$ for PMMA) quite likely makes a more significant contribution to its higher toughness, compared to that of PMMA.

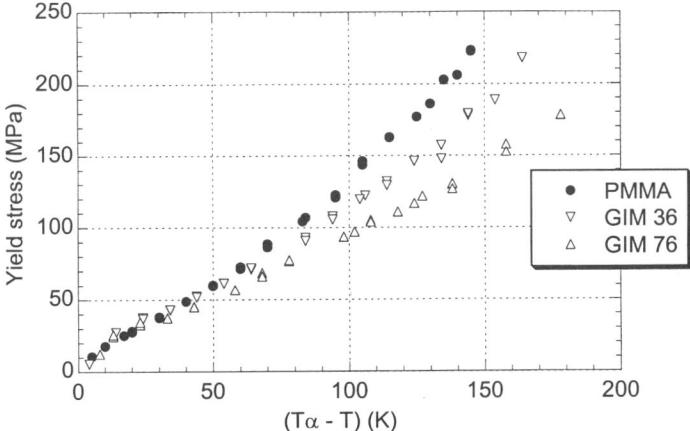

Fig. 65 Yield stress of PMMA, MGIM36 and MGIM76 copolymers, versus $(T_\alpha - T)$ (From [35])

3.4
Conclusion on Poly(methyl methacrylate) and its Maleimide and Glutarimide Copolymers

The investigation of mechanical properties of PMMA and its maleimide (CMIMx) and glutarimide (MGIMx) copolymers reveals several interesting features.

First, in plastic deformation, the strain softening decreases with increasing temperature, particularly in the temperature range corresponding to the high temperature part of the β transition, in which the involved ester group π-flips are coupled to rotation angle changes of main-chain C – C bonds, developing an intramolecular cooperativity. The introduction of CMI units in the PMMA chain hinders the β–α cooperativity occurring above 50 °C and results in larger and larger values of the normalised strain softening amplitude, nSSA. For MGIMx copolymers, nSSA values increase when increasing MGI content to around 60 mol%, then decrease. Such a behaviour is related to the slight hindering of MMA cooperativity at low and moderate MGI contents, and to the occurrence of a new MGI–MGI intramolecular cooperativity at high MGI content. A mechanism has been proposed to account for the relation of strain softening to intramolecular cooperative motions. This mechanism is based, on the one side, on the experimentally observed main-chain conformation change induced by the applied yield stress and, on the other side, on the softening of the polymer medium by the β transition motions. This allows development of local packing adaptations, resulting from the yielding conformation change, in such a way that whole chain motions can occur under a lower stress than that for yielding. In addition to that, in the temperature range where intramolecular cooperative ester group π-flips are active, the conformation changes involved both in yielding and plastic flow are favoured, and occur at stresses close to each other, resulting in a low strain softening. In MGI-rich copolymers, the intramolecular MGI–MGI cooperativity plays an equivalent role.

As regards the micromechanisms of deformation, it appears to be clear that SDZs are controlled by the cooperativity of β transition motions. Occurring at 50 °C for PMMA, they are shifted at temperatures above 100 °C for CMIMx, in which such a cooperativity disappears. However, for MGI-rich copolymers they are present at 10 °C, owing to the MGI–MGI cooperativity.

The fracture results are particularly interesting. In PMMA, the change at the crack tip from multiple crazing (happening at low temperature) to single crazing (above – 20 °C) has been assigned to the occurrence of the β transition and, more precisely, to the high temperature part of this transition, where intramolecular cooperative ester group π-flips occur. However, the most striking behaviour deals with fracture results on MGIMx copolymers. Indeed, whereas an almost constant entanglement density, ν_e, is observed in the series, quite a significant increase in toughness occurs with increasing MGI content, in contradiction to the expected dependence of toughness with ν_e or $\nu_e^{1/2}$. Furthermore, in the case of MGIM58, it has been shown that the reinforcement originates from interaction between SDZs and crazes. Finally, let us point out that CMIMx copolymers become more and more brittle with increasing CMI content, but this behaviour can result from either lower and lower ν_e values, or the hindering of cooperative β transition motions.

4
Bisphenol A (or Tetramethyl Bisphenol A) Polycarbonate

Bisphenol A polycarbonate (BPA-PC), whose the chemical structure is shown in Fig. 66a, has very interesting fracture properties, exhibiting quite a high toughness for a pure amorphous polymer. At a very low temperature (– 100 °C at 1 Hz) it presents a secondary β transition, shown in Fig. 67, which has been analysed in detail in [1] (Sect. 5).

In order to investigate the effect of a methyl substitution of the phenyl hydrogen in *ortho* position to the carbonate group, tetramethyl bisphenol A polycarbonate (TMBPA-PC), whose the chemical structure is shown in Fig. 66b, has also been considered. In the case of TMBPA-PC, the secondary β transition happens at 50 °C, at 1 Hz (Fig. 67), in such a way that it allows one to examine the consequences of such a large shift of the β transition on the mechanical properties.

Basic mechanical behaviours, such as plastic deformation, deformation micromechanisms, and fracture, are successively presented. The characteristics of the studied polymers are gathered in Table 7.

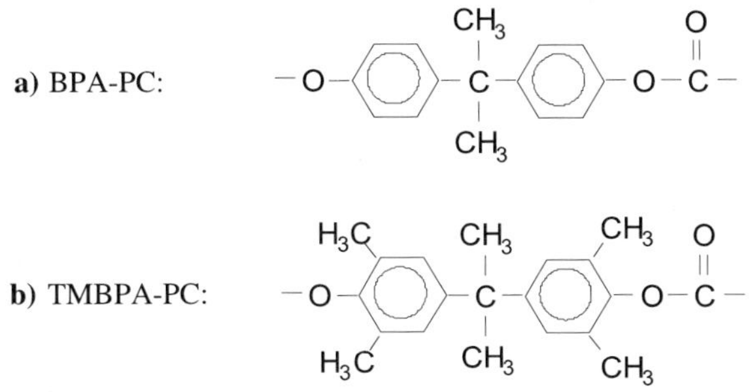

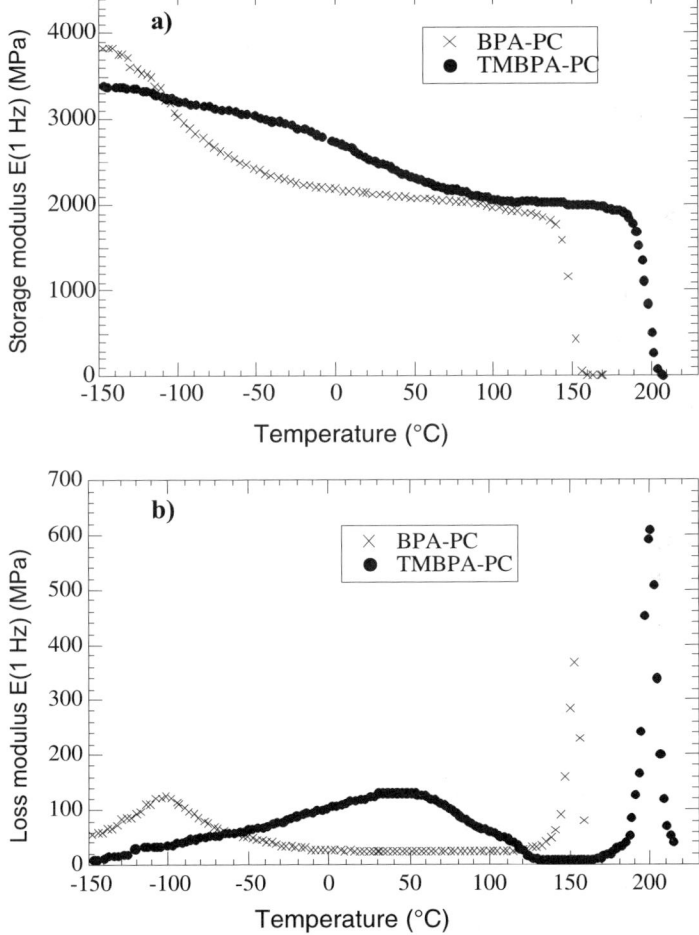

Fig. 66 Chemical structures of **a** BPA-PC and **b** TMBPA-PC

Fig. 67 Temperature dependence for BPA-PC and TMBPA-PC of **a** the modulus, E', at 1 Hz and **b** the loss modulus, E'', at 1 Hz (From [53])

Table 7 Characteristics of BPA-PC and TMBPA-PC

Polymer code	M_w g mol^{-1}	M_n g mol^{-1}	T_α °C	ρ kg m^{-3}	M_e g mol^{-1}	ν_e 10^{24} m^{-3}	M_w/M_e	M_n/M_e
BPA-PC(18)	18 000	8000	150		2100		8.6	3.8
BPA-PC(31)	31 000	12 000	150	1140	2100	330	14.8	5.7
TMBPA-PC(28)	126 000	69 000	200	1100	4000	170	7.0	2.25

4.1
Plastic Deformation

The plastic deformation characteristics: yield stress, σ_y, plastic flow stress, σ_{pf}, and strain softening, have been studied under uniaxial compression at a strain rate of 2×10^{-3} s^{-1} [53] in a temperature range from -110 °C to typically $T_\alpha - 20$ K. Indeed, for temperatures closer to T_α, the experimental results are less reliable, some creep behaviour occurring.

4.1.1
BPA-PC

The stress–strain curves for BPA-PC obtained at various temperatures are shown in Fig. 68. When decreasing temperature, in addition to the increase of

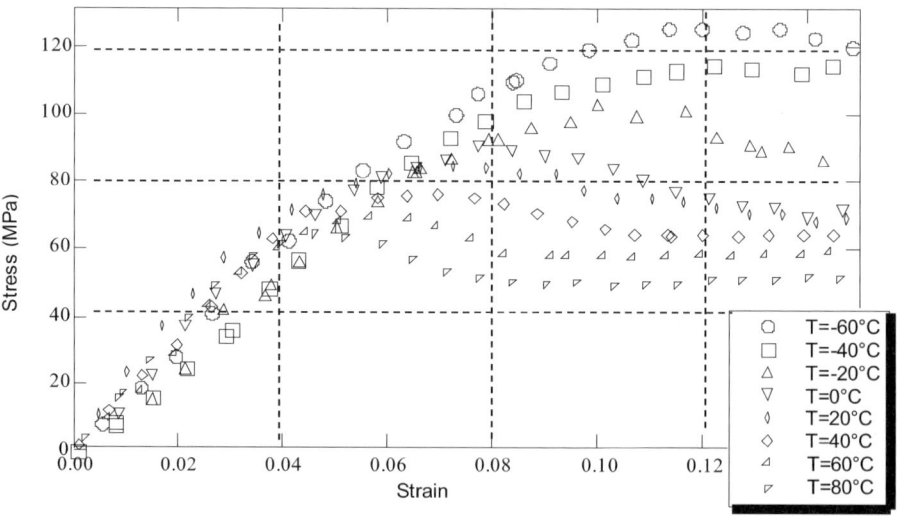

Fig. 68 Compression stress–strain curves for BPA-PC at various temperatures at a strain rate of 2×10^{-3} s^{-1} (From [53])

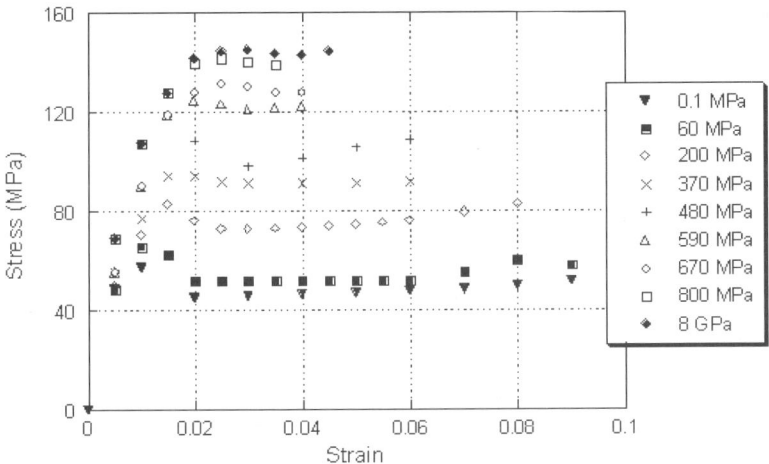

Fig. 69 Tensile stress–strain curves for BPA-PC at various pressures (From [54])

yield stress and yield strain, one observes a decrease in the strain softening. This latter behaviour is opposite to that of poly(methyl methacrylate) and its maleimide and glutarimide copolymers, as described in Sect. 3.

It is interesting to notice that similar effects on yielding characteristics have been reported [54] when increasing pressure to 80 kbar, as shown in Fig. 69. Here too, the decrease of strain softening with increasing pressure is opposite to behaviour observed for PMMA (Sect. 3.1.1.3).

4.1.1.1
Yield Stress

The temperature dependence of yield stress, σ_y, for BPA-PC is shown in Fig. 70. After a rapid decrease of σ_y between -110 and about $-40\,°C$, one observes a smaller, almost linear, decrease to $120\,°C$.

Considering the loss modulus β transition peak (Fig. 67a) and the downward temperature shift of about $20\,°C$ (associated with the strain rate of $2 \times 10^{-3}\,s^{-1}$ used in the experiments) it turns out that the temperature domain (around $-40\,°C$) where the behaviour change of σ_y occurs, corresponds to the end of the β transition.

The strain rate, $\dot{\varepsilon}$, dependence (in the range 2×10^{-4} to $2 \times 10^{-1}\,s^{-1}$) of σ_y at various temperatures yields a linear increase of σ_y with $\log(\dot{\varepsilon})$, as expected from the Eyring relationship (Sect. 2.2.1.1). The activation volume, V_0, determined from this relationship, is equal to $1\,nm^{-3}$ at $-60\,°C$. Above $-40\,°C$ it gradually increases to $3\,nm^{-3}$ at $40\,°C$, as shown in Fig. 71.

Physical ageing slightly increases σ_y by about 10 MPa over the whole temperature range [55].

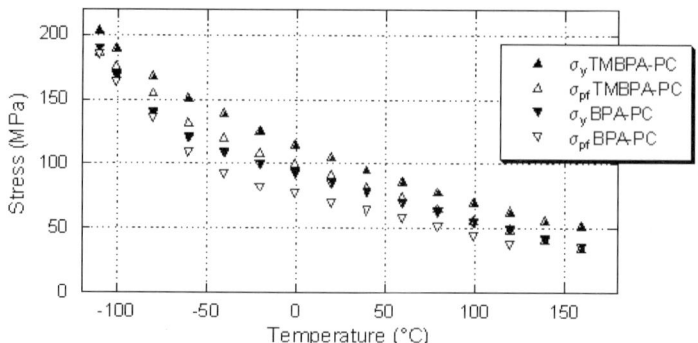

Fig. 70 Temperature dependence of yield stress, σ_y, and plastic flow stress, σ_{pf}, for BPA-PC and TMBPA-PC

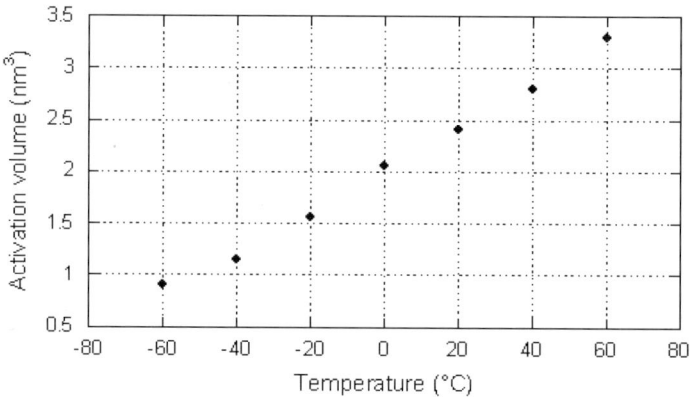

Fig. 71 Activation volume, V_0, as a function of temperature for BPA-PC (From [53])

4.1.1.2
Plastic Flow Stress

The plastic flow stress, σ_{pf}, for BPA-PC is shown in Fig. 70. As can be seen, the general shape of the temperature dependence of σ_{pf} is similar to that of σ_y.

4.1.1.3
Strain Softening

As already mentioned, strain softening develops when temperature increases (Fig. 68). The temperature dependence of the strain softening amplitude, SSA, for BPA-PC is shown in Fig. 72. Actually, at very low temperatures quite a small strain softening amplitude is observed, then a large increase takes place from − 90 to about − 30 °C, followed by a linear decrease at higher temperatures.

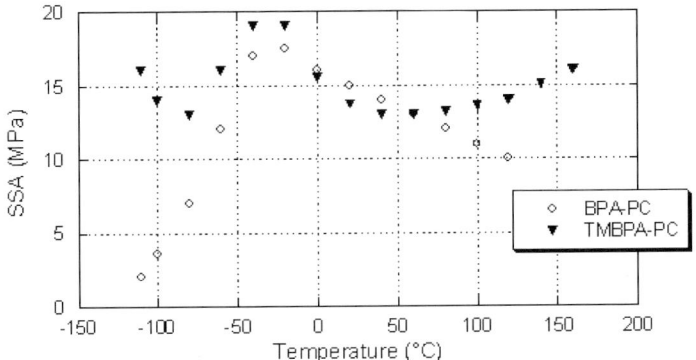

Fig. 72 SSA versus temperature for BPA-PC and TMBPA-PC

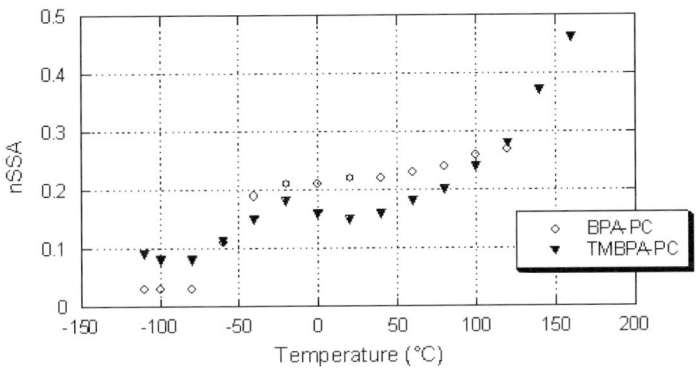

Fig. 73 Temperature dependence of nSSA for BPA-PC and TMBPA-PC

The SSA normalised by the plastic flow stress, σ_{pf}, is shown in Fig. 73.

A completely different behaviour is obtained since in the high temperature range a slight increase of nSSA is observed.

4.1.2
TMBPA-PC

In the case of TMBPA-PC, the large shift of the β transition peak (Fig. 67) provides an opportunity to investigate the role of this transition in plastic deformation.

4.1.2.1
Yield Stress

The temperature dependence of σ_y for TMBPA-PC is shown in Fig. 70. One observes a regular smooth decrease, without any evidence of two different regimes, contrary to the situation for BPA-PC (Fig. 70). Such a behaviour can be related to the very broad β transition of TMBPA-PC (Fig. 67), which extends over almost the whole investigated temperature range.

4.1.2.2
Plastic Flow Stress

Figure 70 presents the results of σ_{pf} for TMBPA-PC. The curve shape is quite similar to that for σ_y.

4.1.2.3
Strain Softening

The strain softening for TMBPA-PC is shown in Fig. 72 as a function of temperature. In the low temperature range, one observes at $-100\,^\circ$C quite a significant SSA value, which increases and goes through a maximum around $-20\,^\circ$C. This behaviour is similar to that obtained for BPA-PC (Fig. 72), except that for the latter there is no strain softening at $-100\,^\circ$C. Above $20\,^\circ$C, the strain softening increases, whereas it decreases in the case of BPA-PC (Fig. 72).

The evolution of nSSA with temperature is shown in Fig. 73. The general behaviour is similar to that of BPA-PC but, in the low and high temperature limits, the differences observed between the two polymers in the case of SSA still remain.

4.1.3
Relation of Plastic Deformation to β Transition

Before discussing the relation of plastic deformation temperature dependence with the β transition motions, it is useful to recall the main results achieved through investigations into the motions involved in this transition (as described in [1], Sect. 5).

First, the temperature dependence of the mechanical loss modulus, E'', is shown in Fig. 67.

The β transition occurring for BPA-PC around $-100\,^\circ$C, at 1 Hz, involves motions of both carbonates and phenyl rings, accompanied by main-chain reorientation. More precisely, for BPA-PC:

- In the low temperature range, carbonate groups perform limited amplitude motions, whereas phenyl rings undergo oscillations of moderate amplitude (around $12°$). In both cases the intermolecular cooperativity is quite weak, due to the small size of the involved motions.
- In the high temperature range, the conformation change of the carbonate group is coupled to motions of the adjacent phenyl rings. This induces

some rearrangement of the surroundings and an intermolecular contribution. However, the most important feature deals with the occurrence of phenyl ring π-flips. Indeed, these latter motions have a strong intramolecular coupling with the motions of the phenyl ring attached to the same isopropylidene unit. In addition to the rearrangement of the surroundings at a scale as large as $10\,\text{Å}$, the phenyl ring π-flips can induce motions or conformation changes of groups as far away as $7\,\text{Å}$. Furthermore, these carbonate and phenyl ring motions lead to main-chain reorientation of about $15°$. All these effects result in quite a high activation entropy ($110\,\text{J}\,\text{K}^{-1}\,\text{mol}^{-1}$).

In the case of TMBPA-PC, the methyl groups in the *ortho* position to the carbonate at first hinder the limited oscillations of the carbonate group at low temperature, and shift to a much higher temperature the carbonate conformation change coupled to phenyl ring motions. As regards the effect of these methyl groups on the phenyl ring motions, the oscillations occurring at low temperature are still active, the phenyl ring π-flips being shifted to much higher temperature. This results in a very broad dynamic mechanical β transition peak. Consequently, even considering the β peak maximum, the activation entropy for TMBPA-PC is quite small, $20\,\text{J}\,\text{K}^{-1}\,\text{mol}^{-1}$, indicating very localised motions.

4.1.3.1
BPA-PC

Concerning the plastic deformation of BPA-PC, the most surprising behaviour, is the strain softening. The above reported results show that the strain softening is higher at high temperature than at low temperature (Fig. 68 and Fig. 72), whereas it decreases with increasing pressure (Fig. 69). So, decreasing temperature or increasing pressure qualitatively lead to the same effect. The effect of pressure is quite interesting for it only concerns intermolecular contributions, without mixing them with available thermal energy, as happens when changing temperature. By increasing pressure, the phenyl ring π-flips are slowed down ([1], Sect. 5.3.4), and the extent and strength of their intermolecular cooperativity is increased. Consequently, it clearly shows that the strain softening amplitude is related to intermolecular effects.

Concerning the temperature-pressure equivalence, a quantitative analysis has been performed [54] on the yield stress. The results for BPA-PC are shown in Fig. 74. For a given volume change, the required yield stress is much higher when the temperature is changed than when the pressure is varied at $25\,°\text{C}$. Such a result can be interpreted by the fact that, in order to get a negative volume change through temperature variation, one has to decrease the temperature quite significantly. Consequently, one shifts from a bulk BPA-PC at $25\,°\text{C}$, where lots of groups are undergoing motions, to a material

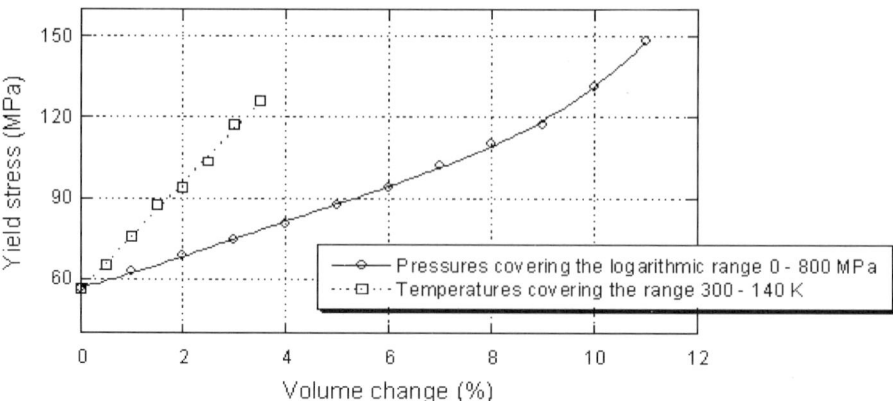

Fig. 74 Cross-plot of yield stress for BPA-PC against volume changes obtained by varying temperature or pressure (From [54])

where motions are slowed down and their amplitude is decreased. At low-enough temperatures, even phenyl ring π-flips are replaced by phenyl ring oscillations, suppressing the surrounding rearrangements associated with the phenyl ring π-flips. At the same value of volume change (e.g. 3%) to reach yielding at a low temperature of $-120\,°C$ in a low-mobility medium, requires a higher stress than in a mobile medium at $25\,°C$ under about 1.4 kbar. Applying a hydrostatic pressure of 1.5 kbar to BPA-PC leads to a decrease of phenyl ring π-flip frequency corresponding to a downward temperature shift of $20\,°C$ ([1], Sect. 5.3.4), in such a way that under 1.4 kbar at $25\,°C$ the polymer medium still keeps a high mobility.

The remaining feature deals with a molecular approach of the strain softening amplitude in BPA-PC and its temperature dependence.

As plastic flow corresponds to large chain displacements analogous to those occurring without applied stress above T_α, it implies the main-chain motions involved in the α transition, whatever the temperature at which the plastic flow develops. Thus, the plastic flow constitutes, in some way, a reference.

The molecular process involved in yielding has been described (Sect. 2.2.1.2) as corresponding to main-chain conformation changes, induced by the applied stress and leading to an increase of high-energy conformations, in a way similar to that induced by increasing temperature above T_α without any applied stress. In the case of BPA-PC, the electron conjugation within the carbonate group yields a planar structure of this unit. However, from a conformation point of view, three conformations are available for the carbonate group: *trans,trans*, *trans,cis* (or *cis,trans*), and *cis,cis* conformation. These are shown in Fig. 75 for the situation encountered in BPA-PC. The *trans,cis* (or *cis,trans*) conformation has an energy higher than that of *trans,trans* conformation by about $5\,kJ\,mol^{-1}$, whereas the *cis,cis* conformation is com-

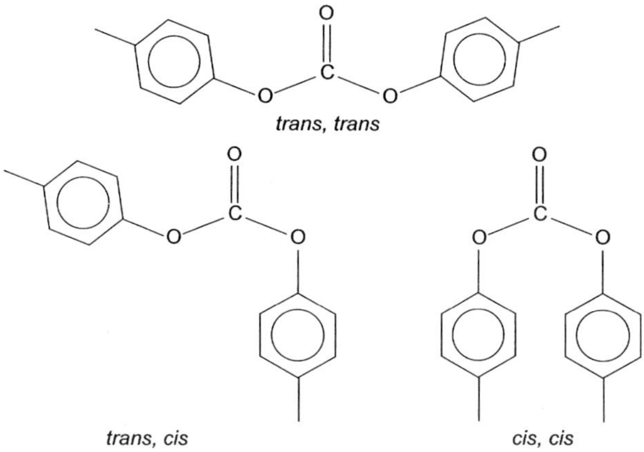

trans, trans

trans, cis   *cis, cis*

Fig. 75 Various conformations of the carbonate group

pletely forbidden because of steric hindrance. In the glassy state, the percentage of *trans,cis* (or *cis,trans*) corresponds to the value at $T_\alpha = 150\,°C$ (i.e. 34%). Under a stress level corresponding to σ_y, *trans,trans* to *trans,cis* (or vice-versa) conformation changes occur, resulting in an increase of the total percentage of the *trans,cis* (or *cis,trans*) conformation. Such conformation change of a single carbonate group, within a BPA-PC chain, implies a drag on the remaining part of the chain, which is unlikely not only in bulk, but also in solution. In order to avoid this effect, compensating counter-rotation motions happen on the neighbouring carbonates, along the chain sequence, on both sides of the carbonate undergoing the conformation change (as mentioned in Sect. 2.2.1.2).

Under conditions (low temperature or high pressure) where a strong intermolecular packing exists and where only small motions occur, each stress-induced carbonate conformation change remains localised and the associated rearrangements are limited to a small surrounding volume, which is softened. Consequently, the absence of an extended medium softening implies that the same stress level as required for yielding has to be applied in order to continue the carbonate conformation changes required for reaching the main-chain displacements involved in plastic flow. This results in a small strain softening, as observed in BPA-PC, through nSSA, below $-80\,°C$ (Fig. 73).

At higher temperatures, the occurrence of motions, in particular phenyl ring π-flips, makes the structure more mobile and carbonate conformation changes can occur under a lower applied stress (as discussed for the comparison between temperature and pressure). In addition, each phenyl ring π-flip induces rearrangements of the chain units within the surroundings at a scale of 10 Å. Furthermore, an intermolecular cooperativity exists in such a way that a phenyl ring can induce a flip of another phenyl ring or a conformation change of a carbonate situated as far away as 7 Å from the initial moving phenyl or from the carbonate group undergoing a stress-induced conformation change. Consequently, the occurrence of the carbonate conformation changes, induced by the yield stress, make it easier to develop the main-chain motions involved in the chain displacements characteristic of plastic flow. The latter will occur at a stress level lower than the yield stress, resulting in strain softening development. Such a behaviour is well observed in BPA-PC at − 80 to − 20 °C, through nSSA (Fig. 73). Above − 20 °C, which corresponds to the end of the BPA-PC β transition (Fig. 67), all the described motions are active and the proposed mechanism for yielding and plastic flow remain valid. Indeed, no new group motion occurs that is able to change the intermolecular cooperativity. The increase in temperature leads to an increase in the frequency of the various motions, as well as a softening of the material and thus to a decrease of σ_y. However, the difference relative to σ_{pf}, expressed through nSSA, should remain since the same mechanism operates. Actually, nSSA only shows a very slight increase from − 20 to 120 °C (Fig. 73).

4.1.3.2
TMBPA-PC

As described, the methyl groups in *ortho* position to the carbonate strongly affect the motions of the carbonate group as well as the phenyl ring π-flips, which remain very localised even at the β peak maximum.

A lower packing is achieved with the TMBPA-PC, relative to BPA-PC, as reflected in the corresponding densities at 25 °C of 1083 and 1208 kg m^{-3}, respectively. The lower density accounts for the lower modulus, E', observed at − 150 °C for TMBPA-PC.

In spite of the lower density, the *trans* to *cis* (or vice-versa) carbonate conformation change under stress, requires a higher value of σ_y relative to BPA-PC, as shown in Fig. 70. Furthermore, the gradual increase in amplitude of the phenyl ring oscillations and the occurrence of phenyl ring π-flips around 50 °C only, leads to a smooth decrease of σ_y over the whole temperature range.

For strain softening, the molecular mechanism proposed for BPA-PC can be considered. Thus, the increase of nSSA (Fig. 73) above 50 °C occurs in the high temperature part of the β transition peak (shifted to lower temperature due to the 10^{-3} s^{-1} strain rate used in the experiments). Phenyl ring π-flips become active, associated with a development of intermolecular cooperativity, similar to that existing in BPA-PC in the temperature range of − 80 to − 20 °C. Under such conditions, according to the proposed mechanism, the carbonate conformation changes induced by the yield stress make it easier to develop the main-chain motions involved in the chain displacements required for plastic flow. The increase of nSSA above 50 °C for TMBPA-PC would have the same molecular origin as the one observed between − 80 and − 20 °C for BPA-PC.

Concerning the strain softening at low temperatures (below $- 80\,°C$), the proposed molecular mechanism would lead to a low nSSA value, as for BPA-PC. The molecular origin of the higher, though small, value observed for TMBPA-PC is not quite clear. It could be related to the lower density of the polymer and the associated decrease of packing, allowing for larger disturbance of the surroundings of the carbonate undergoing *trans* to *cis* (or vice-versa) conformation change under the applied yield stress and, consequently, favouring the further changes required for the main-chain displacement involved in plastic flow.

4.1.3.3
Comment on the Molecular Mechanism for Plastic Deformation of BPA-PC

Detailed analysis of the molecular motions involved in the β transition of BPA-PC (described in [1], Sect. 5 and summarised above) shows that, due to the chemical structure of BPA-PC, an intramolecular cooperativity intrinsically exists, above $- 100\,°C$, between phenyl ring π-flips and carbonate conformation changes. These latter directly concern the main-chain behaviour, its reorientation, and the ease with which it undergoes *trans* to *cis* (or vice-versa) transition of the carbonate groups under an applied stress.

In addition to the intrinsic intramolecular cooperativity, BPA-PC is able to develop efficient, long range, intermolecular cooperativity for phenyl ring π-flips and, to a lesser degree, for carbonate conformation changes. In contrast, oscillation motions of phenyl rings or carbonate units do not lead to such a long range intermolecular cooperativity. Furthermore, this intermolecular cooperativity is sensitive to the surrounding packing and is gradually squeezed by the volume reduction resulting from an applied pressure.

Finally, in BPA-PC, intermolecular cooperativity controls the temperature (and pressure) dependence of the plastic behaviour, in particular the strain softening amplitude reflected in nSSA.

The importance of the intermolecular cooperativity is well supported by the large activation volume (around $1\,nm^3$ at $- 60\,°C$) associated with yielding in BPA-PC.

4.2
Micromechanism of Deformation

In order to understand the fracture behaviour, it is important to determine the types of deformation mechanisms undergone under strain: chain scission craze (CSC), shear deformation zone (SDZ), chain disentanglement craze (CDC), and the temperature range over which each one occurs. Furthermore, it is interesting to wonder whether these micromechanisms are related to β transition motions.

The most convenient technique for performing such a study is the thin film technique described in Sect. 2.3.

4.2.1
BPA-PC

In an early study [56], thin film deformation of un-aged and aged BPA-PC was performed at room temperature and at high temperatures (100 and 125 °C).

At 25 °C, a diffuse shear morphology is observed, without any craze. In aged samples (30 h at 130 °C), fine bands (ca. 100 Å thick) that grow in both the maximum shear directions have a tendency to collect and localise the shear deformation into ca. 3000 Å-wide diffuse shear bands, as indicated by the arrow D in Fig. 76a. In the case of un-aged sample, the fine bands are less distinct and more delocalised, as shown in Fig. 76b.

At high temperatures, in un-aged samples the fine bands collect into diffuse shear bands (Fig. 77a, arrow D). Furthermore, at 100 °C, isolated craze nuclei form at the intersect of diffuse shear bands (Fig. 77a, arrow C), but they do not coalesce in larger crazes, in contrast to what happens at 125 °C (Fig. 77b, arrow C). In the case of aged samples, sharp shear bands develop profusely and, at 100 °C, ca. 5000 Å-wide crazes, consisting of a collection of small craze nuclei 200–300 Å in width and 1000 Å in length, are observed.

A more quantitative study has been performed [57] by considering the temperature dependence of the strain of yielding for the occurrence of SDZs and crazes. The effects of ageing time at 130 °C and of MW were also considered. The results are shown in Fig. 78.

It appears that the strain of yielding for SDZs greatly increases with ageing time and does not depend on MW.

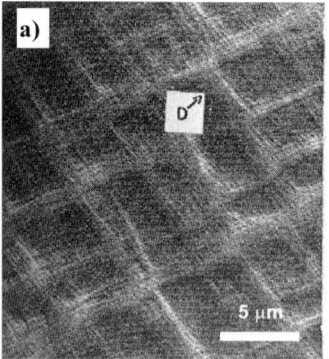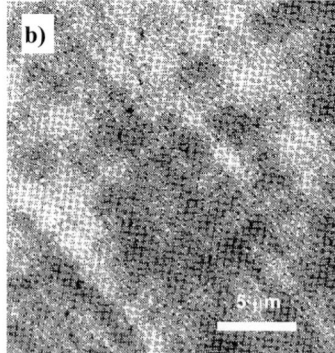

Fig. 76 Shear deformation bands in BPA-PC at 10% strain: **a** aged sample and **b** un-aged sample (From [56])

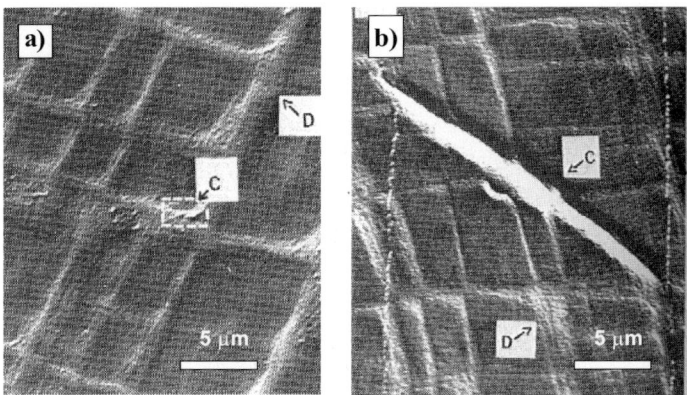

Fig. 77 Localised deformation shear bands and crazes in un-aged BPA-PC at **a** 100 °C and **b** 125 °C (From [56])

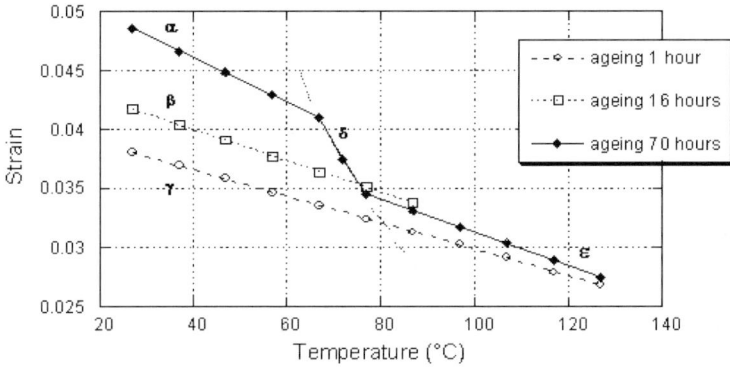

Fig. 78 Temperature dependence of the strain for deformation onset for BPA-PC(PC 0.5) for various ageing times at 130 °C. Points on curves α, β and γ represent SDZs. Points on curves δ and ε represent crazes (From [57])

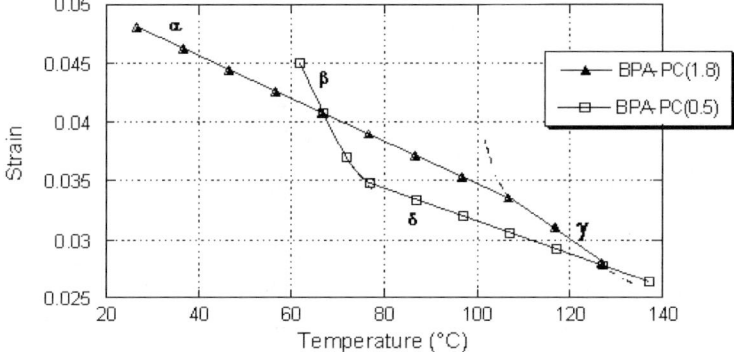

Fig. 79 Effect of MW on temperature dependence of the strain for deformation onset for samples aged 70 h at 130 °C. Points on curve α represent SDZs. Points on curves β, γ and δ represent crazes (From [57])

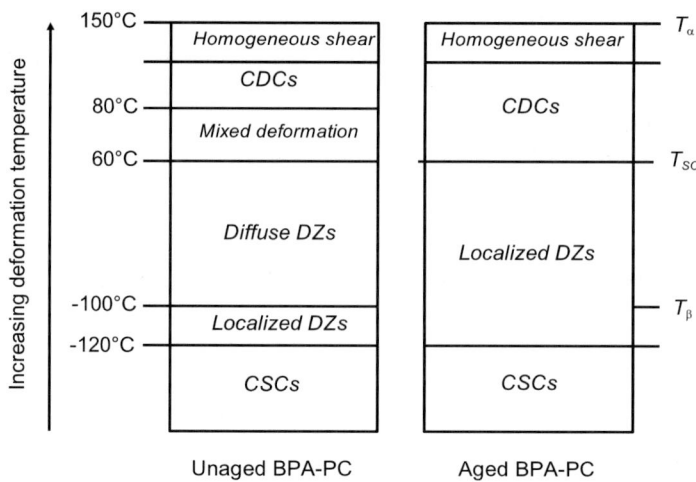

Fig. 80 Diagram of thin film deformation behaviour as a function of temperature for un-aged and aged BPA-PC ($M_w = 31\,000$); strain rate of $4 \times 10^{-4}\,s^{-1}$ (From [58])

When the straining temperature is increased, a transition from SDZs to crazes happens. This transition is characterised by a sharp decrease in the strain to produce deformation mechanisms. In contrast to the SDZ case, the strain to craze is independent of ageing time. As a consequence, the temperature at which shear–craze transition occurs decreases with increasing ageing time, as evidenced in Fig. 78. When dealing with crazes, MW has an influence; a larger MW shifts the craze occurrence to higher temperature. This is shown in Fig. 79 for PC 0.5 ($M_w = 30\,000\,g\,mol^{-1}$), curves β and δ, and for PC 1.8 ($M_w = 150\,000\,g\,mol^{-1}$), curve γ, leading to a shear–craze transition temperature 50 °C higher for PC 1.8 than for PC 0.5.

In the case of BPA-PC, the crazes occurring at high temperatures and showing a MW dependence have led the authors [57] to introduce the mechanism of craze formation by chain disentanglement (CDC).

Chain scission crazes also exist in BPA-PC films, but they are observed only at temperatures below -120 °C.

Finally, close to T_α, an homogeneous shear of the BPA-PC film is observed.

The temperature ranges in which the various deformation mechanisms occur are summarised in Fig. 80 [58].

4.2.2
TMBPA-PC

Thin film deformation study of TMBPA-PC has been carried out from room temperature to T_α [53].

At room temperature, SDZs are observed and, with increasing temperature, CDCs appear at 60 °C, which is comparable to the behaviour of BPA-PC.

However, as T_α for TMBPA-PC is 50 °C higher than that for BPA-PC (Table 7), it means that CDCs occur farther away from T_α in the case of TMBPA-PC, which could come from the lower values of M_w/M_e compared to BPA-PC ($M_w = 30\,000$).

Unfortunately, there is no information about the transition from CSCs to SDZs.

4.2.3
Relation of Deformation Micromechanisms to β Transition Motions

The limited information available in the case of TMBPA-PC, confines the discussion to BPA-PC. However, it is possible to take advantage of the results obtained on poly(methyl methacrylate) and its maleimide and glutarimide copolymers, as described in Sect. 3. Indeed, it has been shown that the development of cooperative motions in the β transition favours SDZs, relative to CSCs.

Thus, in the case of BPA-PC, the experimental and atomistic modelling investigations of the motions involved in the β transition (described in [1], Sect. 5) clearly show that the high temperature part of the β transition peak, corresponding at 1 Hz to temperatures higher than − 100 °C, is associated with phenyl ring π-flips. Such motions have an intramolecular cooperativity with the motions of the phenyl ring attached to the same isopropylidene unit, as well as with the adjacent carbonate conformation change. In addition, they also induce an intermolecular cooperativity with other phenyl rings or carbonate groups, which can be spatially distant by as much as 7 Å, an adjustment of all the group positions and orientations within a sphere of 10 Å radius around the moving phenyl ring.

At the strain rate of the thin film deformation (10^{-4} s^{-1}), the β transition peak is shifted to − 130 °C in such a way that the temperature at which the SDZs appear (− 120 °C) in BPA-PC (Fig. 80) is well located in the high temperature part of the peak, where the cooperativity described above operates.

4.3
Fracture Behaviour

BPA-PC is one of the toughest amorphous pure polymers and that is the reason why many studies have been performed to elucidate the molecular origin of this fracture behaviour, in particular concerning the role of the β transition analysed in detail in [1] (Sect. 5).

In this section, the temperature and MW dependencies of the fracture characteristics, K_{Ic} and G_{Ic}, obtained by three-point bending measurements, are examined [53].

4.3.1
BPA-PC

4.3.1.1
Temperature Dependence

Effect of MW

The fracture characteristics, K_{Ic} and G_{Ic}, as a function of temperature are shown in Fig. 81. Furthermore, on this figure are shown the results, obtained for BPA-PC with two different MWs, BPA-PC(18) and BPA-PC(31), whose the characteristics are given in Table 7.

At first, in the low temperature range, the toughness does not depend on MW until − 40 °C. At higher temperatures (above − 20 °C), there is a continuous decrease of toughness with increasing temperature in the case of the lower MW BPA-PC(18). In contrast, for the higher MW BPA-PC(31), one observes an increase of toughness up to 60 °C, and over this temperature the sample breaks in a ductile way.

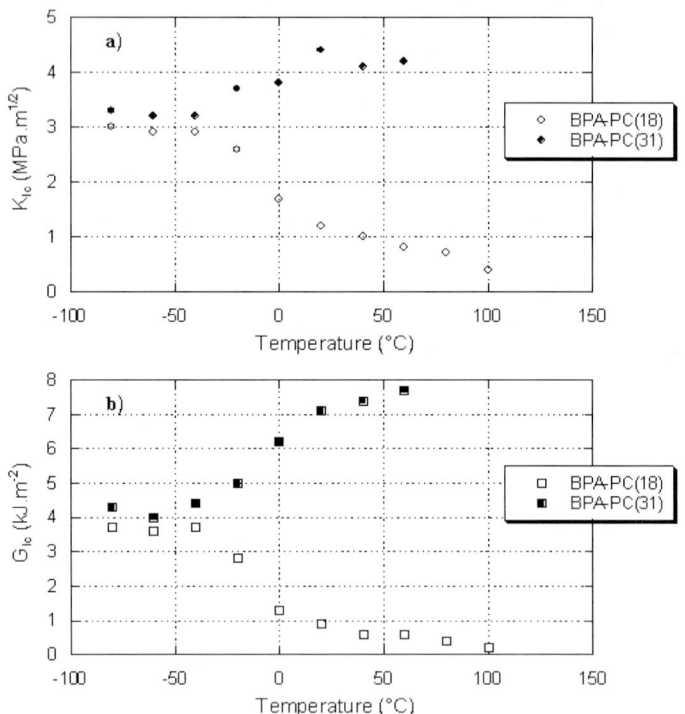

Fig. 81 Temperature dependence for BPA-PC(18) and BPA-PC(31) of **a** K_{Ic} and **b** G_{Ic} (From [53])

Such an effect of MW for BPA-PC has already been reported [59] on fracture measurements performed at $-30\,°C$.

Analysis of the Involved Micromechanisms

The curves shown in Fig. 81 are quite similar to those observed in the case of fracture of aryl-aliphatic polyamides (Sect. 5.3). However, in the latter system, a large range of MWs and chemical structures are available, allowing a detailed analysis of the observed behaviours. For these reasons, the analysis of fracture results for BPA-PC takes advantage of those on aryl-aliphatic polyamides.

Indeed, the effect of MW observed in Fig. 81 is analogous to the results on xT_yI_{1-y} (Fig. 99), for which a low temperature range exists where the toughness is independent of MW, whereas at higher temperatures, the toughness is higher for larger MWs.

In the low temperature range, it is likely that the craze at the crack tip is a chain scission craze (CSC), for chain disentanglement craze (CDC) happens at a much higher temperature ($80\,°C$) in BPA-PC (Sect. 4.2). Furthermore, as this temperature range is still in the β transition domain, there is not enough mobility within the fibrils to get a failing mechanism by chain slippage, of course MW-dependent. Consequently, the fibrils should fail by chain scission. Such a mechanism ends in BPA-PC at $-40\,°C$, a temperature which corresponds to the end at 1 Hz, or the high temperature part at higher frequency, of the β transition, depending on the involved fibril failure time.

The role of β transition motions in the fibril chain slippage is supported by the observation that in the xT_yI_{1-y} series, the temperature range where toughness is independent of MW extends to $-20\,°C/0°C$, which also corresponds to the end of the β transition at 1 Hz, of these compounds.

At higher temperatures, fibrils fail by chain slippage, which is a MW-dependent process. Indeed, even for CSCs, the MW of the chains within the fibrils increases with the initial MW of the sample. Thus, the craze stability is directly related to the polymer MW. When dealing with CDCs, the break of fibril by chain slippage is a direct function of the material MW.

In the case of BPA-PC, the thin film investigation of deformation micromechanisms (Sect. 4.2) shows that CDCs occur around $60\,°C$. So, it is unlikely that the craze at the crack tip occurring at $-20\,°C$, or above, could be a CDC. The observed MW dependence of failure originates from the above described mechanism with CSCs.

For the lower MW BPA-PC(18), increasing temperature makes the chain slippage within the fibrils easier and easier, decreasing their failure time and, by the way, decreasing toughness.

For the higher MW BPA-PC(31), for which an increase of toughness with increasing temperature is observed, an additional mechanism is involved: a contribution of shear deformation zones (SDZ). Indeed, as shown from deformation micromechanism study (Sect. 4.2), in this temperature range SDZ is the preferred mechanism. So, similarly to that observed and described for MGIM58 (Sect. 5.3.3.2), along the craze edges, SDZs are initiated. The SDZs contribute to the energy dissipation within the plastic zone ahead of the crack tip, stabilise microcrazes by stopping their development, and thus increase toughness. However, in order that this SDZ–craze interaction can happen within the plastic zone, it is necessary that the fibrils of the craze ahead of the crack tip can stand long enough. This latter feature requires a sufficiently high MW.

When increasing temperature in the temperature range where the SDZ contribution operates, two opposite effects exist:

1. A decrease of the fibril failure time due to higher chain mobility
2. An easier development of SDZs due to this increase in mobility. The observed increase of toughness would mean that it is the SDZ development which is the more sensitive to the higher mobility

4.3.2
TMBPA-PC

The toughness characteristics, K_{Ic} and G_{Ic}, for TMBPA-PC(28) are shown in Fig. 82.

At low temperature, K_{Ic} values are, in the accuracy range, identical to those of BPA-PC. The G_{Ic} values are lower than for BPA-PC due to the higher modulus, E, in this temperature range for TMBPA-PC relative to BPA-PC (Fig. 67).

However, quite a large difference is observed in the temperature dependence of K_{Ic}. Indeed, for TMBPA-PC(28) the significant decrease of K_{Ic} occurs at 40 °C, instead of − 20 °C for the beginning of the MW-dependent regime for BPA-PC. According to the mechanism described for the MW-independent toughness temperature range, the observed shift of temperature for TMBPA-PC comes from the large shift of the β transition between this polymer and BPA-PC, as evidenced in Fig. 67. So, up to 40 °C, only the motions involved in the low temperature part of the β transition are active and they are not efficient enough to avoiding fibril failure by chain scission.

The further decrease of K_{Ic} at temperatures above 40 °C seems to present quite a significant drop between 60 and 80 °C (Fig. 82). Such a behaviour could be related to the occurrence of the CDC mechanism, which occurs at 60 °C in thin film deformation (Sect. 4.2). The lack of TMBPA-PC samples with various MWs do not allow a deeper analysis of the fracture behaviour in this temperature range.

4.3.3
Conclusion on Fracture

The results obtained on the fracture behaviour of BPA-PC and TMBPA-PC have clearly shown two different temperature ranges.

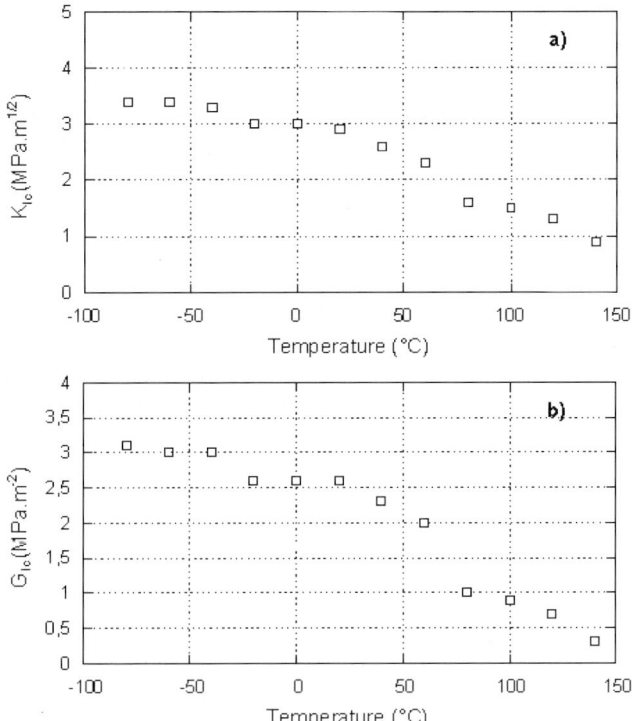

Fig. 82 Temperature dependence for TMBPA-PC(28) of **a** K_{Ic} and **b** G_{Ic} (From [53])

The low temperature range is characterised by a toughness independent of MW. It extends to a temperature where all the β transition motions are not still active. Consequently, the lack of chain mobility within the craze fibrils avoids chain slippage and the craze fibrils fail by chain scission.

The high temperature range shows a MW-dependent toughness associated with chain slippage failure of craze fibrils. It happens in a temperature domain where all the β transition motions (or at least those corresponding to the high temperature part of the β transition) are active. The increase of toughness with increasing temperature originates from a contribution of SDZs and their interaction with crazes. In order to get such a mechanism, a sufficiently high MW is required for increasing the fibril failure time and allowing energy dissipation in the plastic zone ahead of the crack tip.

4.4
Conclusion on Bisphenol A (or Tetramethyl Bisphenol A) Polycarbonate

Investigation of the mechanical properties of bisphenol A polycarbonate (BPA-PC) and tetramethyl bisphenol A polycarbonate (TMBPA-PC) has led to quite unusual results on the role of the β transition motions, compared to the case of PMMA and its maleimide and glutarimide copolymers.

Indeed, whereas in these latter polymers the cooperativity of motions associated with the high temperature part of the β transition is an intramolecular cooperativity, in the case of BPA-PC (besides an intrinsic cooperativity between phenyl ring flips attached to the same isopropylidene unit) there is an intermolecular cooperativity which develops in this high temperature part of the β transition, associated mostly with the occurrence of phenyl ring π-flips.

The normalised strain softening increases with increasing temperature, which is opposite to the PMMA and copolymer behaviour. A mechanism has been proposed to account for this, based on an increasing softening of the medium by the development of intermolecular cooperativity. This makes the yielding and plastic flow conformation changes easier, and increases the extent of perturbation induced by yielding conformation changes, thus leading to a larger strain softening. In the case of TMBPA-PC, the methyl groups in *ortho* position to the carbonate hinder the phenyl ring π-flips and lead to a 150 °C upward shift of the β transition peak maximum temperature. Thus, the medium softening does not occur and disturbance of the surroundings (induced by yielding conformation changes) remains localised, resulting in a smaller strain softening.

The main micromechanism of deformation for BPA-PC is SDZ. They are observed at a very low temperatures (– 120 °C), which correspond to the high temperature part of the β transition of this polymer. These SDZs remain up to about 80 °C.

The fracture behaviour of BPA-PC shows quite a high toughness at room temperature, whereas it is about 30% lower for TMBPA-PC. Furthermore, the temperature dependence of K_{Ic} for BPA-PC exhibits a temperature domain (– 30 to 50 °C) where a large increase of K_{Ic} is observed, followed by a small decrease at higher temperatures. In contrast, a continuous decrease of K_{Ic} with increasing temperature occurs for TMBPA-PC over the whole temperature range. The increase of K_{Ic} for BPA-PC corresponds to the end of its β transition and the easier development of SDZs, which can interact with crazes in the plastic zone ahead of the crack tip. The further K_{Ic} decrease results from the easier chain slippage within the fibrils, owing to the higher chain mobility, reducing the craze stability as well as the interaction with SDZs.

5
Aryl-aliphatic Copolyamides

Investigation of the molecular motions involved in solid-state transitions of amorphous polymers (developed in [1]) has shown that it is interesting to examine both vinyl polymers with side groups (e.g. PMMA and its copolymers) and polymers without side chains, but containing phenyl rings in their chain backbone (e.g. BPA-PC). Indeed, these two types of polymer structures cover most of the situations encountered in amorphous polymers and allows one to reach valuable conclusions, in particular regarding the contribution of cooperative motions of the β transition.

Along this line, the analysis of the plastic deformation, the micromechanisms of deformation, and the fracture behaviour of the series of aryl-aliphatic copolyamides studied in [1] (Sect. 6) is quite suitable.

The polymers considered correspond to two different groups. The first deals with the copolyamides, coded $x T_y I_{1-y}$, whose repeat unit contains x lactam-12 sequences, y and $1 - y$ *tere-* and *iso-*phthalic moieties, respectively, and a 3,3'-dimethylcyclohexyl methane unit in the regular order, as shown in Fig. 83a.

The second group is represented by the copolyamides, coded $M T_y I_{1-y}$, obtained from the condensation of y and $1 - y$ *tere-* and *iso-*phthalic acids, respectively, onto 1,5 diamino-2-methyl pentane, as shown in Fig. 83b.

The characteristics of these polymers are listed in Table 8.

Before considering the various mechanical properties, it is important to notice that the β transition of these copolyamides, as shown by the dynamic mechanical loss modulus, E'', in Figs. 84 and 85 for the $x T_y I_{1-y}$ and $M T_y I_{1-y}$ series, respectively, occurs at quite low temperatures. Indeed, for the first series the β peak maximum occurs at $-60\,^{\circ}C$ at 1 Hz, and at $-110\,^{\circ}C$ for the second series. From this point of view, these copolyamides look more like BPA-PC (Sect. 4) than PMMA (Sect. 3).

Finally, it is worth pointing out that the copolyamides of the $x T_y I_{1-y}$ series present a γ transition, located at 1 Hz around $-140\,^{\circ}C$, with a E'' amplitude larger than the β peak, and an ω transition in the temperature range between 20 and $80\,^{\circ}C$ (Fig. 84).

Owing to the large difference in chemical structures of the two series of copolyamides, they will be considered one after the other.

5.1
$x T_y I_{1-y}$ Copolyamides

5.1.1
Plastic Deformation of $x T_y I_{1-y}$ Copolyamides

For the considered systems, the stress–strain curves and the derived quantities yield stress, σ_y, plastic flow stress, σ_{pf}, and strain softening, have been studied [53, 60] in a temperature range from $-110\,^{\circ}C$ to typically $(T_\alpha - 20)\,^{\circ}C$. Indeed, for temperatures closer to T_α, the experimental results are less reliable as some creep behaviour can to occur.

a)

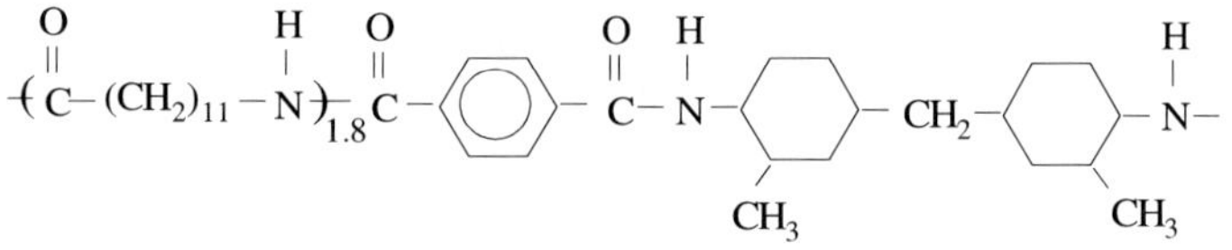

b)

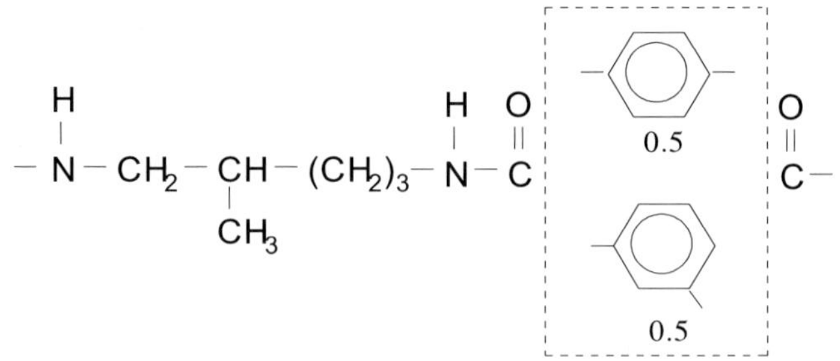

Fig. 83 Chemical formulae of the copolyamide series **a** xT_yI_{1-y} and **b** MT_yI_{1-y}

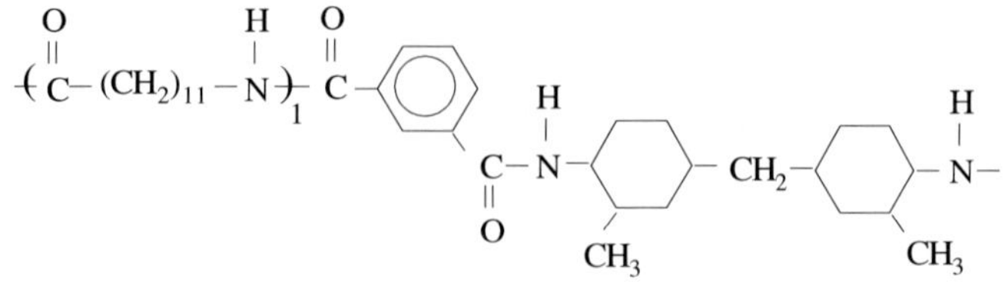

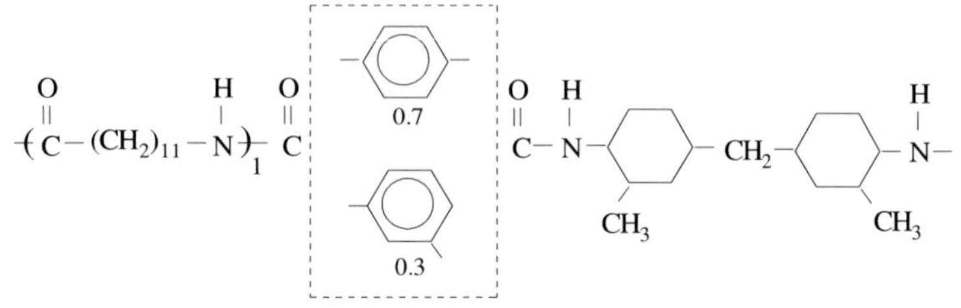

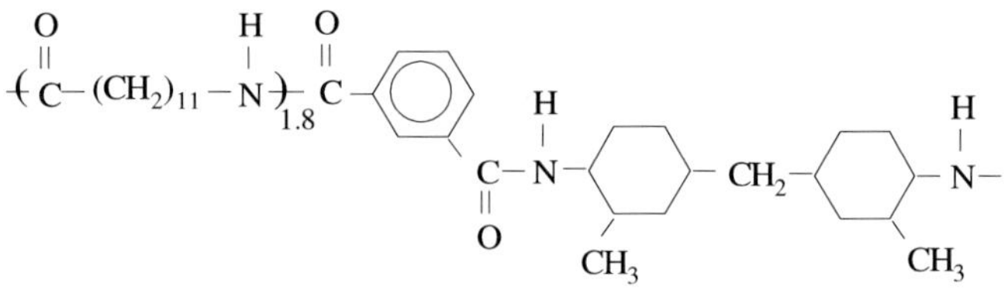

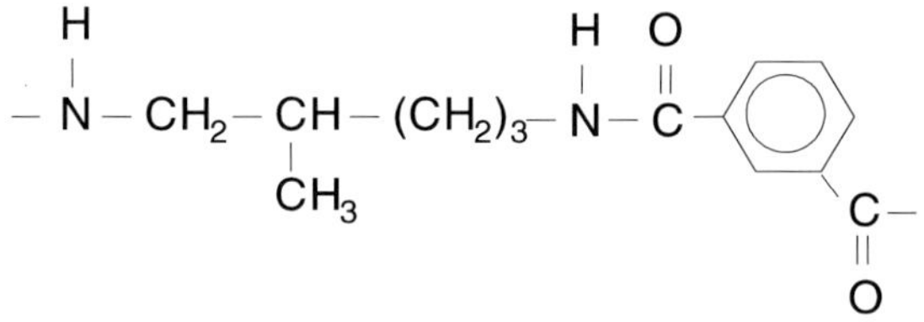

Table 8 Characteristics of copolyamides of xT_yI_{1-y} and MT_yI_{1-y} series

Polymer code	M_w	M_n	ρ	T_g	T_α	M_e	ν_e	M_w/M_e	M_n/M_e	C_∞	$10^{-5} \eta_0$ at $T_\alpha - T = 40$ K	$10^6 \zeta$ at $T_\alpha - T = 40$ K
	g mol^{-1}	g mol^{-1}	kg m^{-3}	°C	°C	g mol^{-1}	10^{24} m^{-3}				Pa s	N s m^{-1}
1.8I(22)	22000	8500	1042	118	130	2700	230	8.1	3.1	5.6	2.4	2.8
1.8T(23)	23000	9000	1042	124	137	3000	210	7.7	3.0	5.8	2.7	3.1
1.8T(39)	39000	12000	1042	124	137	3000	210	13.0	4.0	5.8	21.0	3.5
1I(21)	21000	8000	1055	151	161	2800	225	7.5	2.8	5.3	3.4	3.8
1I(26)	26000	10000	1055	151	161	2800	225	9.2	3.6	5.3	6.1	3.3
$1T_{0.7}I_{0.3}(23)$	23000	8500	1057	159	171	3100	200	7.5	2.8	5.6	3.5	3.3
$1T_{0.7}I_{0.3}(27)$	27000	10000	1057	159	171	3100	200	8.8	3.3	5.6		
$1T_{0.7}I_{0.3}(32)$	32000	11000	1057	159	171	3100	200	10.4	3.6	5.6	11.3	3.4
MI	18000	6000	1194	134	141	2750	260	6.5	2.2	5.5	2.6	2.0
$MT_{0.5}I_{0.5}(23)$	23000	6500	1196	137	145	2900	250	7.9	2.3	5.6	5.5	2.0
$MT_{0.7}I_{0.3}(22)$	22000	5000	1196	139	147	2950	245	7.5	1.7	5.6		

Figures in brackets indicate the MW in kg mol^{-1}

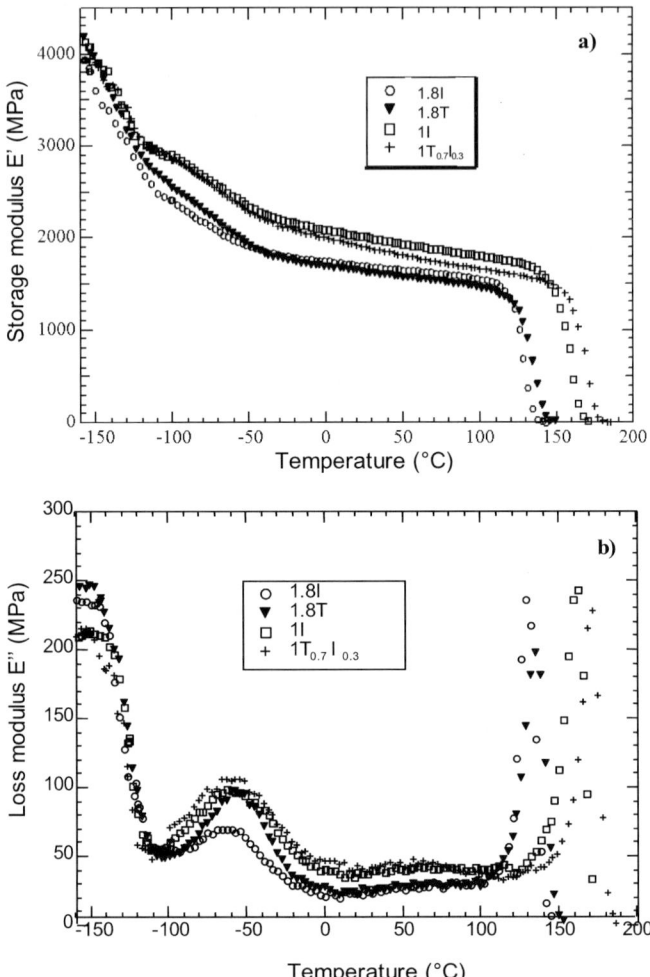

Fig. 84 Temperature dependence at 1 Hz of **a** storage modulus and **b** loss modulus for various xT_yI_{1-y} copolyamides (From [53])

5.1.1.1
Stress–Strain Curves of xT_yI_{1-y} Copolyamides

A typical example of the stress–strain curves is shown in Fig. 86 in the case of $1T_{0.7}I_{0.3}$ copolyamide. Irrespective of whether the variable under consideration is temperature (Fig. 86a) or strain rate (Fig. 86b), all curves present the same profile.

First, the most striking feature deals with the strain softening amplitude. Indeed, it is larger at high temperatures than at low ones, similar to the

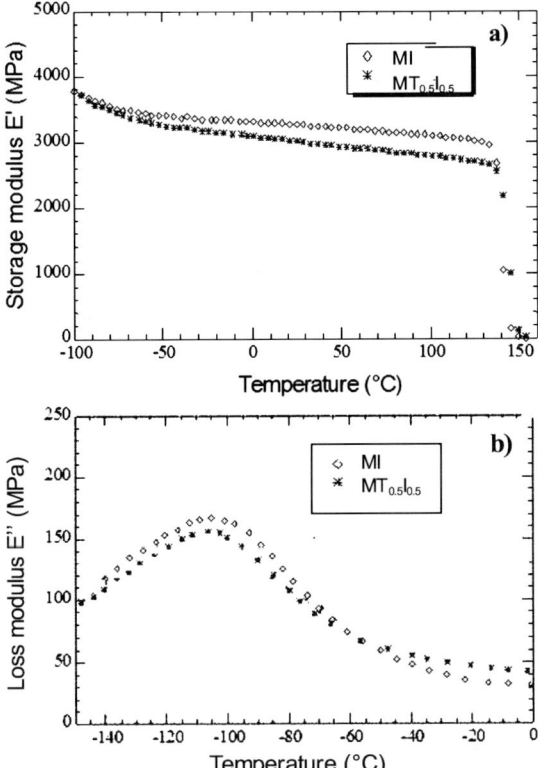

Fig. 85 Temperature dependence at 1 Hz of **a** storage modulus and **b** loss modulus for various MT_yI_{1-y} copolyamides (From [53])

behaviour observed for BPA-PC (Sect. 4) and contrary to that for PMMA (Sect. 3).

5.1.1.2
Yield Stress of xT_yI_{1-y} Copolyamides

The temperature dependence of the yield stress, σ_y, and the plastic flow stress, σ_{pf}, for the various xT_yI_{1-y} copolyamides is shown in Fig. 87. For comparison, the yield stresses of the various copolyamides [61] are shown in Fig. 88.

The effect of physical ageing on σ_y, which is quite significant in PMMA (as described in Sect. 4), does not affect the σ_y values at any temperature.

Investigation of the strain rate dependence of σ_y for the various xT_yI_{1-y} copolyamides has been performed over a large temperature range in order to determine the activation volume, V_0. The results are presented in Fig. 89.

An interesting comment on these results concerns the large values of V_0 (1 nm^3 in order of magnitude) obtained at low temperatures, compared to the

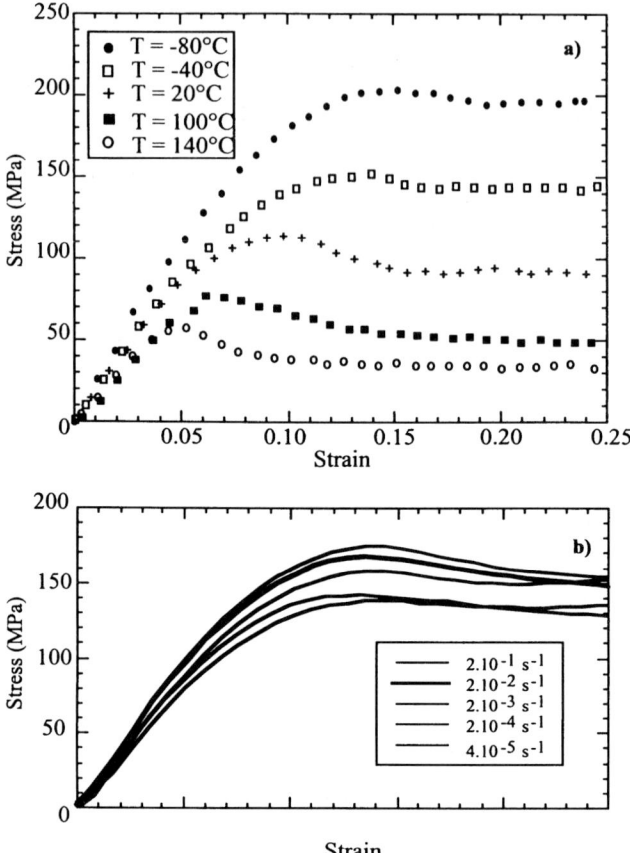

Fig. 86 Typical stress–strain curves relative to the sample $1T_{0.7}I_{0.3}$. **a** Effect of temperature at a strain rate of $2 \times 10^{-3} \text{ s}^{-1}$. **b** Effect of strain rate at a temperature of $-40\,^{\circ}\text{C}$ (From [60])

value of 0.1 nm^3 for PMMA in the same temperature range (see Sect. 3.1.1.3). These values are close to the value derived at $-60\,^{\circ}\text{C}$ for BPA-PC, i.e. 0.9 nm^3 (Sect. 4.1.1.1).

Another feature deals with the marked increase in V_0 observed for all the xT_yI_{1-y} copolyamides at temperatures higher than $-40\,^{\circ}\text{C}$.

5.1.1.3
Plastic Flow Stress of xT_yI_{1-y} Copolyamides

Figure 90 shows the temperature dependence of the plastic flow stress for the various xT_yI_{1-y} copolyamides. As can be seen also from Fig. 87, the general shape of σ_{pf} curves is similar to that of σ_y.

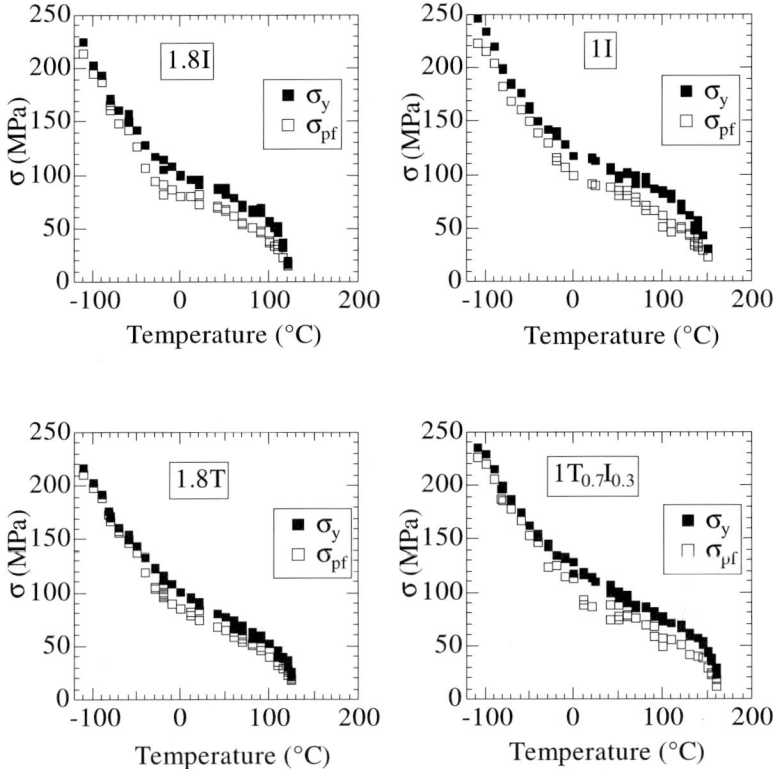

Fig. 87 Temperature dependence of the yield stress, σ_y, and the plastic flow stress, σ_{pf}, for various xT_yI_{1-y} copolyamides (From [60])

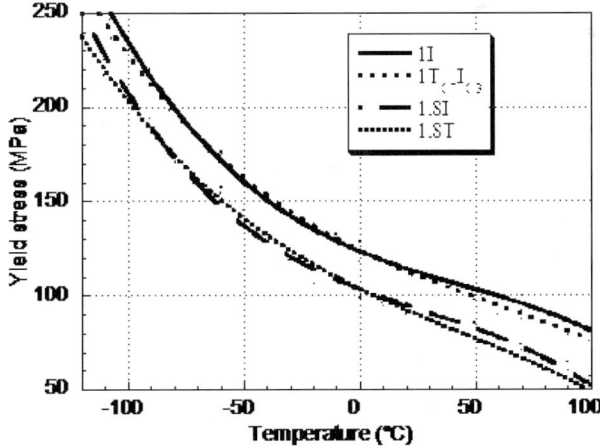

Fig. 88 Yield stress of the xT_yI_{1-y} copolyamides versus temperature at a strain rate of 2×10^{-3} s^{-1} (From [61])

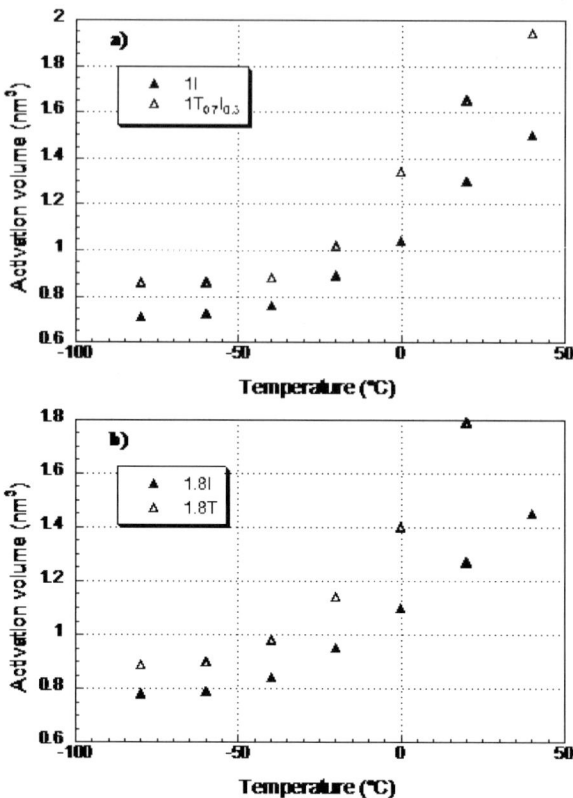

Fig. 89 Temperature dependence of the activation volume, V_0, for **a** $1T_yI_{1-y}$ copolyamides and **b** $1.8T_yI_{1-y}$ copolyamides (From [53])

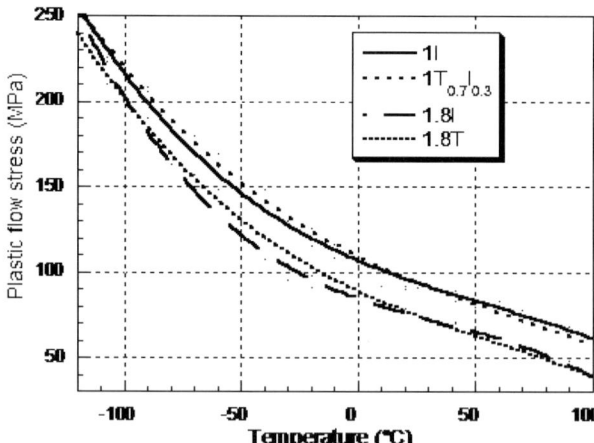

Fig. 90 Temperature dependence of the plastic flow stress for the xT_yI_{1-y} copolyamides (From [61])

5.1.1.4
Strain Softening of xT_yI_{1-y} Copolyamides

The temperature dependence of the strain softening amplitude, SSA, of the various xT_yI_{1-y} copolyamides is shown in Fig. 91. Two different behaviours are observed depending on the lactam-12 content ($x = 1$ or 1.8). However, within each family, quite a similar shape is obtained independently of the *iso-* or *tere-* nature of the phthalic unit.

The results for SSA normalised by the plastic flow stress, nSSA, (defined in Sect. 2.2.3) are shown in Fig. 92. It is striking to see how different are the temperature dependencies of SSA and nSSA. A similar difference is encountered in the case of BPA-PC (Sect. 4.1.1.3). Nevertheless, the observation that the lactam-12 content ($x = 1$ or 1.8) leads to two different behaviours remains (Fig. 91 for 1I and 1.8I), as well as the fact that in each family, similar shapes are obtained independently of the *iso-* or *tere-*nature of the phthalic unit.

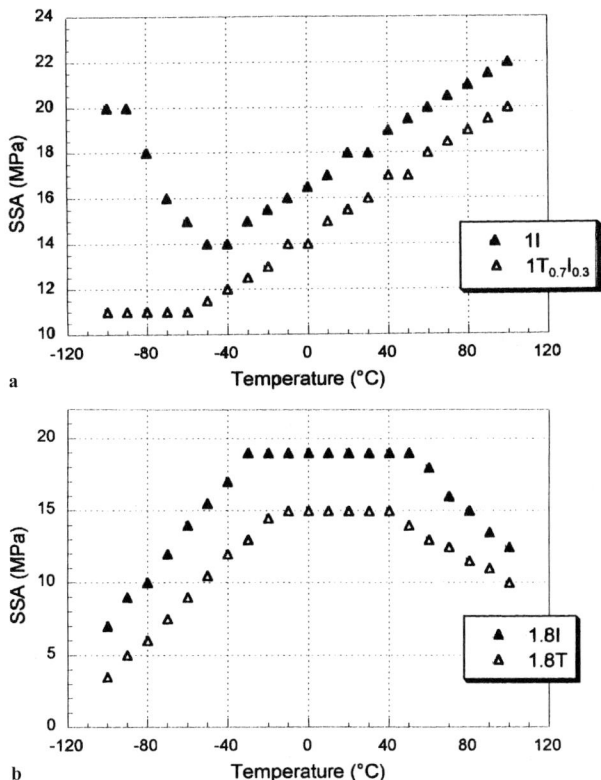

Fig. 91 Temperature dependence of SSA of **a** $1T_yI_{1-y}$ copolyamides and **b** $1.8T_yI_{1-y}$ copolyamides

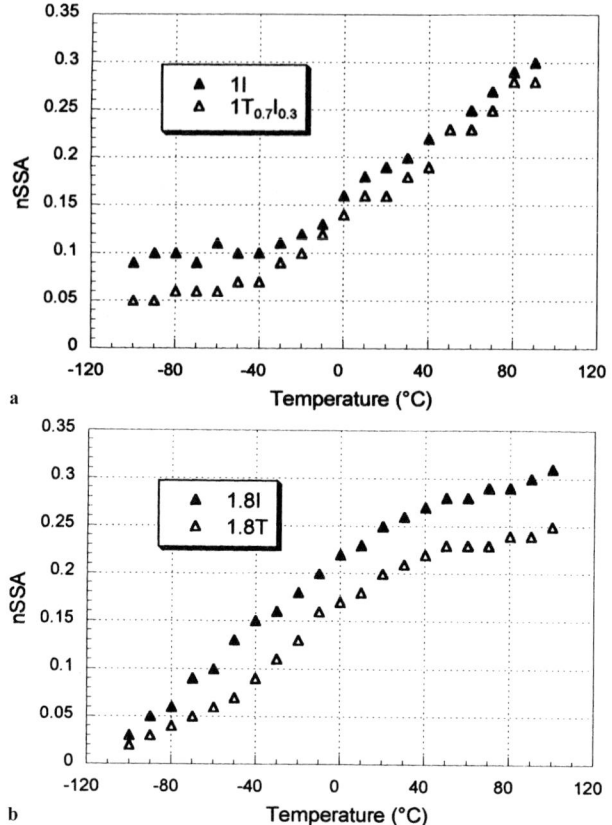

Fig. 92 Temperature dependence of nSSA of **a** $1T_yI_{1-y}$ copolyamides and **b** $1.8T_yI_{1-y}$ copolyamides

In both families, an increase of nSSA with increasing temperature is observed for a large range of the considered temperatures.

5.1.1.5
Effect of the Chemical Structure of xT_yI_{1-y} Copolyamides

Within the xT_yI_{1-y} copolyamide series, it is interesting to analyse the chemical structure dependence of the plastic deformation characteristics.

First, let us look at the effect of the lactam-12 content on σ_y and σ_{pf}. One observes a large decrease, over the whole temperature range, of σ_y and σ_{pf} values when increasing the lactam-12 content per repeat unit from 1 to 1.8, as shown in Fig. 88 and Fig. 90.

The lactam-12 content has also an influence on the shape of the temperature dependence of nSSA. In the case of $1T_yI_{1-y}$ copolyamides, nSSA is constant to about $-40\,°C$, then it regularly increases (Fig. 92a). In contrast, for $1.8T_yI_{1-y}$ copolyamides the increase is continuous from -100 to about $50\,°C$, where a leveling appears.

Replacement of the *iso*-phthalic unit by *tere*-phthalic, at a constant amount of lactam-12, leads to a decrease of strain softening expressed through either SSA (Fig. 91) or nSSA (Fig. 92).

Another effect of the nature of phthalic acid concerns the activation volume, V_0. The results reported in Fig. 89 show that, independently of the lactam-12 content, the replacement of *iso*- by *tere*-phthalic unit leads to higher values of V_0, whatever the temperature. This conclusion remains valid if V_0 values are compared as a function of $(T - T_\alpha)$, in order to take into account the chemical structure dependence of T_α (Table 8).

In contrast, in the same copolyamide series, either *iso*-phthalic or *tere*-phthalic, the lactam-12 content does not have any influence on V_0, whatever the temperature.

It is worth pointing out that the V_0 values obtained for *tere*-phthalic copolyamides are quite similar to those for BPA-PC (Sect. 4.1.1.1) at low temperatures, but they do not increase as much with increasing temperature.

5.1.1.6
Relation of Plastic Deformation of xT_yI_{1-y} Copolyamides to Secondary Transitions

Before developing the role of secondary transitions in the plastic deformation of xT_yI_{1-y} copolyamides, it is useful to recall the main results achieved through the investigation of such transitions (as described in [1], Sect. 6).

First, the temperature dependence of the mechanical loss, E'', at 1 Hz, is shown in Fig. 84 for the various xT_yI_{1-y} copolyamides. Three secondary transitions are observed:

1. γ transition, centred around $-150\,°C$
2. β transition, extending from -100 to $0\,°C$, with a maximum around $-60\,°C$
3. ω transition, wide but weak, centred around $60\,°C$

These various transitions have been assigned precisely and the description will refer to the chemical structures shown in Fig. 83, in particular to the different amide groups, which are specifically denoted depending on their position in the repeat unit. It is worth noticing that, due to the quasi-conjugated character of the CO – NH bond, the amide group is planar.

The γ transition, in the various xT_yI_{1-y} copolyamides, originates from the CO_{aliph} amide group adjacent to a cyclohexyl unit and from motions of CH_2 groups within the lactam-12 sequence. In the case of $1.8T_yI_{1-y}$ copolyamides, the γ transition is broader, due to the existence of an additional γ' transition due to $CO_{aliph\,2}$ amide group situated in the centre of the 1.8 lactam-12 unit.

The β transition, in the low temperature part of the peak, involves only oscillations of the phenyl rings and adjacent amide group near the lactam-12 sequence. In the high temperature part, cooperative π-flips of the *tere*-phthalic rings occur, whereas the motions remain oscillations of larger amplitude for *iso*-phthalic moieties.

The ω transition is assigned to motions of the $CO_{arom\,2}$ amide group situated between the phenyl ring and the cyclohexyl unit.

The plastic flow developing beyond yielding requires main-chain motions similar to those are involved in the glass–rubber transition phenomenon (as mentioned in Sect. 2.2.2). For this reason, the plastic flow can be considered as a reference, since the involved motions are always the same, whatever the considered temperature.

Furthermore, it has been mentioned (Sect. 2.2.1.2) that the yielding molecular process corresponds to main-chain conformation changes, induced by the applied stress, and leads to an increase of high-energy conformations, in a way similar that induced by increasing the temperature above T_α, without any applied stress.

In the case of PMMA (Sect. 3) or BPA-PC (Sect. 4), it is easy to define which main-chain bonds change conformation above T_α, in order to increase the statistical weight of the higher energy conformation (C – C main-chain bond in PMMA, carbonate for BPA-PC). For xT_yI_{1-y} copolyamides, such an assignment is more difficult, due to the presence of several different groups in the repeat unit (as shown in the chemical structures in Fig. 83).

Concerning the lactam-12 unit, it is unlikely that, at temperatures higher than $-110\,°C$, an applied stress could induced C – C bond conformation changes (*trans* to *gauche* or vice-versa). Indeed, such changes are precisely the motions of the C – C bonds involved in the γ transition occurring at $-150\,°C$, at 1 Hz. Thus, thermal energy available above $-110\,°C$ already provides many of these C – C conformation changes.

In the absence of atomistic modelling information, it is not possible to get a definite answer regarding the other groups. In spite of this difficulty, it is nevertheless possible to understand the contribution of the motions involved in the various transitions to plastic deformation, in particular the strain softening, of the various xT_yI_{1-y} copolyamides. Indeed, for a given copolyamide, whatever the temperature, the yield stress will result in the same main-chain conformation change(s), which is(are) the elementary process(es) leading to the chain displacement occurring in the plastic flow.

As the nSSA curves, as a function of temperature, for the various xT_yI_{1-y} copolyamides (Fig. 92) have quite a similar shape to those obtained for BPA-PC and TMBPA-PC (Sects. 4.1.1.3 and 4.1.2.3, respectively), the discussion will be based on the mechanism proposed for the strain softening behaviour of BPA-PC (Sect. 4.1.3) and will emphasise the intermolecular cooperativity of the secondary transition motions.

However, in the case of xT_yI_{1-y} copolyamides, in spite of some differences resulting from the replacement of *iso*-phthalic units by *tere*-phthalic units, similar behaviours occur. In particular, the increase of nSSA with increasing temperature is observed even for xI copolyamides, in which *iso*-phthalic rings cannot undergo π-flips, but oscillations only. Thus, the origin of the intermolecular effects must get a contribution independent of phenyl ring π-flips, in contrast to the BPA-PC case. Such an effect can arise from the hydrogen bonds existing in these copolyamides.

When conformation changes induced by the yielding process occur, they induce a rearrangement of the surrounding medium. At low temperatures, in these copolyamides a rigid medium exists so the surrounding perturbation remains limited, in such a way that the plastic flow requires almost as high a stress as that for yielding.

At higher temperatures, when secondary transitions occur, (γ at first then β), the involved motions, particularly in the case of β transition, result in the weakening or breakdown of some of the hydrogen bonds, softening the polymer medium and, so, allowing for larger rearrangements of the surrounding medium. As a consequence, the plastic flow is able to occur at a lower stress value than that for yielding, and a larger strain softening happens.

It is worth noting that the softening of the polymer medium, resulting from hydrogen bond breakdown due to β transition motions, is well observed in the temperature dependence of the modulus, E', of $x T_y I_{1-y}$ copolyamides (shown in Fig. 84a).

It is in the frame of this mechanism that the plastic deformation results, in particular the strain softening, are considered.

Let us first deal with the $1 T_y I_{1-y}$ copolyamides (Fig. 92a). The low nSSA value plateau extending from -100 to $-40\,^\circ$C indicates that the conformation change induced by σ_y, in this temperature range, happens in too rigid a medium, leading to limited rearrangements in the surroundings, in such a way that the plastic flow requires almost as high a stress as that for yielding. The increase of nSSA starting at $-40\,^\circ$C corresponds to the end of the β transition (taking into account that at the experimental rate of the plastic deformation experiments, $2 \times 10^{-3}\ \text{s}^{-1}$, the β transition is shifted towards lower temperature by about $30\,^\circ$C). Above $-40\,^\circ$C, the β transition motions are completely active and result in an increase of nSSA. At higher temperatures, above $20\,^\circ$C, the occurrence of the ω transition, by providing mobility between the phenyl ring and the 3,3'-dimethylcyclohexyl moiety, still increases the softening of the medium, with the related consequences on the further increase of nSSA.

Concerning the $1.8 T_y I_{1-y}$ copolyamides (Fig. 92b), the main difference to the $1 T_y I_{1-y}$ copolyamide results deals with the low temperature behaviour. Indeed, instead of the plateau of nSSA occurring for the latter series, there is an increase from $-100\,^\circ$C. Such an increase of nSSA corresponds, in the proposed mechanism, to a gradual increase in the softening of the material. Indeed, in the $1.8 T_y I_{1-y}$ copolyamides there is an additional γ' transition, with a maximum around $-90\,^\circ$C at 1 Hz, due to $CO_{\text{aliph 2}}$ amide group motion. The increase of nSSA with increasing temperature can be accounted for in the same way as for $1T_yI_{1-y}$ copolyamides.

Finally, in each copolyamide series, the strain softening is larger for *iso*-phthalic than for *tere*-phthalic copolyamides. Such a behaviour can involve different contributions. At first, the displacement of the kinked *iso*-phthalic structure under the applied yield stress induces a larger disturbance of the surrounding medium than in the case of *tere*-phthalic unit; this would result in a larger nSSA. Another contribution is the intramolecular cooperativity between phenyl ring π-flip and adjacent amide group, which leads to a softer main chain where conformation changes are easier to perform. This intramolecular effect looks like what happens with ester group π-flip and main-chain cooperativity in PMMA (as recalled in Sect. 3.1.1.6), leading to smaller nSSA. The fact that the effect of the nature of the phthalic unit is also observed in the low temperature range, where no ring π-flip occurs, would tend to support the conclusion that the contribution first considered is the one involved in the case of strain softening.

5.1.2
Micromechanisms of Deformation of xT_yI_{1-y} Copolyamides

The micromechanisms of deformation of the various xT_yI_{1-y} copolyamides have been investigated [61] by using the thin film technique (Sect. 2.3). The characteristics of the copolyamides under consideration are gathered in Table 8.

5.1.2.1
Micromechanism Morphologies

All the copolyamides of the xT_yI_{1-y} series present the same micromechanisms of deformation. Due to its chemical structure containing both *iso-* and *tere*-phthalic units, $1T_{0.7}I_{0.3}$ is quite a representative copolyamide for describing the various mechanisms encountered in the explored temperature range (from $-130\,°C$ to $T_\alpha - 20\,K$).

In this temperature range, there are four regimes of micromechanisms, as shown in Fig. 93. At low temperature ($T \leq -100\,°C$), narrow, highly elongated crazes are observed running perpendicular to the tensile axis, with well-defined midribs. In this temperature range, crazes correspond to chain scission (CSC). As the test temperature is raised (-100 to $40\,°C$), both localised and diffuse shear deformation zones (SDZ) become dominant, these latter growing at approximately $45°$ to the tensile axis, as shown in Fig. 94. In particular, Fig. 94b shows part of a typical localised SDZ. At still higher temperatures (40–$80\,°C$), craze-like features and SDZs coexist, as shown in Fig. 95. The SDZs typically develop at the tips of the crazes, again inclined an angle of about $45°$ to the tensile axis. The crazes observed in this tempera-

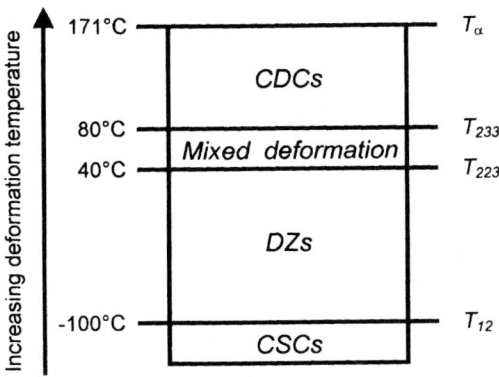

Fig. 93 Deformation map for xT_yI_{1-y} copolyamides (From [61])

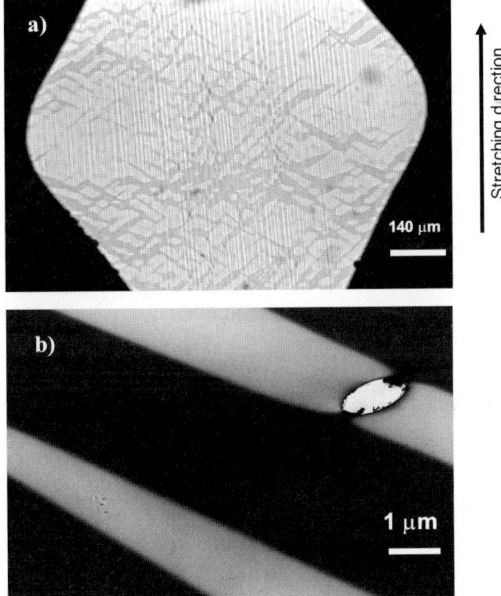

Fig. 94 SDZs in thin films of $1T_{0.7}I_{0.3}$ copolyamide stretched at 25 °C with a strain rate of 2×10^{-3} s^{-1}. **a** Optical microscopy, **b** TEM (From [61])

ture range are consequently relatively short, in a way similar that observed with MGIMx copolymers (Sect. 3.3.2). Finally, between 80 °C and T_α, crazing is again dominant, the craze morphology being very similar to that observed at low temperatures. The crazes originate from chain disentanglement (CDC), as do those observed in the mixed regime from 40 to 80 °C.

Three temperatures, T_{12}, T_{223} and T_{233}, are used to characterise the transitions between the different regimes of deformation micromechanisms, representing the transitions from low temperature CSCs to SDZs, from SDZs to

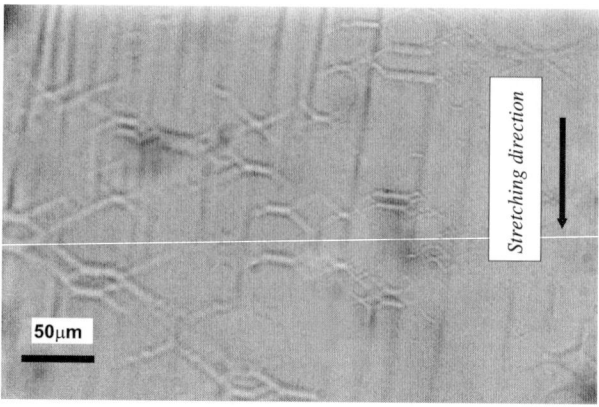

Fig. 95 Optical microscopy of mixed crazes and SDZs for $1T_{0.7}I_{0.3}$ copolyamide stretched at $80\,^{\circ}C$ with a strain rate of $2 \times 10^{-3}\,s^{-1}$ (From [61])

SDZs/CDCs and from SDZs/CDCs to high temperature CDCs only, respectively. Thus, for $1T_{0.7}I_{0.3}$ (Fig. 93), T_{12}, T_{223} and T_{233} are equal to -100, 40 and $80\,^{\circ}C$, respectively.

5.1.2.2
Influence of the Chemical Structure of xT_yI_{1-y} Copolyamides

Studies on the xT_yI_{1-y} copolyamides [53] did not reveal any influence of the molar mass (in the range $M_w/M_e > 7$) on the deformation maps.

The results on T_{12}, T_{223}, T_{233} and $(T_\alpha - T_{233})$ for the considered copolyamides are reported in Table 9.

Occurrence of SDZs (T_{12}) in xT_yI_{1-y} Copolyamides

According to the results in Table 9 at fixed lactam-12 content, the replacement of *tere*-phthalic acid by *iso*-phthalic acid does not influence T_{12}. On the other hand, increasing the lactam-12 content (from 1 to 1.8 mole per repeat unit) results in a decrease of about $10\,^{\circ}C$ in T_{12}.

As described in Sect. 2.3.4, the transition temperature, T_{12}, corresponding to the occurrence of SDZs, can be quantitatively estimated from the temperature dependence of the yield stress, σ_y, the plastic flow stress, σ_{pf} and the critical stress for CSC, σ_{CSC}. However, the absolute value of the latter requires quantities difficult to determine experimentally. To overcome this difficulty, when comparing materials of similar chemical structure, one can take one of them as a reference, using its experimental T_{12}^{ref} value for deriving the T_{12}^i values for the other i polymers of the considered series.

In the case of the xT_yI_{1-y} copolyamides, another difficulty arises from the determination of characteristics of the equivalent chain (Sect. 2.1), namely

Table 9 Experimental and calculated values of the transition temperatures (in °C) for xT_yI_{1-y} copolyamides and experimental values for MT_yI_{1-y} copolyamides

Polymer	T_{12}		T_{223}	T_{233}		$T_\alpha - T_{233}$		$T_{233} - T_{12}$	
	Experimental	Calculated	Experimental	Experimental	Calculated	Experimental	Calculated	Experimental	Calculated
1,8I	−110	−112	20	30	35	100	95	140	137
1,8T	−110	−117	40	70	69	67	68	180	186
1I	−100	−96	40	50	46	111	115	150	142
$1T_{0.7}I_{0.3}$	−100	−100 (ref)	40	80	80 (ref)	91	91 (ref)	180	180 (ref)
MI	−70		110						
$MT_{0.5}I_{0.5}$	−70		120						

the bond length, l_{eq}, of the equivalent chain. Indeed, whereas for the CMIMx and MGIMx copolymers (see Sects. 3.2.2.2 and 3.3.2.2) there is no ambiguity about the calculation of l_{eq}, in the case of xT_yI_{1-y} copolyamides, the occurrence of two cyclohexyl groups within the chemical structure (Fig. 83) excludes a similar calculation due to the chair or boat configurations of the cyclohexyl units. Nevertheless, as the 2,2′-dimethyldicyclohexyl methane residue is present in all the copolyamides of the series, and as the replacement of a *tere*-phthalic unit by a *iso*-phthalic unit leads to a negligible change in the characteristics ($n_{eq}l_{eq}^2$) of the repeat unit, whatever the lactam-12 content, it is possible to consider the molar mass between entanglements, M_e, instead of the rms end-to-end distance between entanglements. This leads to:

$$\sigma_{CSC} \propto \sigma_{pf}^{1/2} M_e^{-1/4} C_\infty^{1/4} U^{1/2} \tag{53}$$

C_∞ is estimated from the empirical relationship [7]:

$$C_\infty = \left(n_v M_e / 3 M_0 \right)^{1/2} \tag{54}$$

where M_0 and n_v are the molar mass and the number of bonds per polymer repeat unit, respectively.

For the reference material at any temperature $\sigma_{CSC}^{ref}(T)$ is expressed as:

$$\sigma_{CSC}^{ref}(T) = \sigma_y^{ref}\left(T_{12}^{ref}\right) \left[\frac{\sigma_{pf}^{ref}(T)}{\sigma_{pf}^{ref}\left(T_{12}^{ref}\right)} \right]^{1/2} \tag{55}$$

For any other copolyamide i of the series, $\sigma_{CSC}^i(T)$ is given by.

$$\sigma_{CSC}^i(T) = \kappa_{CSC}^i(T) \sigma_{CSC}^{ref}(T) \tag{56}$$

where:

$$\kappa_{CSC}^i(T) = \left[\frac{\sigma_{pf}^i(T)}{\sigma_{pf}^{ref}(T)} \right]^{1/2} \left(\frac{M_e^i}{M_e^{ref}} \right)^{-1/4} \left(\frac{C_\infty^i}{C_\infty^{ref}} \right)^{1/4} \tag{57}$$

The quantities, M_e^i and C_∞^i, for the considered copolyamides are given in Table 9.

Taking $1T_{0.7}I_{0.3}$ as a reference, it turns out that terms $\kappa_{CSC}^i(T)$ are close to unity, in such a way that $\sigma_{CSC}^i(T)$ is represented by the corresponding $\sigma_{CSC}^{ref}(T)$.

The calculations for the other copolyamides of the xT_yI_{1-y} series are performed by considering the intercept of $\sigma_y^i(T)$ and $\sigma_{CSC}^{ref}(T)$, as shown in Fig. 96.

The results for T_{12} are reported in Table 9 and show quite a reasonable agreement with the experimental values. Thus, the lower values of T_{12} observed for 1.8 lactam-12 copolyamides, compared to 1 lactam-12, originates entirely from the lower $\sigma_y(T)$ values corresponding to the 1.8 lactam-12 copolyamides. It is worth noting that Fig. 96 leads to T_{12} values differing by

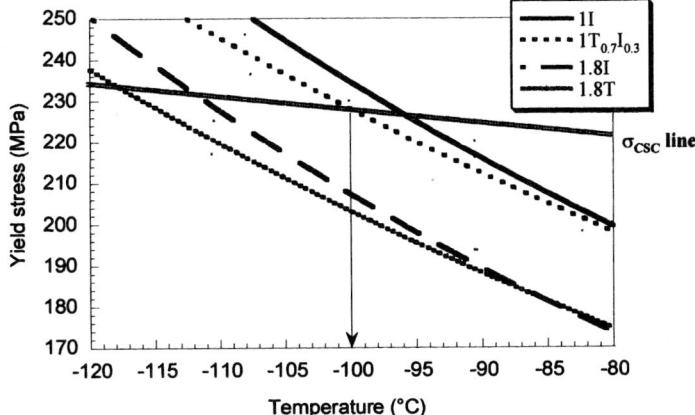

Fig. 96 Graphical determination of the values of T_{12} given in Table 9 (From [61])

about 5 °C only when a *tere*-phthalic unit is replaced by an *iso*-phthalic one, which is too small to be picked up experimentally.

Temperature Range of SDZs ($T_{233}-T_{12}$) in xT_yI_{1-y} Copolyamides

The SDZs are present up to T_{233}, since over this temperature only CDCs are observed. Thus, the extent of SDZs goes from T_{12} to T_{233} and is given in Table 9.

As regards the chemical structure effect, it clearly appears that replacing the *iso*-phthalic residue by a *tere*-phthalic one, considerably increases the temperature range in which SDZs exist. In contrast, the length of the lactam-12 unit does not have any effect. It is worth noting that an opposite influence between the nature of the phthalic group and the length of the lactam-12 unit happens for T_{12}.

Dealing with the temperature range over which the CDCs exist ($T_\alpha-T_{233}$) the values given in Table 9 show that this temperature interval is significantly larger for *iso*- than for *tere*-phthalic units, whatever the length of the lactam-12 moiety. This is a direct consequence of the much more restricted domain of the SDZs in the *iso*-phthalic copolyamides, since the T_α values of these compounds are about 10 °C lower than the corresponding values for the *tere*-phthalic copolyamides.

The temperature T_{233} can be calculated for the various copolyamides by considering $1T_{0.7}I_{0.3}$ copolyamide as a reference (in a way similar to that for T_{12}) except that now the quantity of interest is the critical stress for CDCs, σ_{CDC}, instead of the stress for CSCs, σ_{CSC}.

As indicated in Sect. 2:

$$\sigma_{CDC} \propto (\sigma_{pf})^{1/2} W(M, \zeta) \tag{58}$$

where $W(M, \zeta)$ expression depends on the model, i.e. Kramer or McLeish (Sects. 2.3.1.3 and 2.3.4).

By denoting as "ref" the reference copolyamide and, i, any other copolyamide of the xT_yI_{1-y} copolyamide series, the approach described in Sect. 2.3.4.2 can be applied. It leads to:

$$\sigma_{CDC}^i(T) = \kappa_{CDC}^i(T)\sigma_{CDC}^{ref}(T) \tag{59}$$

where $\kappa_{CDC}^i(T)$ expression depends on the model (see Sect. 2.3.4.2) .

Actually, the $\zeta(T)$ values are identical within the considered series [53] and, thus, they cancel each other. Furthermore, it turns out that, for the various copolyamides of this series, $\kappa_{CDC}^i(T)$ is very close to 1, both with Kramer or McLeish expressions. Consequently, $\sigma_{CDC}^i(T)$ is represented by the corresponding $\sigma_{CDC}^{ref}(T)$.

The CDCs occurring at high temperatures involve molecular motions associated with the glass–rubber transition, in such a way that the temperature interval $(T - T_\alpha)$ plays an important role. Given the differences in T_α values in the xT_yI_{1-y} series (Table 8), the calculations for the various copolyamides are performed by considering the intercept of $\sigma_y(T - T_\alpha)$ and $\sigma_{CDC}^{ref}(T - T_\alpha)$, as shown in Fig. 97.

The results reported in Table 9 show that the agreement between the experimental and calculated values of T_{233} and $(T_\alpha - T_{233})$ is good. It is the same for the temperature range of SDZs, $(T_{233} - T_{12})$. Thus, the differences observed between *iso-* and *tere-*phthalic copolyamides on the one hand, and 1 and 1.8 lactam-12 copolyamides on the other hand, have to be attributed to the higher $\sigma_y(T - T_\alpha)$ (around 7 MPa in the concerned $(T - T_\alpha)$ range) corres-

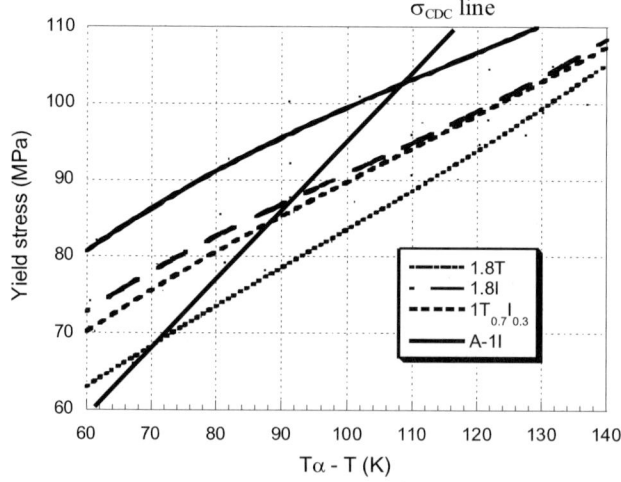

Fig. 97 Graphical determination of the values of $(T_\alpha - T_{233})$ given in Table 9 (From [61])

ponding to the *iso*-phthalic copolyamides at a given lactam-12 content and to the lactam-12 content at a given *iso*- or *tere*-phthalic moiety.

5.1.2.3
Relation of Micromechanisms of Deformation to β Transition in the xT_yI_{1-y} Copolyamides

The secondary transitions of the xT_yI_{1-y} copolyamides, shown in Fig. 84, have been analysed in [1] (Sect. 6) and summarised in Sect. 5.1.1.6.

T_{12} Transition
The SDZs appear at a temperature T_{12}, around $-100\,°C$, which corresponds to the onset of the β transition. This suggests that the motions within the lactam-12 units involved in the γ transition are too localised to lead to the formation of SDZs in this series of copolyamides. Nevertheless, the $10\,°C$ downward shift of T_{12} from 1 to 1.8 lactam-12 copolyamides originates from the higher number of γ motions in the latter polymers, through their lower σ_y values (Fig. 88).

T_{233} Transition
It is striking to notice that the CDCs appear at a temperature, T_{233}, which corresponds to the onset of the ω transition. Furthermore, the $10\,°C$ downward shift of T_{233} observed for 1.8I also occurs for the onset of the ω transition. It would mean that the breakdown of the rigidity of the group constituted by phthalic ring–amide–di-cyclohexyl, due to motion of the amide group, leads to enough flexibility for the chain disentanglement required by the formation of CDCs.

5.1.3
Fracture of xT_yI_{1-y} Copolyamides

The fracture behaviour of the xT_yI_{1-y} copolyamide series has been studied [53, 62, 63] as a function of temperature, and the effect of MW has been also considered. The characteristics of the materials under consideration are given in Table 8. It is worth noting that to identify the various MWs of a given xT_yI_{1-y}, the MW in kg mol^{-1} has been indicated in brackets.

The values of K_{Ic} and G_{Ic}, determined at various temperatures for the considered copolyamide series, are given in Tables 10 and 11, respectively.

Before analysing these results, it is interesting to point out that optical and electron microscopies of the sample region ahead of the crack reveal that a craze develops at the crack tip, whatever the type of fracture regime, i.e. stable or unstable [53].

Table 10 Average values of K_{Ic} (in MPa m$^{1/2}$) for xT_yI_{1-y} and MT_yI_{1-y} copolyamides at various temperatures

	Temperature (°C)											
	−80	−60	−40	−20	0	20	40	60	80	100	120	140
1.8I(22)	2.6	2.55	2.4	2.45	2.2	2.3	2.6	2.5	2.2	1.7		
1.8T(23)	2.8	2.75	2.7	2.65	2.6	2.45	2.5	2.4	2.0	1.9		
1.8T(39)	2.7	2.7	2.7	2.6	2.8	3.0	3.25	3.3	3.1	2.9		
1I(21)	2.7	2.8	2.7	2.35	2.3	2.4	2.3	2.4	2.2	2.05	1.9	1.7
1I(26)	2.8	2.65	2.6	2.35	2.3	2.35	2.6	2.5	2.4	2.15	1.9	1.8
$T_{0.7}I_{0.3}$(23)	2.95	2.85	2.8	2.6	2.55	2.45	2.4	2.2	2.0	1.6	1.3	0.9
$1T_{0.7}I_{0.3}$(27)	2.95	2.9	2.85	2.7	2.55	2.65	2.65	2.6	2.8	2.5	2.35	2.1
$1T_{0.7}I_{0.3}$(32)	2.95	2.85	2.8	2.6	2.5	2.5	2.55	2.6	2.8	2.6	2.35	1.95
MI	3.6	3.25	3.3	3.45	3.6	3.8	3.75	3.6	3.5	3.55		
$MT_{0.5}I_{0.5}$	3.7	3.55	3.5	3.8	3.95	4.0	3.9	3.6	3.4	3.3		
$MT_{0.7}I_{0.3}$	3.95	3.8	3.75	3.7	3.65	3.8	3.4	3.35	3.4	3.2		

Figures in brackets indicate the MW in kg mol^{-1}

Table 11 Average values of K_{Ic} (in KJ m^{-2}) for xT_yI_{1-y} and MT_yI_{1-y} copolyamides at various temperatures

	Temperature (°C)											
	-80	-60	-40	-20	0	20	40	60	80	100	120	140
1.8I(22)	3.0	2.85	2.6	2.6	2.45	2.75	3.3	3.2	2.6	1.75		
1.8T(23)	3.4	3.6	3.4	3.3	3.4	3.0	3.6	2.9	2.6	2.3		
1.8T(39)	3.4	3.5	3.3	3.6	3.8	5.8	6.8	7.0	6.4	6.0		
1I(21)	2.9	3.1	2.9	2.6	2.4	2.6	2.5	2.75	2.6	2.4	2.0	1.7
1I(26)	2.9	2.8	2.75	2.4	2.6	2.8	3.2	3.6	3.3	3.1	2.4	1.7
$1T_{0.7}I_{0.3}$ (23)	3.5	3.4	3.3	3.0	2.9	2.9	2.8	2.6	2.1	1.6	1.1	0.6
$1T_{0.7}I_{0.3}$ (27)	3.4	3.3	3.4	3.3	3.1	3.0	3.4	3.8	4.2	3.7	3.1	2.7
$1T_{0.7}I_{0.3}$ (32)	3.5	3.4	3.2	3.0	2.8	3.05	3.6	3.9	4.3	4.1	3.4	2.6
MI	3.1	2.9	3.0	3.2	3.6	3.9	4.0	3.95	3.75	3.8		
$MT_{0.5}I_{0.5}$	3.3	3.2	3.25	4.0	4.05	4.2	4.3	4.1	4.2	3.9		
$MT_{0.7}I_{0.3}$	3.8	3.6	3.4	3.3	3.2	3.7	3.2	3.5	3.8	4.0		

Figures in brackets indicate the MW in kg mol^{-1}

5.1.3.1
Effect of MW

Figure 98 illustrates the MW dependence of K_{Ic} and G_{Ic} in the case of 1I, $1T_{0.7}I_{0.3}$ and 1.8T. Unfortunately, for 1.8I only one MW was available.

Two important features appear:

1. MW independence of the fracture characteristics over a broad low temperature range. Over this range, K_{Ic} and G_{Ic} exhibit a slight decrease with increasing temperature.
2. MW dependence at higher temperatures. Whereas K_{Ic} and G_{Ic} present a continuous decrease with increasing temperature for the lower MW samples, these quantities go through a maximum for the higher MW samples.

Conventionally, the plots of K_{Ic} and G_{Ic} versus temperature for the samples of sufficient high MW can be divided into three successive temperature ranges, a, b and c. Toughness slightly decreases with increasing temperature over the range a, which extends from low temperatures to a temperature denoted as T_1. Toughness increases over the range b, until reaching a maximum at a temperature denoted as T_m. Finally, toughness decreases again over the range c, on the upper temperature side. The values corresponding to T_1 and T_m for the various xT_yI_{1-y} copolyamides are reported in Table 12.

It is likely that the craze at the crack tip over the low temperature range a corresponds to CSC, in agreement with the observation of CSCs during the thin film deformation of the same samples (as described in Sect. 5.2.1). As the strength required to break a chemical bond is temperature-independent and the CSC stress presents a weak decrease with increasing temperature, toughness would decrease weakly with temperature, as actually observed.

In the high temperature range c, extending close to T_α, the deformation of thin films shows the occurrence of CDCs over a temperature T_{233}. The craze at the tip of the crack in this temperature range is a CDC whose the stability of the fibrils decreases with increasing the mobility of the polymer chains when temperature rises (Sect. 2.3.1.4). It is striking to see that the decrease of toughness happens over a temperature T_m which agrees reasonably with T_{233}, as shown in Table 12.

Concerning the temperature range b, where a MW dependence is observed, as well as an increase of toughness with increasing temperature, the deformation of thin films shows that at the temperature T_1 where it starts, the micromechanisms are mostly SDZs (which are present to T_{233}) with some CSCs. In thin film experiments, CDCs appear at T_{223}; the T_{223} values for the various xT_yI_{1-y} copolyamides (given in Table 9) are higher than the T_1 values. Thus, it is unlikely that CDCs could be involved in this MW dependence of toughness. It is more likely that it deals with chain slippage within CSC fibrils, the MW of chain fragments being a function of the bulk polymer MW

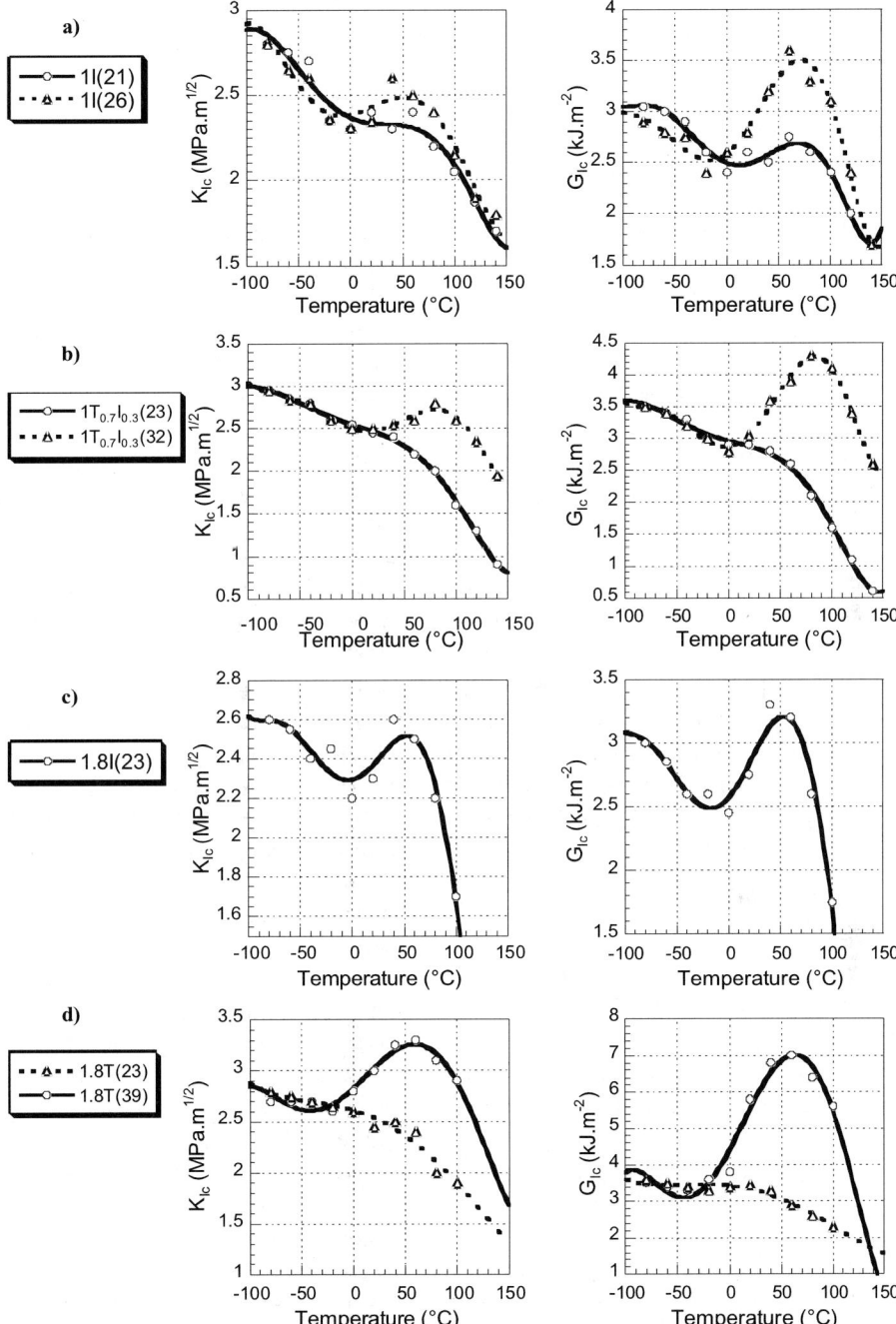

Fig. 98 K_{Ic} (*left*) and G_{Ic} (*right*) versus temperature for **a** 1I, **b** $1T_{0.7}I_{0.3}$, **c** 1.8I and **d** 1.8T copolyamides (From [53])

Table 12 Characteristic temperatures (in °C) for the fracture behaviour T_1 and T_m, and for the micromechanisms of deformation T_{223} and T_{233}

Polymer	T_1	T_m	T_{223}	T_{233}
1I	20	60	40	50
$1T_{0.7}I_{0.3}$	0	80	40	80
1.8I		40	20	30
1.8T	0	60	40	70
MI	(−40)	30	110	None
$MT_{0.5}I_{0.5}$	(−40)	0	120	None

(as described in Sect. 2.3.1.4). The increase of toughness observed in the temperature range b would result:

- From more stable fibrils for longer chains
- From a contribution of the SDZs mixed with CSC, in a way similar to that observed with MGIM 58 (Sect. 3.3.3.3), where SDZs are initiated along the craze edges

Such interaction between crazes and SDZs is evidenced by the short fat morphology of the crazes observed in thin film deformation in this temperature range, up to T_{233}. The SDZs contribute to the energy dissipation within the plastic zone ahead of the crack tip and increase toughness. According to this description, the toughness–temperature profiles of the low MW samples can be explained by the disappearance of the range b. As the chains become short, not enough stabilisation of the craze fibrils occurs for taking advantage of the interaction with the SDZs.

Thus, from a qualitative point of view, the above analysis yields a comprehensive description of the temperature dependence of toughness. However, some peculiar features cannot be accounted for. This is the case for the huge increase in toughness (from 3.3 up to 7.0 kJ m^{-2}, Table 11) that is observed for 1.8T(39) from the range a to the range b, and which is followed by a modest decline over the range c (down to 6.0 kJ m^{-2}). This behaviour, unusual from a quantitative point of view, has been shown to result from a crack deviation around the very stable plastic zone [53].

5.1.3.2
Influence of the Chemical Structure of xT_yI_{1-y} Copolyamides and Relation to β Transition

A careful inspection of Tables 10 and 11 shows that changes, even limited, in the xT_yI_{1-y} chemical structure may noticeably affect their fracture behaviour.

Owing to the change of toughness mechanisms over the investigated temperature range, the influence of the chemical structure has to be considered separately for the MW-independent temperature range *a* and for the MW-dependent temperature ranges *b* and *c*.

MW-Independent Temperature Range

In this temperature range, the replacement of *iso*-phthalic units by *tere-* ones leads systematically to tougher materials, as shown at − 40 °C in Fig. 99, irrespective of the lactam-12 content. At this temperature, neither MW (which does not affect the toughness characteristics) nor the yield stress value (which is slightly dependent on the *iso-* or *tere*-phthalic unit in this temperature range, as shown in Fig. 88) can explain the observations. Changes in entanglement density are also unlikely to be invoked: their effects would go in the opposite direction, since the most entangled chains are the *iso*-phthalic ones (Table 8). Therefore, the unique factor appropriate for explaining the observed results seems to be the molecular mobility, whose characteristics at this temperature are governed by the β relaxation process. As described in [1] (Sect. 6) and summarised in Sect. 5.1.1.6, these β motions present, in the high temperature part of the β peak, a marked cooperative character in the *tere*-phthalic copolyamides as a result of phenyl ring π-flips. The β motions remain isolated in the *iso*-phthalic copolyamides because the phenyl rings cannot flip and oscillate only. These intramolecular cooperative motions would favour SDZ development instead of CSCs, in this temperature range where both are present, and in which craze stability does not contribute to toughness.

The influence of the lactam-12 content on the toughness in this temperature range (shown in Fig. 99) goes in the opposite direction to what would be expected from the higher σ_y values observed for $1T_yI_{1-y}$ than for $1.8T_yI_{1-y}$ copolyamides (Fig. 88). The origin of such a behaviour is yet unexplained.

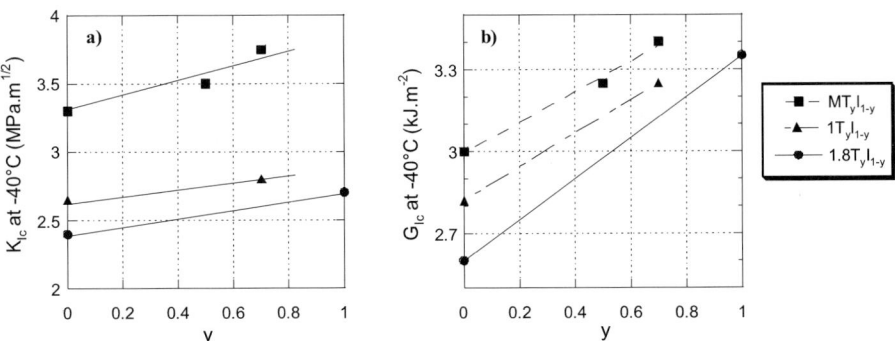

Fig. 99 Dependence of K_{Ic} (*left*) and G_{Ic} (*right*) at − 40 °C on the fraction of *tere*-phthalic units for $1.8T_yI_{1-y}$, $1T_yI_{1-y}$ and MT_yI_{1-y} copolyamides (From [62])

MW-Dependent Toughness Regime

In this regime, corresponding to temperature ranges b and c, analysis of the chemical structure effect requires taking into account the ratios M_w/M_e and M_n/M_e. In all cases, only polyamide samples with a large enough MW to present the temperature range b are discussed.

Furthermore, as chain disentanglements are involved in this regime, it is more appropriate to consider $(T - T_\alpha)$ instead of T for examining the temperature dependence of toughness.

Effect of iso- or tere-phthalic units

Within the $1T_yI_{1-y}$ copolyamide series, the partial replacement of *iso-* by *tere*-phthalic units leads to an increase of toughness, as shown in Fig. 100 for 1I(26) and $1T_{0.7}I_{0.3}$(27). This behaviour is in agreement with the lower σ_y values, in this temperature range, of $1T_{0.7}I_{0.3}$ compared to 1I (Fig. 88). To check whether such a difference in σ_y is able to account for the change of G_{Ic}, it is interesting to apply Brown's model [27], which predicts that the product $G_{Ic}\sigma_y$ does not depend on the chemical structure. As shown in Fig. 101, the two considered copolyamides satisfy this relationship.

As regards the $1.8T_yI_{1-y}$ copolyamide series, the lower MW samples only available make the discussion less liable.

Effect of the lactam-12 content

Due to the sensitivity of toughness to the *iso-* or *tere-* nature of the phthalic unit, analysis of the lactam-12 content can only be performed on 1I and 1.8I copolyamides. The values of K_{Ic} and G_{Ic} are plotted versus $(T - T_\alpha)$ in Fig. 102 for the MWs available for 1I and, unfortunately, for the only MW for 1.8I. It appears that 1.8I(22) has slightly higher K_{Ic} values than 1I(26). However, the ratios M_w/M_e and M_n/M_e are lower for 1.8I(22) than for 1I(26), as indicated in Table 8. Considering the effect on the toughness of 1I of the increase of MW from 1I(21) to 1I(26), one can expect that for the sample of 1.8I with the same ratios M_w/M_e and M_n/M_e as those of 1I(26), its toughness would be significantly higher than that of 1I(26). Such an increase of toughness with increasing the lactam-12 content agrees with the corresponding decrease of σ_y observed in this temperature range (Fig. 88).

5.2
MT_yI_{1-y} Copolyamides

The MT_yI_{1-y} copolyamides have chemical structures shown in Fig. 83b, quite different from those of the xT_yI_{1-y} copolyamides. Indeed, there is no di(methyl cyclohexane) methane unit in the structure, which constitutes a very bulky group, right in the chain backbone. Another important feature is the increase of amide group content, leading to a larger density of hydrogen bonds between the $C = O$ and $N - H$ units of different amide moieties. This latter effect is reflected in the much higher modulus, E', of the MT_yI_{1-y} copolyamides compared to that of xT_yI_{1-y} copolyamides, as evidenced in

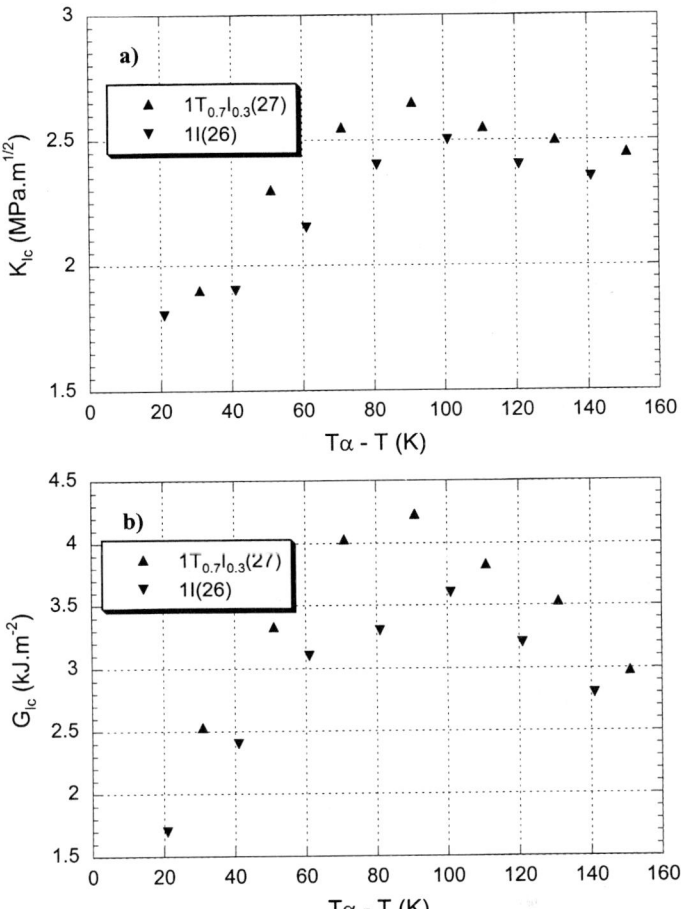

Fig. 100 **a** K_{Ic} and **b** G_{Ic} versus $(T_\alpha - T)$ for 1I(26) and $1T_{0.7}I_{0.3}$(27) copolyamides

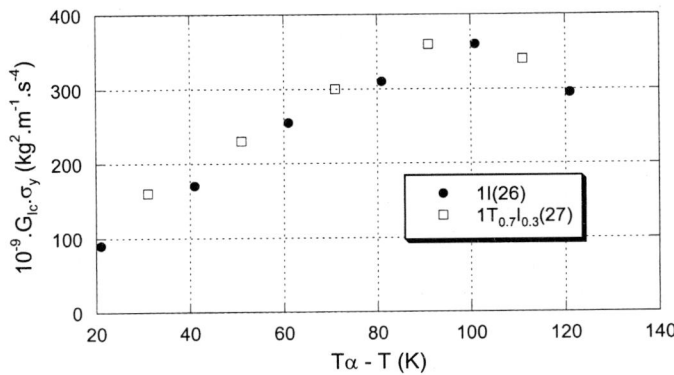

Fig. 101 Product $G_{Ic}\sigma_y$ for 1I(26) and $1T_{0.7}I_{0.3}$(27) copolyamides versus $(T_\alpha - T)$ (From [53])

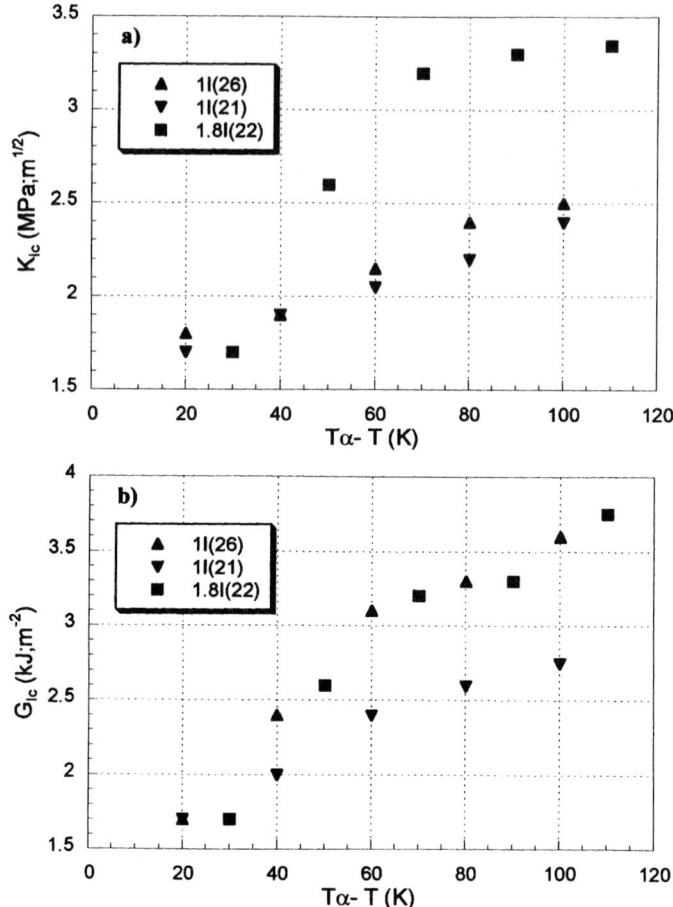

Fig. 102 a K_{Ic} and **b** G_{Ic} versus $(T_\alpha - T)$ for 1I(21), 1I(26) and 1.8I(22) copolyamides

Figs. 84a and 85a. Thus, at − 100 °C, the E' value for MI is 3800 MPa, whereas it is only 2900 MPa for 1I and 2400 MPa for 1.8I. Similarly at 0 °C the values are 3300, 2100 and 1750 MPa, respectively.

5.2.1
Plastic Deformation of MT$_y$I$_{1-y}$ Copolyamides

5.2.1.1
Stress–Strain Curves

Typical stress–strain curves at various temperatures from − 80 to 120 °C, have been obtained for MI under compression at a strain rate of 2×10^{-3} s^{-1} [64]

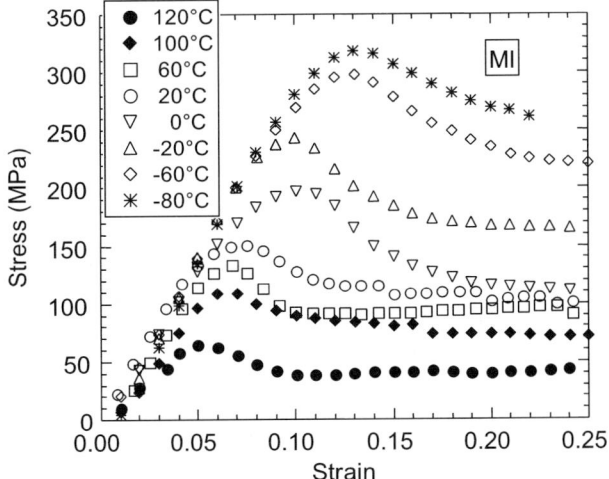

Fig. 103 Stress–strain curves of MI copolyamide at various temperatures. Strain rate 2×10^{-3} s^{-1} (From [64])

(see Fig. 103). A large strain softening is observed in the whole temperature range, in contrast to the behaviour of xT_yI_{1-y} copolyamides (Fig. 86).

5.2.1.2
Yield Stress

The temperature dependence of the yield stress, σ_y, is shown in Fig. 104 for MI and $MT_{0.5}I_{0.5}$ copolyamides. The σ_y values are close to each other, the faster decrease observed between 80 and 100 °C for MI comes from the lower T_α value of this copolyamide relative to that of $MT_{0.5}I_{0.5}$ (as reported in Table 8).

It is interesting to notice that in the whole temperature range, σ_y values of the MT_yI_{1-y} copolyamides are much higher than those of for xT_yI_{1-y}. This results from the higher hydrogen bond density.

5.2.1.3
Plastic Flow Stress

The temperature dependence of the plastic flow stress, σ_{pf}, is shown in Fig. 104 for MI and $MT_{0.5}I_{0.5}$ copolyamides. In the same way as for σ_y, the σ_{pf} values for the two copolyamides are close to each other and the shape of the $\sigma_{pf}(T)$ curves is quite similar to that of $\sigma_y(T)$ curves.

Much larger values are observed for σ_{pf} of MT_yI_{1-y} copolyamides than for xT_yI_{1-y} copolyamides.

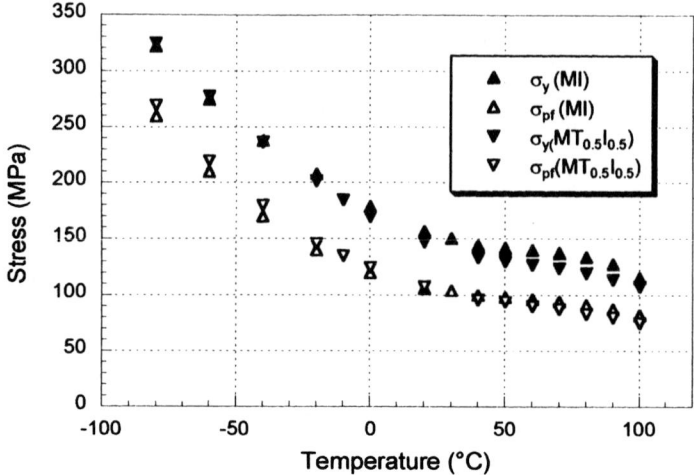

Fig. 104 Temperature dependence of σ_y and σ_{pf} for MI and $MT_{0.5}I_{0.5}$

5.2.1.4
Strain Softening

The $\sigma_y(T)$ and $\sigma_{pf}(T)$ curves in Fig. 104 immediately show that, over the whole temperature range, there is a large difference between σ_y and σ_{pf}, leading to an important strain softening.

More quantitatively, SSA is shown as a function of temperature in Fig. 105. As expected from the reported results, similar behaviours are observed for

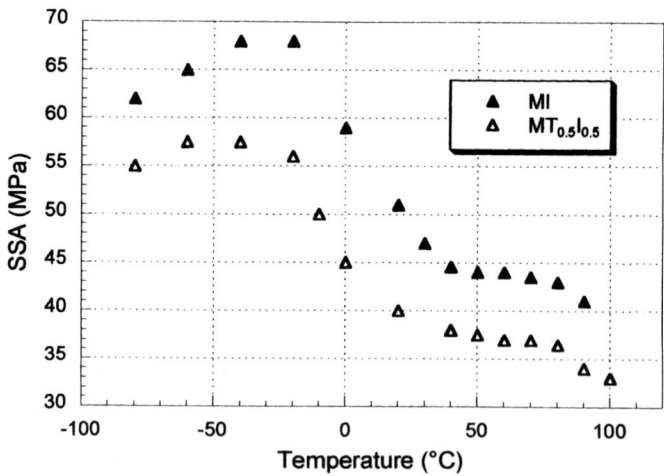

Fig. 105 Temperature dependence of SSA for MI and $MT_{0.5}I_{0.5}$

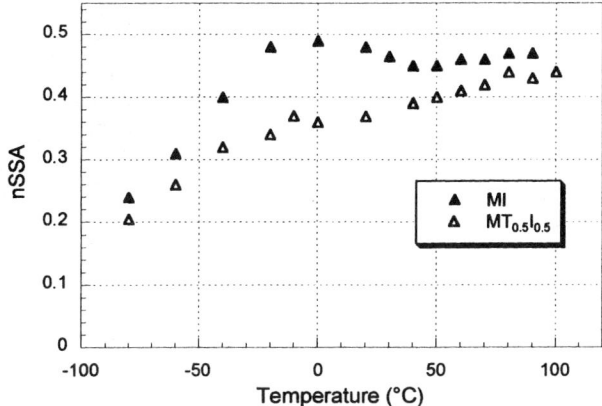

Fig. 106 Temperature dependence of nSSA for MI and $MT_{0.5}I_{0.5}$

both polymers. However, it is worth noting that at any temperature, the strain softening amplitude is larger for MI copolyamide than for $MT_{0.5}I_{0.5}$.

The temperature dependence of nSSA is shown in Fig. 106. As in the case of xT_yI_{1-y} copolyamides, it is striking to see how different are the temperature dependencies of SSA and nSSA. The latter descriptor, taking into account the variation of σ_{pf} with temperature, is more relevant. The two MI and $MT_{0.5}I_{0.5}$ copolyamides have very similar nSSA temperature dependence, showing a large increase from – 80 to about – 20 °C, then very small changes at higher temperatures. However, one always observes lower values of nSSA for $MT_{0.5}I_{0.5}$ relative to MI.

The results shown in Fig. 106 and Fig. 92 reveal that, at any temperature, nSSA is significantly higher in the case of MT_yI_{1-y} copolyamides. Thus, at – 80 °C, the values are 10, 8 and 22% for 1I, 1.8I and MI, respectively. Similarly, at 50 °C for the same copolyamides the nSSA values are 23, 30 and 45%.

5.2.1.5
Relation of Plastic Deformation of MT_yI_{1-y} Copolyamides to β Transition

As can be seen from Figs. 85b and 104, the investigated temperature range for the MT_yI_{1-y} copolyamides corresponds to the higher part of the β transition.

Analysis of the β transition of MT copolyamides (described in [1], Sect. 5) shows that in the mechanical loss β transition of MT_yI_{1-y} copolyamides, in the low-temperature range part of the involved motions are those of CH_2 units. At higher temperatures phenyl ring motions (oscillations then π-flips), coupled with amide group motions, occur.

Concerning the strain softening and its temperature dependence, the proposed mechanism described for BPA-PC, and applied to xT_yI_{1-y} copolyamides, allows one to account for the large values of nSSA and its variation with temperature.

According to the proposed mechanism, the large nSSA values over the whole temperature range, relative to those observed for xT_yI_{1-y} copolyamides (Fig. 92), indicate that the main-chain conformation changes induced by the applied yield stress are much more efficient for softening the polymer medium in the case of MT_yI_{1-y} copolyamides. Such a result is not surprising, considering that these latter copolyamides do not contain in their main chain the bulky and rather stiff sequence phenyl-amide-3,3'-dimethylcyclohexyl methane-amide-phenyl, which limits the softening of the xT_yI_{1-y} copolyamides. It is interesting to notice that the effect of this stiff sequence dominates the opposite effect arising from the higher hydrogen bond density of the MT_yI_{1-y} copolyamides, discussed above in relation to the mechanical modulus, E, and the yield stress, σ_y.

The increase of nSSA in the low temperature part to about $-20\,^\circ C$ corresponds to the softening of the medium by the β transition motions. The leveling observed at higher temperatures is consistent with the fact that, for MT_yI_{1-y} copolyamides, there are no new motions until the glass–rubber transition temperature is reached, in contrast to the case of xT_yI_{1-y} copolyamides for which there is still an ω transition in the range 20–80 $^\circ C$ (Fig. 84b).

Finally, as observed for xT_yI_{1-y} copolyamides, replacing a *tere*-phthalic unit by a *iso*-phthalicunit leads to an increase of strain softening. The same argument, dealing with the larger disturbance introduced by the displacement of the kinked *iso*-phthalic structure under the applied yield stress, can be considered.

5.2.2
Micromechanisms of Deformation of MT_yI_{1-y} Copolyamides

The two copolyamides of this series (i.e. MI and $MT_{0.5}I_{0.5}$) present CSCs at low temperatures to a temperature of about $-70\,^\circ C$, corresponding to the transition T_{12} over which SDZs appear in addition to CSCs and gradually become dominant. Figure 107 shows an electron micrograph corresponding to this mixed regime. At temperatures higher than T_{223} a mixed regime, where SDZs and CDCs coexist, is observed up to T_α. Indeed, with the MT_yI_{1-y} copolyamides, a temperature range with CDCs only is not observed, so that T_{233} cannot be determined. The values of the transition temperatures for the considered copolyamides are given in Table 9.

From the data in Table 9, it appears that the SDZs occur for MT_yI_{1-y} at a higher value of T_{12} than for the xT_yI_{1-y} copolyamide series, but they are present until T_α. It is tempting to apply to MT_yI_{1-y} copolyamides the approach used for calculating T_{12} in the xT_yI_{1-y} copolyamide series. Unfortunately, MI data on σ_y and σ_{pf} are available only above 0 $^\circ C$, due to the too-high brittleness of the samples at lower temperatures. Nevertheless, as

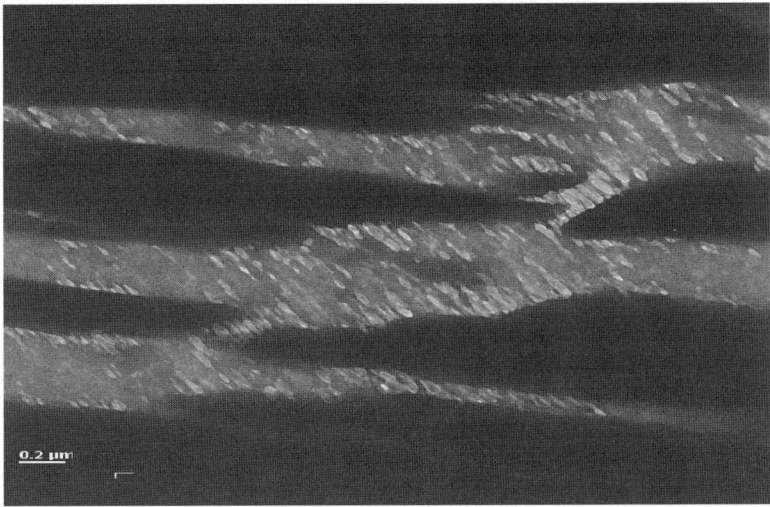

Fig. 107 TEM of mixed crazes and SDZs in thin films of $MT_{0.5}I_{0.5}$ copolyamide stretched at $120\,°C$ with a strain rate of $2 \times 10^{-3}\ s^{-1}$ (From [61])

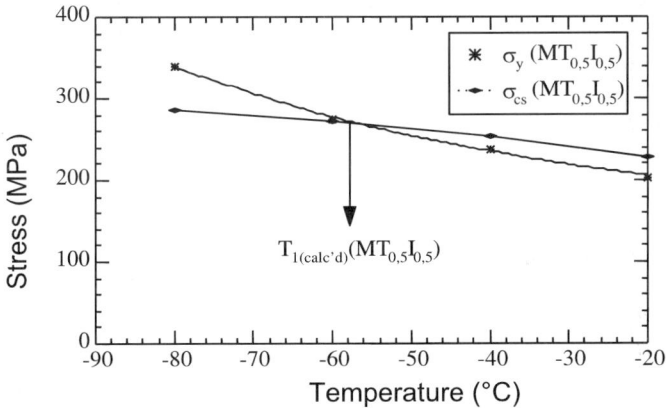

Fig. 108 Intersect of σ_{CSC} (calculated) with experimental $\sigma_y(T)$ for $MT_{0.5}I_{0.5}$ (From [53])

this information exists for $MT_{0.5}I_{0.5}$, a crude estimate of T_{12} for this polymer can be achieved [53], making the assumption that the proportionality constant involved in σ_{CSC} (Sect. 2.3.4.2) keeps the value found for the xT_yI_{1-y} copolyamide series for the $MT_{0.5}I_{0.5}$ copolyamide. T_{12} is determined as shown in Fig. 108. It is worth noting that, in spite of this assumption, the estimated value of $-58\,°C$ for T_{12} compares reasonably with the experimental value of $-70\,°C$.

As regards the onset of CDCs, T_{233}, quite a large temperature shift is observed between the two copolyamide series. Unfortunately, in the MT_yI_{1-y}

series, the lack of data on the monomeric friction coefficient excludes any estimate concerning the CDCs.

5.2.3
Fracture of MT$_y$I$_{1-y}$ Copolyamides

5.2.3.1
Temperature Dependence

The values of K_{Ic} and G_{Ic} determined at various temperatures for MT$_y$I$_{1-y}$ copolyamides are given in Tables 10 and 11, respectively.

It is worth noticing that in this copolyamide series, unfortunately, the MW dependence of toughness cannot be studied due to a lack of samples with different MWs in each of these copolyamides.

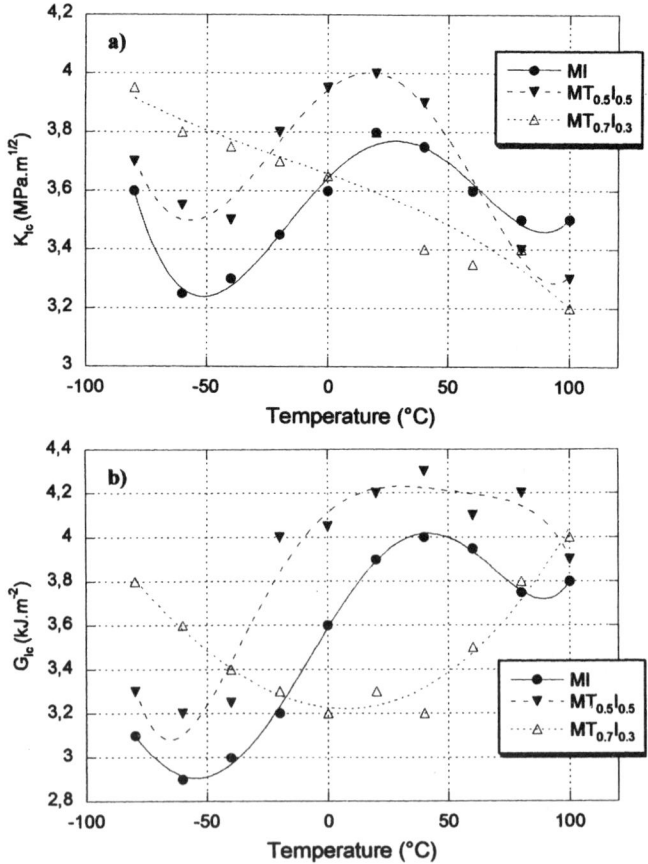

Fig. 109 a K_{Ic} and **b** G_{Ic} versus $(T_\alpha - T)$ for MI, MT$_{0.5}$I$_{0.5}$ and MT$_{0.7}$I$_{0.3}$ copolyamides

Figure 109 presents the results of K_{Ic} and G_{Ic} as a function of temperature for the MT_yI_{1-y} copolyamides under consideration.

The difference in the temperature dependence of K_{Ic} and G_{Ic}, which appears in the high temperature range for $MT_{0.5}I_{0.5}$ and $MT_{0.7}I_{0.3}$ copolyamides, comes from the effect of the larger decrease of the modulus, E, which enters in the relationship of K_{Ic} to G_{Ic}, as indicated in Eq. 39. For this reason, K_{Ic} is preferred for discussing the curve shapes.

For the MT_yI_{1-y} copolyamide series, it is not possible to differentiate the various temperature ranges on the basis of their MW dependence, as was done for the xT_yI_{1-y} copolyamide series. Nevertheless, one can compare the behaviours of these two copolyamide series. So, $MT_{0.5}I_{0.5}$ clearly shows a temperature dependence of K_{Ic}, which presents the three temperature ranges a, b and c defined for xT_yI_{1-y} copolyamides, with a well-pronounced decrease of K_{Ic} in the temperature range c. In the case of MI, the three ranges appear, but with a small decrease of K_{Ic} in the temperature range c. Finally, for $MT_{0.7}I_{0.3}$ copolyamide, the monotonous decrease of K_{Ic} suggests that only temperature ranges a and c occur, as observed with low MW xT_yI_{1-y} copolyamides.

Looking at the molecular characteristics of the MT_yI_{1-y} copolyamides reported in Table 8, the only consistent explanation for the particular behaviour of $MT_{0.7}I_{0.3}$ copolyamide deals with the quite large polydispersity of this sample, leading to a much lower value of the ratio M_n/M_e and, thus, emphasising the influence of the low MW chains.

Investigation of the micromechanisms of deformation (Sect. 5.2.2) shows that the SDZs appear around $-70\,°C$ in the MI and $MT_{0.5}I_{0.5}$ considered copolyamides (Table 9) and a mixed regime with CSCs exists to temperatures (around $10–20\,°C$) where CDCs develop, mixed with SDZs.

On the basis of these micromechanism results, it appears that over the whole temperature range considered in Fig. 109, only CSCs are involved. The increase of toughness occurring above $-50\,°C$ would correspond to the contribution of SDZs in a way similar to that described for xT_yI_{1-y} copolyamides, except that in the present case SDZs interact with CSCs instead of CDCs for the latter copolyamide series. The decrease of toughness at high temperature deals with the fibril stability with increasing chain mobility, due to easier chain slippage. The effect of the low MW species observed for $MT_{0.7}I_{0.3}$ copolyamide originates from quite low MWs present in the corresponding CSC fibrils, which allow fibril rupture at low temperature (typically since $-40\,°C$) and forbid taking advantage of the interaction with SDZs.

5.2.3.2
Influence of the Chemical Structure of MT_yI_{1-y} Copolyamides

Owing to the different mechanisms of deformation involved, it is more appropriate to examine separately the effect of chemical structure in the low temperature range (to about $-40\,°C$) and at higher temperatures.

Low Temperature Range

The effect of replacing the *iso*- by *tere*-phthalic units is shown in Fig. 99. In the same way as for the xT_yI_{1-y} copolyamide series, toughness increases with the *tere*-phthalic content.

It is worth pointing out that the MW data reported in Table 8 do not fit the toughness evolution. The entanglement density, ν_e, decreases with increasing the amount of *tere*-phthalic units, which goes in the opposite direction to the observed toughness. Furthermore, in this temperature range, the yield stress, σ_y, is the same for MI and $MT_{0.5}I_{0.5}$ (Fig. 104). So, it cannot explain the change in toughness.

In the same way as for xT_yI_{1-y} copolyamides, the unique factor able to explain the observed results seems to be related to the motions involved in the high temperature part of the β transition (summarised in Sect. 5.2.1.5). Indeed, a marked cooperativity exists for the *tere*-phthalic copolyamides, related to the π-flips of the *tere*-phthalic phenyl rings, whereas the β motions of the *iso*-phthalic copolyamides remain isolated, the corresponding phenyl rings not being able to undergo π-flips. These intramolecular cooperative motions would favour SDZ development, already existing in this temperature range, instead of CSCs.

Higher Temperature Range

As regards the temperature range above $-40\,°C$, the effect of chemical structure is limited to the results of toughness of MI and $MT_{0.5}I_{0.5}$ copolyamides, due to the meaningless data obtained for $MT_{0.7}I_{0.3}$ copolyamide (as explained above). Furthermore, the temperature dependence is expressed in terms of $(T - T_\alpha)$, since the chain mobility plays an important role in fibril stability in this temperature range. The corresponding data for K_{Ic} and G_{Ic} are plotted as a function of $(T - T_\alpha)$ in Fig. 110. Furthermore, the yield stress, σ_y, for the two copolyamides is shown as a function of $(T - T_\alpha)$ in Fig. 111.

The difference in the relative behaviour of MI and $MT_{0.5}I_{0.5}$ copolyamides obtained between K_{Ic} and G_{Ic} comes from the difference in modulus, E, and its temperature dependence (Fig. 85). Indeed, in this temperature range, $MT_{0.5}I_{0.5}$ has a lower modulus than MI, an effect which enlarges with increasing temperature since the ratio $E(MT_{0.5}I_{0.5})/E(MI)$ goes from 0.94 at $-20\,°C$ to 0.90 at $100\,°C$.

In the $(T - T_\alpha)$ range from -160 to $-90\,°C$, the larger increase in toughness observed for $MT_{0.5}I_{0.5}$ than for MI does not agree with the difference in ν_e (it would go in the opposite direction). Furthermore, it cannot be explained from the yield stress values since they are identical in this temperature range, as shown in Fig. 111. The effect of the intramolecular cooperativity of the π-flip motions of the *tere*-phthalic rings, already invoked for the results in the low temperature range, appears to be the most likely reason: these cooperative

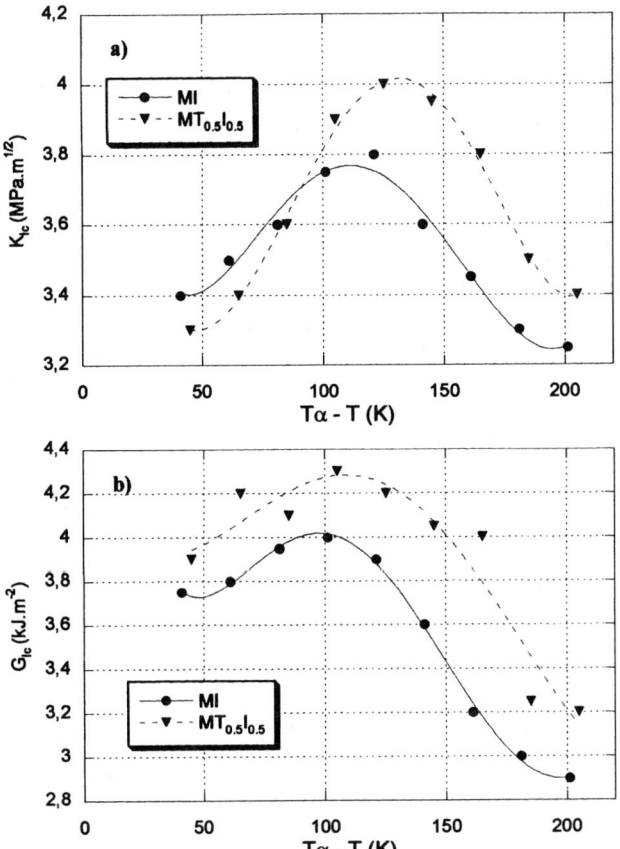

Fig. 110 a K_{Ic} and **b** G_{Ic} versus $(T_\alpha - T)$ for MI and $MT_{0.5}I_{0.5}$ copolyamides

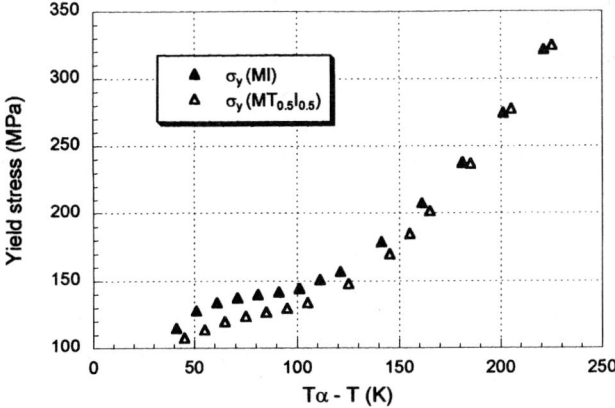

Fig. 111 Yield stress, σ_y, versus $(T_\alpha - T)$ for MI and $MT_{0.5}I_{0.5}$ copolyamides

motions favour SDZs more and more, interacting with CSCs when mobility increases with increasing temperature.

The inversion of toughness, occurring in the high temperature part, between $MT_{0.5}I_{0.5}$ and MI is completely opposite to the expectation from the σ_y values, $\sigma_y(MI)$ being larger than $\sigma_y(MT_{0.5}I_{0.5})$ in this temperature range (Fig. 111). Furthermore, the MWs of the corresponding copolyamides do not explain the lower craze stability of $MT_{0.5}I_{0.5}$ compared to MI, the opposite is expected. The only remaining argument would be that the intramolecular cooperativity of the β motions present in the $MT_{0.5}I_{0.5}$ copolyamide enhances the increase of chain mobility generated by the increase in temperature and, consequently, makes chain slippage easier, leading to craze fibril breakdown.

5.2.3.3
Comparison of the Fracture Behaviour of xT_yI_{1-y} and MT_yI_{1-y} Copolyamides

Due to the different types of micromechanisms of deformation involved, the comparison is made separately for the low temperature range and for higher temperatures.

Low Temperature Range
In this temperature range (to $-40\,°C$), no MW dependence has been observed for the xT_yI_{1-y} copolyamides, so there is no need to mention it for the compounds being considered.

The smaller difference in G_{Ic} compared to K_{Ic} (Tables 10 and 11) reflects only the effect of the higher modulus of the MT_yI_{1-y} copolyamides owing to their higher hydrogen bond density.

The results reported in Fig. 99 show that a similar effect of the replacement of iso- by tere-phthalic units is obtained in the two copolyamide series.

The most important feature is the much higher K_{Ic} values observed for MT_yI_{1-y} copolyamides than for xT_yI_{1-y} copolyamides. Such a difference completely disagrees with the yield stress values, which are about twice as large for MT_yI_{1-y} copolyamides as for xT_yI_{1-y}. As regards the entanglement density, ν_e, the values given in Table 8 favour a higher toughness for MT_yI_{1-y} copolyamides (average $\nu_e = 2.5$) than for xT_yI_{1-y} (average $\nu_e = 2.15$). However, it is quite unlikely that this small difference could account for the quite large observed difference in K_{Ic}.

To account for the difference in toughness between the two copolyamide series, one has to consider, once again, the corresponding β transitions. Indeed, there is quite a large difference in the mechanical β transitions, as shown in Fig. 84 and Fig. 85 and described in [1] (Sect. 6):

– For xT_yI_{1-y} copolyamides, the β peak corresponds to the temperature range from -100 to $-10\,°C$, with a maximum around $-60\,°C$

– For the MT_yI_{1-y} copolyamides, the β peak is very broad, from temperatures lower than – 150 to about – 20 °C

Thus, whereas the low temperature range (– 80 to – 40 °C) considered here for the toughness data is located right in the middle of the β transition in the case of xT_yI_{1-y} copolyamides, for the MT_yI_{1-y} copolyamides, it corresponds to the end of the β peak. The higher mobility existing in this latter case could favour the dissipation mechanisms existing in this temperature range for both copolyamide series (CSCs and SDZs) or their interaction. The absence, in MT_yI_{1-y} copolyamides, of the bulky (and rigid in this temperature range) sequence constituted by phenyl–amide– di-cyclohexyl methane, also contributes to their higher mobility.

Higher Temperature Range

As in this temperature range (above – 40 °C) the toughness of xT_yI_{1-y} copolyamides is MW-dependent, for comparison the data corresponding to the highest MW have been taken for each composition.

The results of K_{Ic} and G_{Ic} are shown for 1I(26) and MI copolyamides in Fig. 112, and those for $1T_{0.7}I_{0.3}(32)$ and $MT_{0.5}I_{0.5}$ in Fig. 113.

As for the low temperature range, the toughness of MT_yI_{1-y} copolyamides is much larger than that of $1T_yI_{1-y}$ copolyamides.

Whereas neither the consideration of the MW values, nor that of the σ_y values (higher for MT_yI_{1-y} than for $1T_yI_{1-y}$ copolyamides in this temperature range) nor that of ν_e (whose difference goes in the appropriate direction, but is not large enough) can account for the observed toughness difference, it is worth pointing out that in the highest temperature range, where there is a decrease of toughness of xT_yI_{1-y} copolyamides, the involved mechanisms of deformation are different. Indeed, instead of the only CDCs occurring for the xT_yI_{1-y} copolyamide series, in the case of MT_yI_{1-y} copolyamides SDZs are still present. This favours the toughness of these latter copolyamides in the highest temperature range.

However, such an argument is no longer valid when considering the results at lower temperatures for both copolyamide series present SDZs mixed either with CDCs (for xT_yI_{1-y} copolyamides) or with CSCs (for MT_yI_{1-y} copolyamides). Consequently, one is led to take into account the secondary transitions corresponding to each copolyamide series.

In the MT_yI_{1-y} copolyamide series, there is only the β transition, which ends up around – 20 °C, and thus is already performed in the discussed temperature range (typically – 20 to + 40 °C). In contrast, in the $1T_yI_{1-y}$ copolyamide series, in addition to the β transition, which ends up around – 10 °C, there is the ω transition occurring in the temperature range from 20–30 to 80 °C. This transition involves the motion of the amide group between the phthalic unit and the di-cyclohexyl methane group (as described in [1],

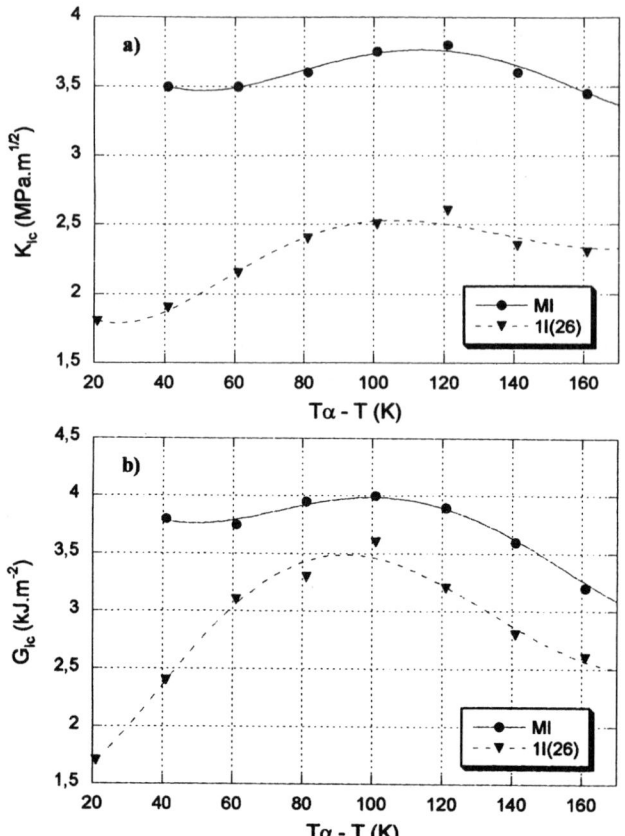

Fig. 112 a K_{Ic} and **b** G_{Ic} versus $(T_\alpha - T)$ for 1I(26) and MI copolyamides

Sect. 6). So, in this temperature range these ω motions do not yet occur, resulting in a rigid linkage (phthalic ring–amide group–cyclohexyl unit), which could favour CSCs over SDZs and decrease their interaction. This hindering of SDZs and their interaction with CSCs does not exist for the MT_yI_{1-y} copolyamides; this fact could contribute to the difference in toughness between the two copolyamide series.

5.3
Conclusion on Aryl-aliphatic Copolyamides

The mechanical properties of aryl-aliphatic copolyamides, both xT_yI_{1-y} and MT_yI_{1-y} series, present interesting features.

At first, the normalised strain softening increases with increasing temperature, opposite behaviour to that observed for PMMA and its copolymers, but similar that of BPA-PC. Owing to this similarity, the mechanism proposed for

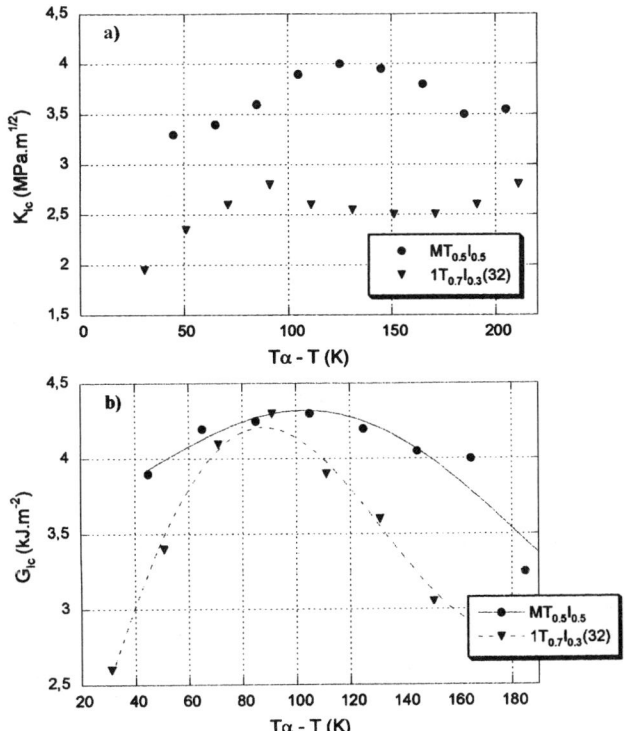

Fig. 113 a K_{Ic} and **b** G_{Ic} versus $(T_\alpha - T)$ for $1T_{0.7}I_{0.3}(27)$ and $MT_{0.5}I_{0.5}$ copolyamides

BPA-PC has been used to account for the results. This mechanism is based on the fact that the β transition motions lead to a softening of the surrounding medium. As a result, the rearrangements induced by the conformation changes associated with yielding are able to develop, allowing plastic flow processes to occur at a lower stress than yielding. In the case of aryl-aliphatic copolyamides, the medium softening arises from a breakdown of some of the hydrogen bonds under the action of amide group motions involved in the β transition.

For copolyamides containing *tere*-phthalic units, in the high temperature part of the β transition, intramolecular cooperativity of phenyl ring π-flips with adjacent amide groups exists, whatever the xT_yI_{1-y} and MT_yI_{1-y} series considered.

Such an intramolecular cooperativity plays quite an important role, in particular in fracture behaviour. Indeed, when considering the effect of replacing *iso*- by *tere*-phthalic units, in most cases it has not been possible to account for the results by considering the entanglement density, nor the yield stress. Only the occurrence of intramolecular cooperativity leads to consistent explanations. Furthermore, this cooperativity leads to K_{Ic} values that increase with the amount of *tere*-phthalic units in the chemical structure.

In the xT_yI_{1-y} copolyamide series (where various MWs were available for several compounds) at temperatures below about $-40\,°C$, a regime where toughness is independent of MW was observed. At higher temperatures, for sufficiently high MW, an increase of toughness with increasing temperature was achieved up to 60–80 °C for the xT_yI_{1-y} copolyamide series, but only to 40 °C for MT_yI_{1-y} copolyamides. It was followed by a decrease of K_{Ic} values at higher temperatures. The increase of K_{Ic} values results from an interaction between SDZs and crazes, when these latter are stable enough. The mechanism is similar to that observed for MGIMx copolymers with high MGI content. The decrease of K_{Ic} values at higher temperatures comes from a decrease of craze stability due to easier chain slippage and leading to fibril breakdown.

As regards the micromechanisms of deformation, in the xT_yI_{1-y} copolyamide series, by taking one of the compounds as a reference, the various CSC–SDZ, SDZ–CDC transition temperatures have been calculated and a satisfactory agreement with the observed values was achieved.

Finally, it is important to mention that MT_yI_{1-y} copolyamides are characterised by quite a high modulus, E'. For copolyamides of this series with high *tere*-phthalic unit content (50%) fairly high K_{Ic} values are obtained, in such a way that, around room temperature, they are only slightly lower than those of BPA-PC.

6
Conclusion

The goal of this investigation of the mechanical properties of amorphous polymers (plastic deformation, micromechanisms of deformation, fracture) was to analyse the influence of secondary transition motions on these properties.

A first approach was to consider amorphous polymers with quite different chemical structures, either containing side groups responsible for the β transition (PMMA and its copolymers), or only a chain backbone without side chains (BPA-PC, aryl-aliphatic copolyamides) in which the β transition results from motions of some of the main-chain units (typically phenyl ring, carbonate, or amide groups).

In order to achieve a more accurate analysis of the chemical structure effect, gradual changes of chemical structure were performed for each polymer type, e.g. MMA-*co*-maleimide (CMIMx) and MMA-*co*-glutarimide (MGIMx) copolymers, BPA-PC and TMBPA-PC, and the content and nature (*iso*-(I) or *tere*(T)) of phthalic units in aryl-aliphatic copolyamides (xT_yI_{1-y} and MT_yI_{1-y}).

From this large set of results, several important features appear.

The mechanical properties are related to β transition motions, more precisely the cooperativity associated with the motions occurring in the high temperature part of the β transition plays quite an important role. Depending on the chemical structure, this cooperativity can be intra- or intermolecular. Thus, in PMMA and MGIMx copolymers (except for MGI-rich copolymers) the ester group π-flips have an intramolecular cooperativity with rotation angle changes of main-chain C – C bonds (for MGI-rich copolymers, there is an intramolecular cooperativity in MGI–MGI sequences). In contrast, even at low CMI content (around 10 mol%) the intramolecular cooperativity is hindered in CMIMx copolymers, due to the rigid CMI unit. In xT$_y$I$_{1-y}$ and MT$_y$I$_{1-y}$ copolyamides, there are the phenyl ring π-flips of the *tere*-phthalic units which develop a cooperativity with adjacent amide groups. Finally, in BPA-PC, in addition to the intramolecular cooperativities for the phenyl ring motions (intrinsic cooperativity with the motion of the phenyl ring attached to the same isopropylidene unit, cooperativity of the phenyl ring π-flips with conformation change of adjacent carbonate groups) there is a long-range intermolecular cooperativity of the phenyl ring π-flips.

For strain softening, in the case of PMMA and its copolymers, the amplitude decreases with the occurrence of the intramolecular cooperativity of the ester group π-flips. This is due to the effect on the main-chain conformation rotation, which favours the conformation changes involved in the whole chain motions characteristic of plastic flow. In contrast, for BPA-PC and aryl-aliphatic copolyamides, it is an increase of strain softening which arises when increasing temperature. A mechanism based on the softening of the surrounding medium, consecutive to β transition motions, has been proposed. In the case of BPA-PC, such a softening results from the long range intermolecular cooperativity of phenyl ring π-flips, whereas in aryl-aliphatic copolyamides the softening originates from the breakdown of hydrogen bonds (present in these compounds) to β transition motions, particularly those involving amide groups.

For micromechanisms of deformation, it clearly appears that the temperature at which SDZs appear is directly related to the high temperature part of the β transition and, consequently, to the occurrence of intramolecular cooperativity of the involved motions. Thus, SDZs appear at very low temperatures ($-120\,^{\circ}$C) in BPA-PC, and at around $-100\,^{\circ}$C or $-70\,^{\circ}$C for xT$_y$I$_{1-y}$ and MT$_y$I$_{1-y}$ copolyamides, respectively. For PMMA, SDZs appear at $50\,^{\circ}$C, but at $10\,^{\circ}$C for MGI-rich copolymers. It is interesting to note that the calculated transition temperatures for CSC–SDZ and SDZ–CDC, obtained by considering one of the compounds in a polymer series as a reference, lead to a satisfactory agreement with the experimentally determined ones.

The most challenging feature deals with the fracture behaviour. Indeed, in Sect. 2.4.7 it has been indicated that, according to empirical or theoretical approaches, K_{Ic} values should increase with the entanglement density,

ν_e, proportionally to ν_e or $\nu_e^{1/2}$, depending on the model. If these dependencies look reasonable when considering quite different chemical structures (as shown in Fig. 11) they completely fail when performing gradual changes of the chemical structure in a given polymer series, e.g. for MGIMx copolymers, or xT$_y$I$_{1-y}$ and MT$_y$I$_{1-y}$ copolyamides. In the first case, ν_e is constant but K_{Ic} increases with MGI content, in relation with the development of intramolecular MGI–MGI cooperativity. In both copolyamide series, when replacing *iso*-phthalic units by *tere*-phthalic units, K_{Ic} changes cannot be accounted for by considering either ν_e, or σ_y values. Only the intramolecular cooperativity associated with the phenyl ring π-flips leads to consistent explanations.

As a general comment, one can say that the empirical statement made a long time ago by Heijboer [65], "secondary transition originating from a motion within the main chain can, very likely, lead to a major increase of impact strength, whereas if the transition originates from a motion of side chains only, no major effect on the impact strength can be expected", appears valid when considering the fracture behaviour of amorphous polymers. However, the whole set of results reported in this paper provide a considerably deeper understanding of the way the type of β transition motions, and in particular their cooperativity in the high temperature part of the transition, controls the plastic deformation, the micromechanisms of deformation, and the fracture behaviour. Thus, a gap is filled between polymer chemical structure and mechanical properties, which constitutes quite a significant progress.

Acknowledgements The authors are greatly indebted to the PhD students B. Brûlé, O. Julien, L. Tézé and P. Tordjeman, and to their colleagues P. Béguelin, L. Canova, S. Choe, A.M. Donald, A. Dubault and C.J.G. Plummer for their fruitful collaboration.

References

1. Monnerie L, Lauprêtre F, Halary J-L (2005) Adv Polym Sci 187:35
2. Williams JG (1981) Stress analysis of polymers, 2nd edn. Wiley, New York
3. Ward IM (1983) Mechanical properties of solid polymers, 2nd edn. Wiley, New York
4. Kinloch AJ, Young RJ (1983) Fracture behaviour of polymers. Applied Science, London
5. Williams JG (1984) Fracture mechanics of polymers. Ellis Horwood, Chichester
6. Flory PJ (1989) Statistical mechanics of chain molecules. Hanser, New York
7. Wu S (1989) J Polym Sci Polym Phys 27:723
8. Graessley WW (1971) J Polym Sci 54:5143
9. Robertson RE (1966) J Chem Phys 44:3950
10. Robertson RE (1968) Appl Polym Symp 7:201
11. Theodorou M, Jasse B, Monnerie L (1985) J Polym Sci Polym Phys 23:445
12. Xu Z, Jasse B, Monnerie L (1989) J Polym Sci Polym Phys 27:355
13. Helfand E, Wasserman ZR, Weber TA (1980) Macromolecules 13:526

14. Weber TA, Helfand E (1983) J Phys Chem 87:2881
15. Bahar I, Erman B, Monnerie L (1992) Macromolecules 25:6315
16. Lauterwasser BD, Kramer EJ (1979) Philos Mag A 39:469
17. Kausch H-H (ed)(1983) Crazing in polymers. Adv Polym Sci 52/53
18. Kausch H-H (ed)(1990) Crazing in polymers. Adv Polym Sci 91/92
19. Kramer EJ (1983) Adv Polym Sci 52/53:1
20. Kramer EJ, Berger LL (1990) Adv Polym Sci 91/92:1
21. Berger LL, Kramer EJ (1987) Macromolecules 20:1980
22. McLeish TCB, Plummer CJG, Donald AM (1989) Polymer 30:1651
23. Trassaert P, Schirrer R (1983) J Mater Sci 18:3004
24. Donald AM, Kramer EJ (1981) J Mater Sci 16:2967
25. ISO 13586-1, Determination of fracture toughness for plastics, an LEFM approach.
26. Wu S (1992) Polym Eng Sci 32:823
27. Brown HR (1991) Macromolecules 24:2752
28. Hui CY, Ruina A, Creton C, Kramer EJ (1992) Macromolecules 25:3948
29. Sha Y, Hui CY, Ruina A, Kramer EJ (1995) Macromolecules 28:2450
30. Kausch H-H (1987) Polymer fracture, 2nd edn. Springer, Heidelberg Berlin New York
31. Jud K, Kausch H-H, Williams JG (1981) J Mater Sci 16:204
32. Tordjeman P, Tézé L, Halary J-L, Monnerie L (1997) Polym Eng Sci 37:1621
33. Julien O (1995) PhD Thesis, Université Pierre et Marie Curie, Paris
34. Haussy J, Cavrot J-P, Escaig B, Lefebvre J-M (1980) J Polym Sci Polym Phys 18:311
35. Tézé L, Halary J-L, Monnerie L, Canova L (1999) Polymer 40:971
36. Rabinowitz S, Ward IM, Parry J SC (1970) J Mater Sci 5:29
37. Tordjeman P (1992) PhD Thesis, Université Pierre et Marie Curie, Paris
38. Tordjeman P, Halary J-L, Monnerie L, Donald AM (1995) Polymer 36:1627
39. Levett RJ, Donald AM (1994) In: Proceedings of 9th international conference on deformation, yield and fracture in polymers. 11–14 April 1994, Cambridge, UK :45
40. Marshall GP, Coutts LH, Williams JG (1974) J Mater Sci 9:1409
41. Morgan GP, Ward IM (1977) Polymer 18:87
42. Mizutani K (1987) J Mater Sci Lett 6:915
43. Balzano M, Ravi-Chandar K (1991) J Mater Sci 26:1387
44. Döll W, Könczöl L (1990) Adv Polym Sci 91/92:137
45. Schirrer R (1990) Adv Polym Sci 91/92:215
46. Schirrer R, Goett C (1981) J Mater Sci 16:2563
47. Schirrer R, Goett C (1982) J Mater Sci Lett 1:355
48. Lousteau B (2001) PhD Thesis, Université Pierre et Marie Curie, Paris
49. Bordes B (1999) PhD Thesis, Université Pierre et Marie Curie, Paris
50. Wu S (1989) J Polym Sci Polym Phys 27:723
51. Plummer CJG, Kausch H-H, Tézé L, Halary J-L, Monnerie L (1996) Polymer 37:4299
52. Tézé L (1995) PhD Thesis, Université Pierre et Marie Curie, Paris
53. Brûlé B (1999) PhD Thesis, Université Pierre et Marie Curie, Paris
54. Christiansen AW, Baer E, Radcliffe SV (1971) Phil Mag 24:451
55. Bauwens-Crowet C, Bauwens J-C (1983) J Phys C 44:185
56. Wellinghoff ST, Baer E (1978) J Appl Polym Sci 22:2025
57. Plummer CJG, Donald AM (1989) J Polym Sci Polym Phys 27:325
58. Plummer CJG, Soles CL, Xiao C, Wu J, Kausch H-H, Yee AF (1995) Macromolecules 28:7157
59. Pitman GL, Ward IM (1979) Polymer 20:895
60. Brûlé B, Halary J-L, Monnerie L (2001) Polymer 42:9073
61. Brûlé B, Kausch H-H, Monnerie L, Plummer CJG, Halary J-L (2003) Polymer 44:1181

62. Brûlé B, Monnerie L, Halary J-L (2003) In: Blackman BRK, Pavan A, Williams JG (eds) Fracture of polymers, composites and adhesives II. Elsevier and ESIS

63. Halary J-L, Monnerie L (2003) In: Proceedings of the 12th international conference on deformation, yield and fracture of polymers. IOM communications ed: 25

64. Choe S, Brûlé B, Bisconti L, Halary J-L, Monnerie L (1999) J Polym Sci Polym Phys 37:1131

65. Heijboer J (1968) J Polym Sci C 16:3755

Author Index Volumes 101–187

Subject Index

Printing: Krips bv, Meppel
Binding: Stürtz, Würzburg